AF361965

Selected Papers from IEEE ICKII 2019

Selected Papers from IEEE ICKII 2019

Special Issue Editors

Teen-Hang Meen
Wenbing Zhao
Cheng-Fu Yang

MDPI • Basel • Beijing • Wuhan • Barcelona • Belgrade • Manchester • Tokyo • Cluj • Tianjin

Special Issue Editors

Teen-Hang Meen	Wenbing Zhao	Cheng-Fu Yang
National Formosa University	Cleveland State University	National University of Kaohsiung
Taiwan	USA	Taiwan

Editorial Office
MDPI
St. Alban-Anlage 66
4052 Basel, Switzerland

This is a reprint of articles from the Special Issue published online in the open access journal *Energies* (ISSN 1996-1073) (available at: https://www.mdpi.com/journal/energies/special_issues/ICKII_2019).

For citation purposes, cite each article independently as indicated on the article page online and as indicated below:

LastName, A.A.; LastName, B.B.; LastName, C.C. Article Title. *Journal Name* **Year**, *Article Number*, Page Range.

ISBN 978-3-03936-160-1 (Hbk)
ISBN 978-3-03936-161-8 (PDF)

Contents

About the Special Issue Editors

Teen-Hang Meen was born in Tainan, Taiwan, in 1967. He received his BSc from the Department of Electrical Engineering of National Cheng Kung University (NCKU), Tainan, Taiwan, in 1989, and his MSc and PhD from the Institute of Electrical Engineering, National Sun Yat-Sen University (NSYSU), Kaohsiung, Taiwan, in 1991 and 1994, respectively. He was the chairman of the Department of Electronic Engineering of National Formosa University, Yunlin, Taiwan, from 2005 to 2011. He recevied the Excellent Research Award from National Formosa University in 2008 and 2014. Currently, he is a Distinguished Professor with the Department of Electronic Engineering, National Formosa University, Yunlin, Taiwan. He is also the president of International Institute of Knowledge Innovation and Invention (IIKII) and the chair of the IEEE Tainan Section Sensors Council. He has published more than 100 SCI, SSCI and EI papers in recent years.

Wenbing Zhao is a Full Professor of Electrical Engineering and Computer Science (EECS) at Cleveland State University (CSU), Cleveland, Ohio, USA. He obtained his BSc and MSc in Physics from Peking University, Beijing, China, in 1990 and 1993, respectively, and his MSc and PhD in Electrical and Computer Engineering from the University of California, Santa Barbara, in 1998 and 2002, respectively. Prior to joining Cleveland State University in 2004, Dr. Zhao worked as a post-doctoral researcher at the University of California, Santa Barbara, and as Senior Research Engineer/Chief Architect at Eternal Systems, Inc. (now dissolved), which he co-founded in 2000. Dr. Zhao has conducted research in several different areas, including fault tolerance computing, computer and network security, smart and connected healthcare, machine learning, Internet of Things, quantum optics and superconducting physics. Currently, his research focuses on smart and connected healthcare. Dr. Zhao's recent research was funded by the National Science Foundation, the Ohio Bureau of Workers' Compensation, the Ohio Department of Higher Education, the Ohio Advancement Office (via the Ohio Third Frontier Program), the US Department of Transportation (via CSU Transportation Center), Cleveland State University, and private companies.

Cheng-Fu Yang was born in Taiwan in 1964. Yang received his MSc and PhD in 1988 and 1993, respectively, from the Department of Electrical Engineering of Cheng Kung University, Tainan, Taiwan. Yang entered professional academic life in 1993 with the Department of Electronic Engineering, Chinese Air Force Academy, and then, in 2000, as a professor. In 2004 he joined the faculty of the National University of Kaohsiung (NUK). Currently, he is a Professor of Chemical and Materials Engineering at NUK. He received the Outstanding Contribution Awards of the Chinese Ceramic Society in 2009. In 2010, he was the first (and only) person to receive the title of Distinguished Professor from NUK. In 2014, he became the Fellow of Taiwanese Institute of Knowledge Innovation (TIKI) and in 2015 the Fellow of the Institution of Engineering and Technology (IET). He was also labeled the Mingjiang Scholar and Chair Inviting Professor of Jimei University, Xiamen, Fujian, China.

Editorial

Special Issue on Selected Papers from IEEE ICKII 2019

Teen-Hang Meen [1,*], Wenbing Zhao [2] and Cheng-Fu Yang [3,*]

[1] Department of Electronic Engineering, National Formosa University, Yunlin 632, Taiwan
[2] Department of Electrical Engineering and Computer Science, Cleveland State University, Cleveland, OH 44011, USA; w.zhao1@csuohio.edu
[3] Department of Chemical and Materials Engineering, National University of Kaohsiung, Kaohsiung 811, Taiwan
* Correspondence: thmeen@nfu.edu.tw (T.-H.M.); cfyang@nuk.edu.tw (C.-F.Y.)

Received: 23 March 2020; Accepted: 3 April 2020; Published: 14 April 2020

Abstract: This Special Issue on "Selected papers from IEEE ICKII 2019" selected 13 excellent papers from 260 papers presented in IEEE ICKII 2019 on topics in energies. The fields include: energy fundamentals, energy sources and energy carriers, energy exploration, intermediate and final energy use, energy conversion systems, and energy research and development. The main goal of this Special Isue is to discover new scientific knowledge relevant to the topic of energies.

Keywords: energy sources and energy carriers; energy conversion systems; energy research and development

---

The 2nd IEEE International Conference on Knowledge Innovation and Invention 2019 (IEEE ICKII 2019) was held in Seoul, South Korea on 12–15 July 2019. It provided a unified communication platform for researchers in the topics of information technology, innovation design, communication science and engineering, industrial design, creative design, applied mathematics, computer science, electrical and electronic engineering, mechanical and automation engineering, green technology and architecture engineering, material science, and other related fields. This Special Issue on "Selected papers from IEEE ICKII 2019" selected 13 excellent papers from 260 papers presented in IEEE ICKII 2019 on topics in energies. The fields include: energy fundamentals, energy sources and energy carriers, energy exploration, intermediate and final energy use, energy conversion systems, and energy research and development. The main goal of this Special Issue is to discover new scientific knowledge relevant to the topic of energies.

The Topic of Energies and its Applications

This Special Issue on "Selected papers from IEEE ICKII 2019" selected 13 excellent papers from 260 papers presented in IEEE ICKII 2019 on topics in energies. The published papers are introduced as follows:

Hwang et al. reported on "Optimization and Application for Hydraulic Electric Hybrid Vehicle" [1]. In this research, the rule-based control strategy was implemented as the energy distribution management strategy first, and then the genetic algorithm was utilized to conduct global optimization strategy analysis. The results from the genetic algorithm were employed to modify the rule-based control strategy to improve the electricity economic performance of the vehicle. The simulation results show that the electricity economic performance of the designed hydraulic hybrid vehicle was improved by 36.51% compared to that of a pure electric vehicle. The performance of energy consumption after genetic algorithm optimization was improved by 43.65%.

Liao reported "A Step Up/Down Power-Factor-Correction Converter with Modified Dual Loop Control" [2]. In this study, A step up/down AC/DC converter with a modified dual loop control is proposed. The step up/down AC/DC converter features the bridgeless characteristic which can reduce bridge–diode conduction losses. Based on the step up/down AC/DC converter, a modified dual loop control scheme is proposed to achieve input current shaping and output voltage regulation. Fewer components are needed compared with the traditional bridge and bridgeless step up/down AC/DC converters. In addition, the intermediate capacitor voltage stress can be reduced. Furthermore, the top and bottom switches still have a zero-voltage turn-on function during the negative and positive half-line cycle, respectively. Hence, the thermal stresses can also be reduced and balanced. Simulation and experimental results are provided to verify the validity of the proposed step up/down AC/DC converter and its control scheme.

Wu et al., reported "The Optimal Control of Fuel Consumption for a Heavy-Duty Motorcycle with Three Power Sources Using Hardware-in-the-Loop Simulation" [3]. This study presents a simulation platform for a hybrid electric motorcycle with an engine, a driving motor, and an integrated starter generator (ISG) as three power sources. This platform also consists of the driving cycle, driver, lithium-ion battery, continuously variable transmission (CVT), motorcycle dynamics, and energy management system models. Two Arduino DUE microcontrollers integrated with the required circuit to process analog-to-digital signal conversion for input and output are utilized to carry out a hardware-in-the-loop (HIL) simulation. A driving cycle called the worldwide motorcycle test cycle (WMTC) is used for evaluating the performance characteristics and response relationship among subsystems. Control strategies called rule-based control (RBC) and equivalent consumption minimization strategy (ECMS) are simulated and compared with the purely engine-driven operation. The results show that the improvement percentages for equivalent fuel consumption and energy consumption for RBC and ECMS using the pure software simulation were 17.74%/18.50% and 42.77%/44.22% respectively, while those with HIL were 18.16%/18.82% and 42.73%/44.10%, respectively.

Tsai et al., reported "Optimal Configuration with Capacity Analysis of a Hybrid Renewable Energy and Storage System for an Island Application" [4]. This study uses a Philippine offshore island to optimize the capacity configuration of a hybrid energy system (HES). A thorough investigation was performed to understand the operating status of existing diesel generator sets and load power consumption, and to collect the statistics of meteorological data and economic data. Using the Hybrid Optimization Models for Energy Resources (HOMER) software we simulated and analyzed the techno-economics of different power supply systems containing stand-alone diesel systems, photovoltaic (PV)-diesel HES, wind-diesel HES, PV-wind-diesel HES, PV-diesel-storage HES, wind-diesel-storage HES, and PV-wind-diesel-storage HES. In addition to the lowest cost of energy (COE), capital cost, fuel saving and occupied area, the study also uses entropy weight and the Technique for Order Preference by Similarity to an Ideal Solution (TOPSIS) method to evaluate the optimal capacity configuration. The proposed method can also be applied to design hybrid renewable energy systems for other off-grid areas.

Li et al., reported "Integrated Analysis of Influence of Multiple Factors on Transmission Efficiency of Loader Drive Axle" [5]. In this study, a loader drive axle digital model was built using 3D commercial software. On the basis of this model, the transmission efficiency of the main reducing gear, the differential planetary mechanism, and the wheel planetary reducing gear of the loader drive axle were studied. The functional relationship of the transmission efficiency of the loader drive axle was obtained, including multiple factors: the mesh friction coefficient, the mesh power loss coefficient, the normal pressure angle, the helix angle, the offset amount, the speed ratio, the gear ratio, and the characteristic parameters. This revealed the influence law of the loader drive axle by the mesh friction coefficient, mesh power loss coefficient, and speed ratio. The research results showed that the transmission efficiency of the loader drive axle increased with the speed ratio, decreased when the mesh friction coefficient and the mesh power loss coefficient increased, and that there was a greater influence difference in the transmission efficiency of the loader drive axle.

Yu et al., reported "Management and Distribution Strategies for Dynamic Power in a Ship's Micro-Grid System Based on Photovoltaic Cell, Diesel Generator, and Lithium Battery" [6]. This study examines the stable parallel operation of a ship's micro-grid system through a dynamic power management strategy involving a step change in load. With cruise ships in mind, the authors construct a micro-grid system consisting of photovoltaics (PV), a diesel generator (DG), and a lithium battery, and establish a corresponding simulation model. The authors analyze the system's operating characteristics under different working conditions and present the mechanisms that influence the power quality of the ship's micro-grid system. Based on an analysis of the power distribution requirements under different working conditions, the authors design a power allocation strategy for the micro-grid system. Next the authors propose an optimization allocation strategy for dynamic power based on fuzzy control and a load current feed-forward method, and finally, the authors simulate the whole system. Through this study, the authors prove that the proposed power management strategy not only verifies the feasibility and correctness of the ship's micro-grid structure and control strategy, but also greatly improves the reliability and stability of the ship's operation.

Lin et al., reported "Analysis of Energy Flux Vector on Natural Convection Heat Transfer in Porous Wavy-Wall Square Cavity with Partially-Heated Surface" [7]. This study utilizes the energy-flux-vector method to analyze the heat transfer characteristics of natural convection in a wavy-wall porous square cavity with a partially heated bottom surface. The effects of the modified Darcy number, modified Rayleigh number, modified Prandtl number, and length of the partially heated bottom surface on the energy-flux-vector distribution and mean Nusselt number are examined. The results show that when a low modified Darcy number with any value of modified Rayleigh number is given, the recirculation regions are not formed in the energy-flux-vector distribution within the porous cavity. Therefore, a low mean Nusselt number is presented. The recirculation regions still do not form, and thus the mean Nusselt number has a low value when a low modified Darcy number with a high modified Rayleigh number is given.

Luo et al., reported "Performance Enhancement of Hybrid Solid Desiccant Cooling Systems by Integrating Solar Water Collectors in Taiwan" [8]. In this study, a solar-assisted hybrid Solid Desiccant Cooling System (SDCS) was developed, in which solar-heated water is used as an additional heat source for the regeneration process, in addition to recovering heat from the condenser of an integrated heat pump. A solar thermal collector sub-system is used to generate solar regenerated water. Experiments were conducted in the typically hot and humid weather of Taichung, Taiwan, from the spring to fall seasons. The experimental results show that the overall performance of the system in terms of power consumption can be enhanced by approximately 10% by integrating a solar-heated water heat exchanger, in comparison to the hybrid SDCS system. The results show that the system performs better when the outdoor humidity ratio is high. In addition, regarding the effect of ambient temperature on the coefficient of performance (COP) of the systems, a critical value of outdoor temperature exists. The COP of the systems gradually rises with the increase in ambient temperature. However, when the ambient temperature is greater than the critical value, the COP gradually decreases with the increase in ambient temperature. The critical outdoor temperature of the hybrid SDCS is from 26 to 27 °C, and the critical temperature of the solar-assisted hybrid SDCS is from 27 to 30 °C.

Hsueh et al. reported "Condition Monitor System for Rotation Machine by CNN with Recurrence Plot" [9]. In this paper, the authors introduce an effective framework for the fault diagnosis of 3-phase induction motors. The proposed framework mainly consists of two parts. The first part explains the preprocessing method, in which the time-series data signals are converted into two-dimensional (2D) images. The preprocessing method generates recurrence plots (RP), which represent the transformation of time-series data such as 3-phase current signals into 2D texture images. The second part of the paper explains how the proposed convolutional neural network (CNN) extracts the robust features to diagnose the induction motor's fault conditions by classifying the images. The generated RP images are considered as input for the proposed CNN in the texture image recognition task. The proposed framework is tested on the dataset collected from different 3-phase induction motors working with different failure modes. The experimental results of the proposed framework show its competitive performance over traditional methodologies and other machine learning methods.

Chen et al., reported "Design of a Logistics System with Privacy and Lightweight Verification" [10]. This study designs a secure logistics system, with anonymous and lightweight verification, in order to meet the following requirements: mutual authentication, non-repudiation, anonymity, integrity, and a low overhead for the logistics environment. A buyer could check the goods and know if the parcel has been exchanged by a malicious person. Moreover, the proposed scheme not only presents a solution to meet the logistics system's requirements, but also to reduce both computational and communication costs.

Chang et al. reported on the "Current Control of the Permanent-Magnet Synchronous Generator Using Interval Type-2 T-S Fuzzy Systems" [11]. In this study, the current control of the permanent-magnet synchronous generator (PMSG) using an interval type-2 (IT2) Takagi-Sugeno (T-S) fuzzy systems is designed and implemented. PMSG is an energy conversion unit widely used in wind energy generation systems and energy storage systems. Its performance is determined by the current control approach. IT2 T-S fuzzy systems are implemented to deal with the nonlinearity of a PMSG system in this paper. First, the IT2 T-S fuzzy model of a PMSG is obtained. Second, the IT2 T-S fuzzy controller is designed based on the concept of parallel distributed compensation (PDC). Next, the stability analysis can be conducted through the Lyapunov theorem. Accordingly, the stability conditions of the closed-loop system are expressed in Linear Matrix Inequality (LMI) form. The AC power from a PMSG is converted to DC power via a three-phase six-switch full bridge converter. The six-switch full bridge converter is controlled by the proposed IT2 T-S fuzzy controller. The analog-to-digital (ADC) conversion, rotor position calculation and duty ratio determination are digitally accomplished by the microcontroller. Finally, the simulation and experimental results verify the performance of the proposed current control.

Lin et al., reported "The Optimal Energy Dispatch of Cogeneration Systems in a Liberty Market" [12]. This paper investigates the cogeneration systems of industrial users and collects fuel consumption data and data concerning the steam output of boilers. On the basis of the relation between the fuel enthalpy and steam output, the Least Squares Support Vector Machine (LSSVM) is used to derive boiler and turbine Input/Output (I/O) operation models to provide fuel cost functions. The CO_2 emission of pollutants generated by various types of units is also calculated. The objective function is formulated as a maximal profit model that includes profit from steam sold, profit from electricity sold, fuel costs, costs of exhausting carbon, wheeling costs, and water costs. By considering Time-of-Use (TOU) and carbon trading prices, the profits of a cogeneration system in different scenarios are evaluated. By integrating the Ant Colony Optimization (ACO) and Genetic Algorithm (GA), an Enhanced ACO (EACO) is proposed to come up with the most efficient model. The EACO uses a crossover and mutation mechanism to alleviate the local optimal solution problem and to seek a system that offers an overall global solution using competition and selection procedures. The results show that these mechanisms provide a good direction for the energy trading operations of a cogeneration system. This approach also provides a better guide for operation dispatch to use in determining the benefits accounting for both cost and the environment in a liberty market.

Hsu et al. reported "Article Numerical Simulation of Crystalline Silicon Heterojunction Solar Cells with Different p-Type a-SiOx Window Layer" [13]. In this study, p-type amorphous silicon oxide (a-SiOx) films are deposited using a radio-frequency, inductively coupled plasma chemical vapor deposition system. Effects of the CO_2 gas flow rate on film properties and crystalline silicon heterojunction (HJ) solar cell performance are investigated. The experimental results show that the band gap of the a-SiOx film can reach 2.1 eV at CO_2 flow rate of 10 standard cubic centimeters per minute (sccm), but the conductivity of the film deteriorates. In the device simulation, the transparent conducting oxide and contact resistance are not taken into account. The electrodes are assumed to be perfectly conductive and transparent. The simulation result shows that there is a tradeoff between the increase in the band gap and the reduction in conductivity at an increasing CO_2 flow rate, and the balance occurs at the flow rate of six sccm, corresponding to a band gap of 1.95 eV, an oxygen content of 34%, and a conductivity of 3.3 S/cm. The best simulated conversion efficiency is 25.58%, with an open-circuit voltage of 741 mV, a short-circuit current density of 42.3 mA/cm^2, and a fill factor of 0.816%.

Author Contributions: Writing and reviewing all papers, T.-H.M.; English editing, W.Z.; checking and correcting the manuscript, C.-F.Y. All authors have read and agreed to the published version of the manuscript.

Funding: This research received no external funding.

Acknowledgments: The guest editors would like to thank the authors for their contributions to this Special Issue and all the reviewers for their constructive reviews. We are also grateful to Chloe Wu, the Assistant Editor of *Energies*, for her time and efforts in the publication of this special issue for Energies.

Conflicts of Interest: The authors declare no conflict of interest.

References

1. Hwang, H.Y.; Lan, T.S.; Chen, J.S. Optimization and Application for Hydraulic Electric Hybrid Vehicle. *Energies* **2020**, *13*, 322. [CrossRef]
2. Liao, Y.H. A Step Up/Down Power-Factor-Correction Converter with Modified Dual Loop Control. *Energies* **2020**, *13*, 199. [CrossRef]
3. Wu, C.H.; Xu, Y.X. The Optimal Control of Fuel Consumption for a Heavy-Duty Motorcycle with Three Power Sources Using Hardware-in-the-Loop Simulation. *Energies* **2020**, *13*, 22. [CrossRef]
4. Tsai, C.T.; Beza, T.M.; Wu, W.B.; Kuo, C.C. Optimal Configuration with Capacity Analysis of a Hybrid Renewable Energy and Storage System for an Island Application. *Energies* **2020**, *13*, 8. [CrossRef]
5. Li, J.Y.; Wu, T.L.; Chi, W.M.; Hu, Q.C.; Meen, T.H. Integrated Analysis of Influence of Multiple Factors on Transmission Efficiency of Loader Drive Axle. *Energies* **2019**, *12*, 4540. [CrossRef]
6. Yu, W.N.; Li, S.W.; Zhu, Y.H.; Yang, C.F. Management and Distribution Strategies for Dynamic Power in a Ship's Micro-Grid System Based on Photovoltaic Cell, Diesel Generator, and Lithium Battery. *Energies* **2019**, *12*, 4505. [CrossRef]
7. Lin, Y.T.; Cho, C.C. Analysis of Energy Flux Vector on Natural Convection Heat Transfer in Porous Wavy-Wall Square Cavity with Partially-Heated Surface. *Energies* **2019**, *12*, 4456. [CrossRef]
8. Luo, W.J.; Faridah, D.; Fasya, F.R.; Chen, Y.S.; Mulki, F.H.; Adilah, U.N. Performance Enhancement of Hybrid Solid Desiccant Cooling Systems by Integrating Solar Water Collectors in Taiwan. *Energies* **2019**, *12*, 3470. [CrossRef]
9. Hsueh, Y.M.; Ittangihala, V.R.; Wu, W.B.; Chang, H.C.; Kuo, C.C. Condition Monitor System for Rotation Machine by CNN with Recurrence Plot. *Energies* **2019**, *12*, 3221. [CrossRef]
10. Chen, C.L.; Lin, D.P.; Chen, H.C.; Deng, Y.Y.; Lee, C.F. Design of a Logistics System with Privacy and Lightweight Verification. *Energies* **2019**, *12*, 3061. [CrossRef]
11. Chang, Y.C.; Tsai, C.T.; Lu, Y.L. Current Control of the Permanent-Magnet Synchronous Generator Using Interval Type-2 T-S Fuzzy Systems. *Energies* **2019**, *12*, 2953. [CrossRef]
12. Lin, W.M.; Yang, C.Y.; Tu, C.S.; Huang, H.S.; Tsai, M.T. The Optimal Energy Dispatch of Cogeneration Systems in a Liberty Market. *Energies* **2019**, *12*, 2868. [CrossRef]
13. Hsu, C.H.; Zhang, X.Y.; Lin, H.J.; Lien, S.Y.; Cho, Y.S.; Ye, C.S. Article Numerical Simulation of Crystalline Silicon Heterojunction Solar Cells with Different p-Type a SiOx Window Layer. *Energies* **2019**, *12*, 2541. [CrossRef]

Article

Optimization and Application for Hydraulic Electric Hybrid Vehicle

Hsiu-Ying Hwang [1], Tian-Syung Lan [2,*] and Jia-Shiun Chen [1]

[1] Department of Vehicle Engineering, National Taipei University of Technology, Taipei 10608, Taiwan; hhwang@mail.ntut.edu.tw (H.-Y.H.); chenjs@mail.ntut.edu.tw (J.-S.C.)

[2] College of Mechatronic Engineering, Guangdong University of Petrochemical Technology, Maoming 525000, China

* Correspondence: tslan888@gmail.com

Received: 27 November 2019; Accepted: 7 January 2020; Published: 9 January 2020

Abstract: Targeting the application of medium and heavy vehicles, a hydraulic electric hybrid vehicle (HEHV) was designed, and its energy management control strategy is discussed in this paper. Matlab/Simulink was applied to establish the pure electric vehicle and HEHV models, and backward simulation was adopted for the simulation, to get the variation of torque and battery state of charge (*SOC*) through New York City Cycle of the US Environmental Protection Agency (EPA NYCC). Based on the simulation, the energy management strategy was designed. In this research, the rule-based control strategy was implemented as the energy distribution management strategy first, and then the genetic algorithm was utilized to conduct global optimization strategy analysis. The results from the genetic algorithm were employed to modify the rule-based control strategy to improve the electricity economic performance of the vehicle. The simulation results show that the electricity economic performance of the designed hydraulic hybrid vehicle was improved by 36.51% compared to that of a pure electric vehicle. The performance of energy consumption after genetic algorithm optimization was improved by 43.65%.

Keywords: hydraulic hybrid vehicle; NYCC driving cycle; optimization; genetic algorithm

---

1. Introduction

The increasing demand for fossil fuels in different fields since the Industrial Revolution has led to increasing global CO_2 emission and worsening global warming. Among all CO_2 emission, the emission of means of transportation is only second to the industry. Now, the passenger vehicles all develop toward alternative energy, whereas the medium and heavy vehicles for goods transportation are still using gasoline or diesel engines as the main power source. With global warming and increasing stringent laws and regulations, they will definitely develop toward the same clean energy as the passenger vehicles. According to Navigant Research, the market survey company, hydraulic hybrid vehicles seldom known and underestimated in significance will gain a position in the heavy-duty truck market, and even can be expected to apply to the next generation of vehicles. Therefore, hydraulic electric hybrid vehicles (HEHV) will be the first choice for medium vehicles, heavy vehicles, and common carriers. With the DSHplus software simulation, Sokar [1] compared the fuel economy of the hydraulic transmission vehicles and hydraulic hybrid vehicles in urban and highway driving cycles. Chen [2] compared the energy consumption of different hydraulic hybrid configurations, and it showed the HEHV could have better energy efficiency over the pure EV system. The energy optimization can be divided to hardware optimization and control strategy optimization. As for hardware optimization, Ramakrishnan et al. [3] proposed the study on influence of system parameters in hydraulic system on the overall system power and established the series hydraulic hybrid power vehicle with LMS AMESim software. Change of size of accumulator and hydraulic motor/pump and internal pressure greatly improves the output power of the whole system, which also reduces the fuel consumption and pollution of the hydraulic hybrid vehicles. The energy control strategy can be divided into two categories [4]: (1) rule-based strategy and (2) optimization-based strategy. For optimization strategy, Lu et al. [5] introduced the weighted-sum method and no-preference method to solve the multi-objective optimization problem of plug-in electric vehicles, and it was validated with ADVISOR software. Zeng et al. [6] proposed a different strategy, Equivalent Consumption Minimum Strategy (ECMS), to solve the optimization problem of PHEV, and the Simplified-ECMS strategy could effectively shorten the calculation time. Wang et al. [7] applied the Dynamic Programming for PHEV and received an approximately 20% improvement in fuel economy.

The rule-based control, featuring a smaller amount of calculation, is adopted by many studies, to design the energy management strategy. Yu et al. [8] developed a simulation model and rule-based control strategy for extended-range electric vehicle (E-REV) and showed that a small engine can be used to reduce the weight of vehicle and batteries of E-REV. Gao et al. [9] proposed two control strategies, thermostat and power follower. With dynamic programming, it showed that the thermostat control strategy optimized the operation of the internal combustion engine, and the power follower control strategy minimizes the battery-charging and -discharging operations. Konev et al. [10] developed a control strategy for series hybrid vehicle. The control strategy was to ensure gradual operation of the motor along the steady-state Optimal Operating Points Line (OOP-Line) in the engine speed–torque map, which could improve the efficiency of series hybrid vehicle. Liu et al. [11] developed a control strategy for a series hybrid vehicle which included two parts, constant SOC control, and driving-range optimization. Comparing to thermostat control strategy, the constant SOC control could have a longer driving range. Li et al. [12] proposed a fuzzy logic energy-management system, using the battery working state, which ensured that the engine would operate in the vicinity of its maximum fuel-efficiency region. The rule-based design is fast and easy and can be readily applied to real vehicle-control strategy. However, the rule-based control strategy is simple, so it cannot provide optimal power management to HEV in real time. Therefore, an optimization algorithm is required for rule-based control to improve the energy efficiency. Ho and Klong [13] introduced an optimization algorithm for series plug-in hybrid electric vehicles by utilizing the genetic algorithm (GA), which could determine the optimal driving patterns offline. Xu et al. [14] developed a fuzzy control strategy for parallel hybrid electric vehicle. The control strategy was adjusted with GA. It was verified that GA could effectively improve the efficiency of the engine and fuel consumption. Kaur et al. [15] proposed a control strategy to control the speed of a hybrid electric vehicle. The controller, which was using GA, could improve fuel economy and reduce pollution. Hu and Zhao [16] applied an adaptive based hybrid genetic algorithm to optimize the energy efficiency of parallel hybrid electric vehicles and presented the effectiveness of the hybrid genetic algorithm.

Therefore, global optimization, together with rule-based control method, are selected in this paper for medium and heavy vehicles in fixed driving route, to adjust the rule-based control strategy and improve the electricity economic performance of vehicles. The optimization approach selected in this paper is genetic algorithm (GA). With global optimization ability and probability optimization approach, GA can automatically obtain and instruct the optimized search space and adaptively adjust the search direction without the need of clear rules.

2. Modeling

In this study, Matlab/Simulink serves as the main simulation program, and backward simulation is used to establish the model. In order to compare the difference between an HEHV and a pure electric vehicle, subsystem models of the electric system are established, including models of electric motor, generator, and lithium ion battery. The subsystem models of hydraulic system include variable hydraulic motor/pump and accumulator models. The whole vehicle model can be divided into following subsystem models: (1) driver model; (2) vehicle dynamic model; (3) tyre and drive model; (4) power component element; and (5) energy storage component model. Driving cycle of the EPA

NYCC is employed in this study to get the vehicle driving force, and then gear ratio of the transmission system is adopted to calculate the torque and speed needed for the motor. In HEHV, the electric motor does not function as the regenerative brake; rather the hydraulic pump is used for energy recovery. This is introduced in the following.

2.1. Driver Model

The EPA NYCC driving cycle for testing, as shown in Figure 1, is employed in this model. The total driving time is 599 s. The stop time accounts for 35.08% of the total driving time. The maximum speed and the average speed are 44.6 and 11.4 km/h, respectively.

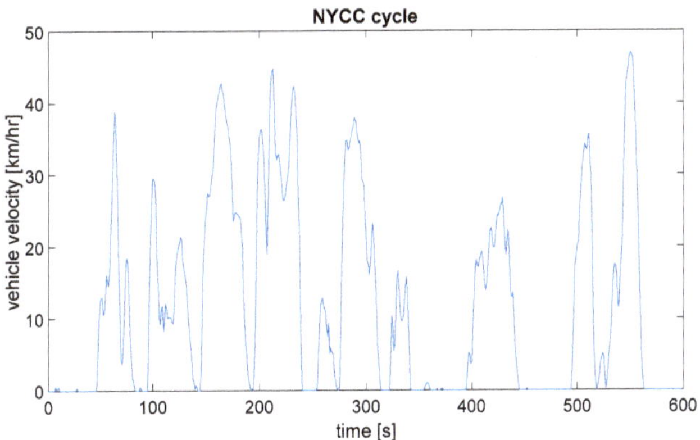

Figure 1. United States Environmental Protection Agency New York City Cycle (US EPA NYCC) driving cycle.

2.2. Vehicle Dynamic Model

A vehicle dynamic model is applied to respond to the driving tractive effort and resistance needed for the simulation vehicle. The resistance included rolling resistance (R_r), aerodynamic resistance (R_a), grading resistance (R_c), and acceleration resistance R_s. The tractive effort for driving needed by the vehicle can be obtained with a vehicle dynamic model, which can be represented by Equation (1). The detailed calculation of resistance will be introduced in the following:

$$F_t = R_r + R_a + R_c + R_s \tag{1}$$

2.2.1. Rolling Resistance

During vehicle traveling, interaction force is produced in both radial and axial directions in the area where the wheels make contact with the ground, and there is also deformation between the tyre and the ground. The deformation process will be accompanied by a certain energy loss, regardless of whether or not it is in tyre or ground. This energy loss is the cause of rolling resistance during wheel turning. The rolling resistance can be represented by Equation (2), where μ_r is the rolling resistance coefficient, and W is the vehicle weight:

$$R_r = R_{r,A} + R_{r,B} = \mu_r \cdot W \tag{2}$$

2.2.2. Aerodynamic Resistance

The aerodynamic resistance can be represented by Equation (3) as follows, where C_D is the aerodynamic resistance coefficient, ρ is the air density, A_f is the front area of the vehicle, v is the vehicle speed, and v_w is the wind speed.

$$R_a = C_D \cdot \frac{\rho}{2} \cdot A_f \cdot (v - v_w)^2 \tag{3}$$

2.2.3. Grading Resistance

During climbing, grading resistance is produced due to the influence of the vehicle weight. During downhill, this resistance becomes the driving force instead. It can be represented by Equation (4), where θ is the slope angle:

$$R_c = W sin(\theta) \tag{4}$$

2.2.4. Acceleration Resistance

The vehicle driving state covers the acceleration and deceleration for most of the time, except on highways, where it is fixed-speed driving. The required force for acceleration can be represented by Equation (5), where a is the acceleration, and g is the gravity acceleration:

$$R_s = W \cdot \frac{a}{g} \tag{5}$$

2.3. Tyre and Drive Model

Vehicle dynamics is used to analyze the vehicle tyre model. The angular speed (ω_{drive}) and the torque (T_{drive}) of its transmission shaft can be represented by Equations (6) and (7), where GR is the final transmission gear ratio, r is the tyre radius, η_{fd} is the transmission efficiency, and F_{tire} is the tyre force.

$$\omega_{drive} = GR \cdot \frac{60}{2\pi \cdot r} \cdot v \tag{6}$$

$$T_{drive} = F_{tire} \cdot r \cdot \frac{\eta_{fd}}{GR} \tag{7}$$

2.4. Electric Motor Model

A 150 kW permanent magnetic motor was applied in this study. An efficient map of the motor was reproduced from Autonomie simulation software. In simulation, the motor efficiency can be obtained from a 2D look-up table through the efficiency curve shown in Figure 2, based on the motor torque and speed.

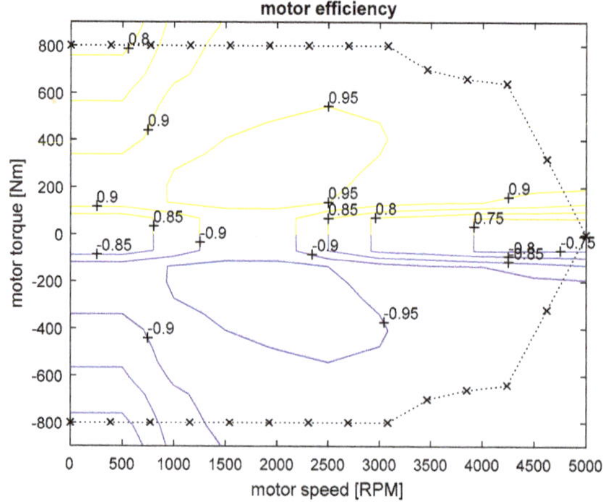

Figure 2. Efficiency of electric motor.

2.5. Variable Hydraulic Motor/Pump Model

Axial slope plate plunger type of hydraulic motor/pump is applied in this study, and its efficiency is obtained through a look-up table, as shown in Figure 3. The fluid flow rate ($Q_{P/M}$) and output torque ($T_{P/M}$) are calculated according to Equations (8) and (9), where $D_{P/M}$ is the maximum hydraulic motor displacement, N is the hydraulic motor speed, S_p is the plate angular position, $\eta_{vP/M}$ is the volumetric efficiency, $\Delta P_{P/M}$ is the pressure difference at the entry and exit, and $\eta_{tP/M}$ is the mechanical efficiency.

$$Q_{P/M} = D_{P/M} N S_p / (1000 \eta_{vP/M}) \tag{8}$$

$$T_{P/M} = (S_P \Delta P_{P/M} D_{P/M} \eta_{tP/M}) / 63 \tag{9}$$

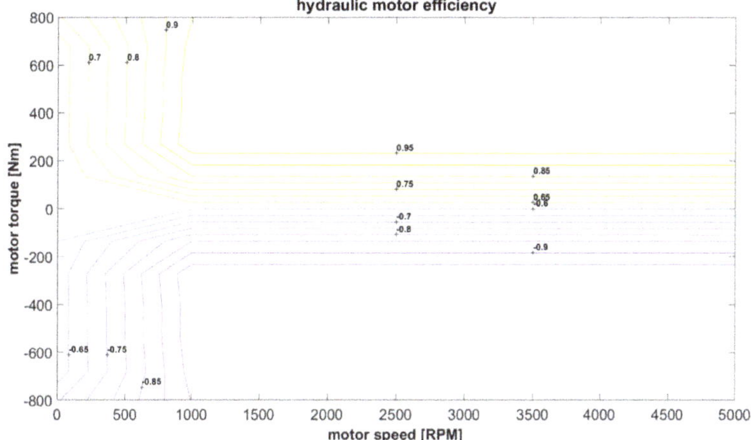

Figure 3. Efficiency of hydraulic motor/pump [1].

2.6. Battery Model

The battery used in this study was a lithium ion battery. An RC circuit design was applied, as shown in Equation (10), where V_t is the battery terminal voltage, V_{oc} is the battery open-circuit voltage, I_{bat} is the output current, and R_{int} is the internal resistance.

$$V_t = V_{oc} - I_{bat} \cdot R_{int} \tag{10}$$

Since the terminal voltage and the current can be measured, output of battery power P_{bat} can be received from Equation (11).

$$P_{bat} = I_{bat} \cdot V_{oc} \tag{11}$$

Equation (12) can be obtained by substituting Equation (10) to Equation (11).

$$I_{bat} = \frac{V_t - \left(V_t^2 - 4 \cdot R_{int} \cdot P_{bat}\right)^{0.5}}{2R_{int}} \tag{12}$$

The battery *SOC* is expressed by the capacity ampere hour. Since *SOC* changes with the current charging and discharging, *SOC* can be obtained from Equation (13), where SOC_{int} is the initial value of the battery.

$$SOC = SOC_{int} - \frac{\int_0^t I_{bat} dt}{Ah} \tag{13}$$

2.7. Accumulator Model

For the accumulator model, the influence due to temperature change was not considered in this study, so the gas-state change is set to be adiabatic process (rapid change, $n = 1.4$). The relationship between the pressure and volume is shown in Equations (14) and (15), during actual expansion and compression of gas.

$$PV^n = C \tag{14}$$

where P is pressure, V is volume of container area, and C is a constant value.

$$P_0 V_0^n = P_1 V_1^n = P_2 V_2^n = C \tag{15}$$

where P_0 is initial pressure, P_1 is the maximum activate pressure of accumulator, P_2 is the minimum activate pressure of accumulator, V_0 is the total volume of accumulator, V_1 is the volume of air in accumulator when the pressure is P_1, and V_2 is the volume of air in accumulator when the pressure is P_2.

The boundary movement work, W_b, of the accumulator can be expressed by Equation (16).

$$W_b = \int_1^2 PdV = P_1 V_1 \ln \frac{P_1}{P_2} \tag{16}$$

The inlet/outlet fluid, V_f, of the accumulator can be expressed by Equation (17).

$$V_f = V_1' - V_2'$$
$$= (V_0 - V_1) - (V_0 - V_2) = V_2 - V_1 = P_0^{\frac{1}{n}} V_0 \left\{ \left(\frac{1}{P_2} \right)^{\frac{1}{n}} - \left(\frac{1}{P_1} \right)^{\frac{1}{n}} \right\} = \left(\frac{P_0}{P_1} \right)^{\frac{1}{n}} V_0 \left\{ \left(\frac{P_1}{P_2} \right)^{\frac{1}{n}} - 1 \right\} \tag{17}$$

The accumulator *SOC* is expressed by the volume. Since the *SOC* changes with the volume flow rate, the accumulator *SOC* can be expressed by Equation (18).

$$SOC = SOC_{int} - \frac{\int_0^t Qdt}{V_f} \tag{18}$$

2.8. Vehicle Configurations

Two vehicles configurations were applied in this study for energy-efficiency comparison. The electric vehicle (EV) configuration is shown in Figure 4, and the HEHV configuration is presented in Figure 5.

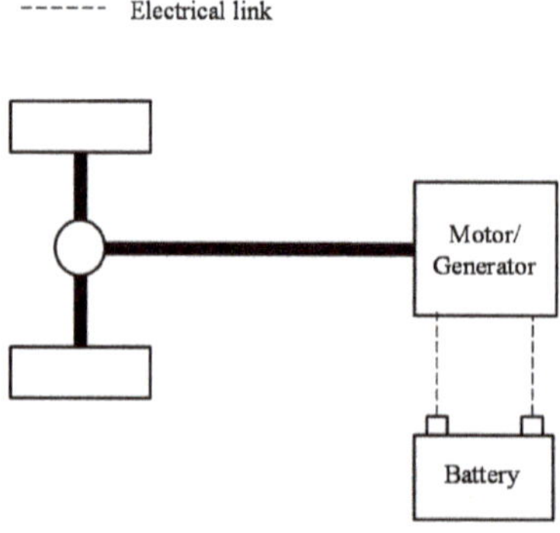

Figure 4. Electric vehicle (EV) configuration.

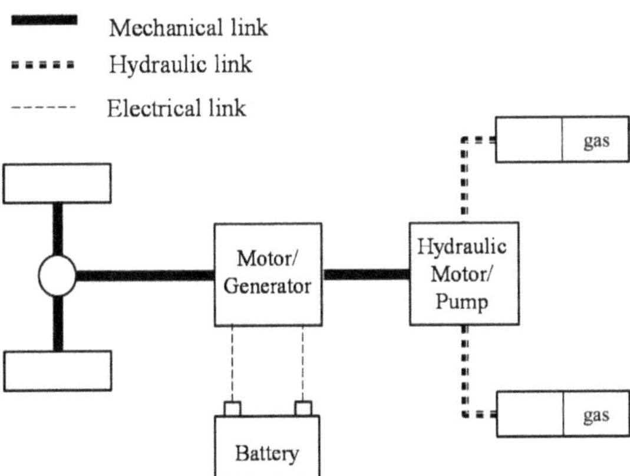

Figure 5. Hydraulic electric hybrid vehicle (HEHV) configuration.

3. Optimization Control Strategy

In this study, genetic algorithm (GA) was applied as the optimization function. The rule-based control was taken as the energy management strategy of HEHV first, and the simulation result was compared with the pure-electric-vehicle model. Then, the selected optimization approach was implemented for global optimization. From those results, together with a rule-based control approach, the optimal electricity economic performance was obtained.

The global optimization calculation was made by genetic algorithm. The objective of GA optimization was to minimize electricity consumption, and the objective function was set to be the reciprocal of the lithium ion battery's state of charge, SOL Li, as shown in Equation (19). The setting of objective function in GA could correspond to the fitness function, as shown in Equation (20). Parameters of GA set in this study are shown in Table 1.

$$\text{cost} = \min \left(1/\sum (SOC \text{ Li})\right) \tag{19}$$

$$\text{Fitness} = 1/\text{cost} \tag{20}$$

Table 1. Parameter settings of genetic algorithm (GA).

Gene Length	20 Bits (10 Bits for Both Acceleration and Accumulator *SOC*)
Group number	50
Algebra	40
Mating rate	0.9
Mutation probability	0.01

Two design variables (vehicle acceleration and accumulator *SOC*) were applied to judge the time to use the hydraulic system in control strategy. The thresholds of vehicle acceleration and accumulator *SOC* were set as selected variables *x* and *y* for optimization, respectively. If the vehicle acceleration was higher than the acceleration threshold and the accumulator *SOC* was higher than the accumulator threshold, the hydraulic pump provided the required power for vehicle acceleration. If the vehicle acceleration was lower than the acceleration threshold and the accumulator *SOC* was lower than the accumulator threshold, the electric motor provided the required acceleration power. If the vehicle acceleration was higher than the acceleration threshold and the accumulator *SOC* was lower than the accumulator threshold, the electric motor provided the major portion of required acceleration power. Some other detailed judgements of applying hydraulic pump and the overall control flow are shown in Figure 6. To prevent the calculation of variables x and y from exceeding the maximum component scope, the setting constraints of the variables are shown in the constraint Equations (21) and (22).

$$0 < x \leq 1 \text{(vehicle acceleration constraint)} \tag{21}$$

$$0 \leq y \leq 0.4 \text{(accumulator constraint)} \tag{22}$$

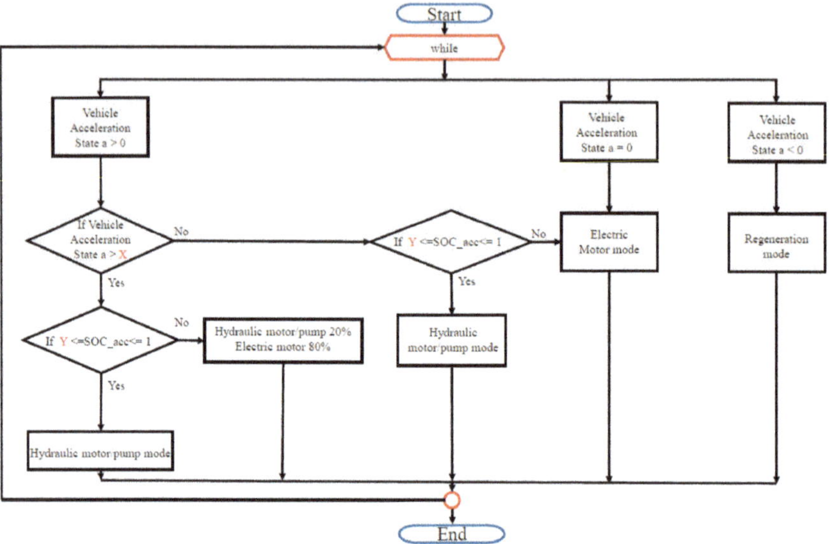

Figure 6. Control flow of genetic algorithm.

The fitness function was adapted to judge whether the solution of GA was suitable for the overall response of the hydraulic system. Values of acceleration threshold x and accumulator threshold y were recorded each time the GA was simulated. After the algorithm completed the iteration set of simulation, its fitness performance was looked up to ensure the value of fitness function was reasonable. The number of mutations and whether the optimization was convergence were checked during the operation of GA. In this study, the convergence of GA was judged by the difference of fitness values between the final four generations. If each difference was smaller than 1%, the optimization reached the convergence. From the solution of optimal fitness value, the recorded variables x and y were selected as the optimal set threshold. This set of variables could be implemented in rule-based control algorithm for real-time simulation and improve the energy consumption. With the implement of genetic algorithm, the rule-based control algorithm for real-time simulation could achieve the energy performance close to optimization.

The thresholds of vehicle acceleration and accumulator *SOC* calculated from the genetic algorithm were 0.9 and 0.1, respectively. The diagram of control strategy was modified, as shown in Figure 7.

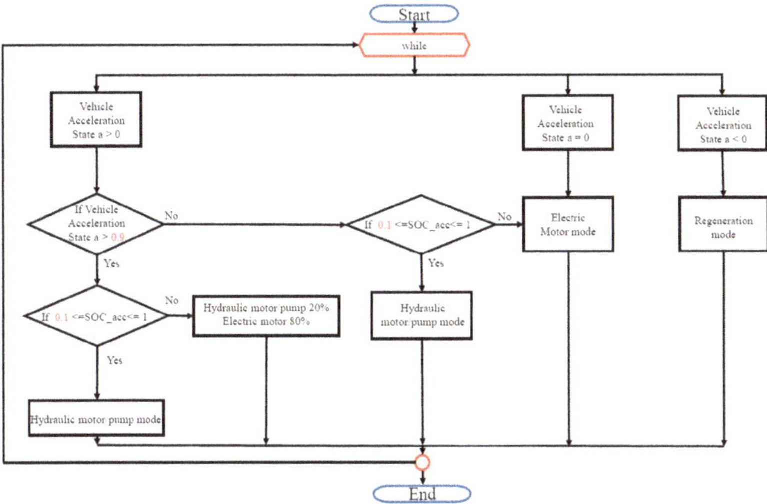

Figure 7. Updated control strategy from genetic algorithm.

4. Results

The vehicle parameters of the simulated vehicle are presented in Table 2. The mass of vehicle includes the gross weight, which is 7200 kg, and 20 passengers, which is 1600 kg.

Table 2. Vehicle parameters.

Parameter	Symbol	Unit	Value
Vehicle mass	W	kg	8800
Wheelbase		cm	378
Wheel track		cm	168
Front area	A_f	m^2	5.4
Aerodynamic drag coefficient	C_D		0.28
Rolling resistance coefficient	μ_r		0.008
Tire radius	r	m	0.334
Air density	ρ	kg/m^3	1.225
Gravitational acceleration	g	m/s^2	9.81
Final reduction gear ratio	GR		11.5
Hydraulic system weight		kg	200

In this section, simulation results of the pure electric vehicle and HEHV are compared, and that of the HEHV with optimized energy management strategy is explored. With the energy consumption of the NYCC driving cycle as the analysis basis, the difference of component performance is discussed. Firstly, the pure electric vehicle was established based on the set subsystem model, and it was taken as the basic model. Then the HEHV model was established based on the hydraulic components (hydraulic motor/pump and accumulator), and the rule-based control strategy was applied for the energy management of the power system. Finally, the rule-based control strategy was improved based on the results from the genetic algorithm, to get the HEHV with optimization energy management strategy.

4.1. EV vs. HEHV (Rule-Based)

This section compares the difference between the EV and HEHV and presents the causes of the differences. The operating points of the EV electric motor are presented in Figure 8, and those of HEHV electric motor are show in Figure 9. It is obvious that the HEHV electric motor does not work at heavy load and low speed, so it was replaced with a hydraulic motor/pump. Therefore, the operating efficiency points concentrate on the high-efficiency region, and the HEHV features better electricity economic performance than the pure electric vehicle.

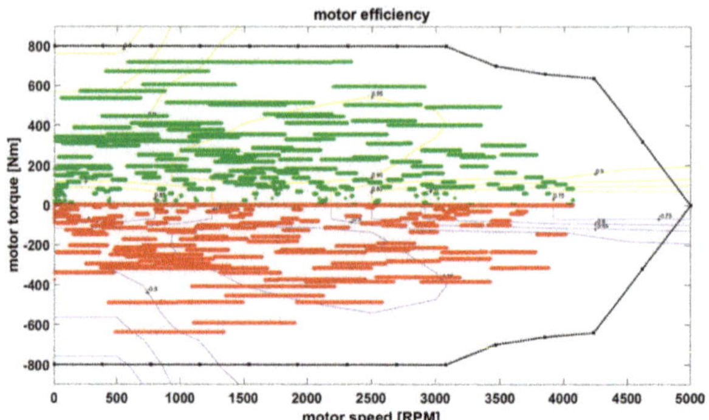

Figure 8. Electric-motor operating points of EV.

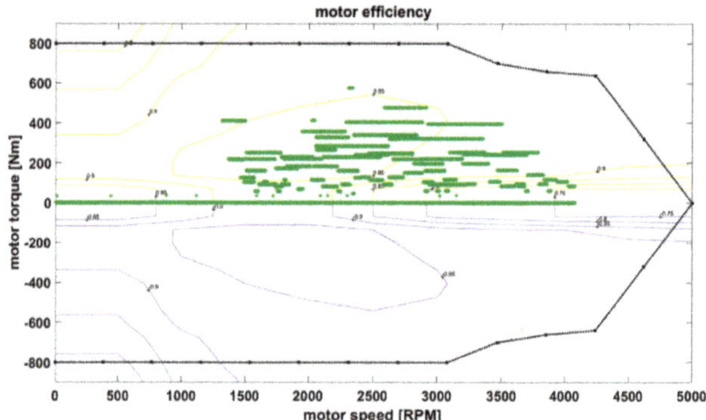

Figure 9. Electric motor operating points of HEHV (rule-based).

In order to better understand the reason that the HEHV has a better economic performance than the pure electric vehicle, the power of electric motors is compared. As shown in Figure 10, the power of HEHV electric motor is smaller than that of pure electric vehicle. Figure 11 shows the comparison of battery *SOC*, where the electricity economic performance of HEHV is greatly improved.

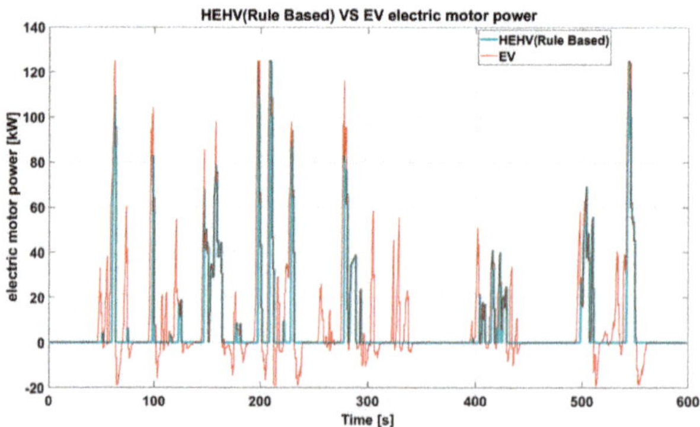

Figure 10. Electric motor power comparison of EV and HEHV.

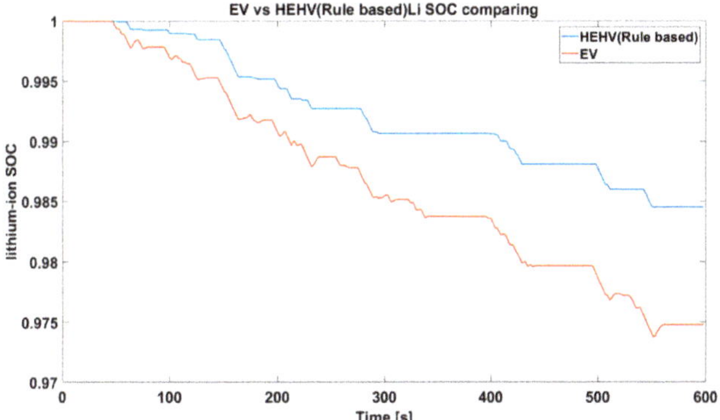

Figure 11. Battery *SOC* comparison of EV and HEHV.

Table 3 shows the electricity economic performance of the EV and HEHV (rule-based). Since a hydraulic motor functions as the drive at the HEHV's low speed, the use of an electric motor in the low-efficiency zone is reduced, and the electricity is optimized. The electricity economy of the HEHV has 36.5% improvement over that of EV.

Table 3. Comparison of electricity economic performance between the EV and HEHV (rule-based).

	Energy Consumption (kWh)	Electricity Economy (kWh /km)
Electric Vehicle (EV)	0.63	0.334
Hydraulic Electric Hybrid Vehicle, HEHV (Rule-Based)	0.40	0.212
Percent difference	+36.5%	+36.5%

4.2. HEHV (Rule-Based) vs. HEHV (GA)

This research had taken the genetic algorithm (GA), together with rule-based control, to perform global optimization, and it got the optimal electricity economic performance. In this section, the HEHV with original rule-based control is compared with the HEHV with modified rule-based control based on the genetic algorithm optimization. The distribution of operating points of the HEHV (rule-based) and HEHV (GA) electric motors is presented in Figures 12 and 13, respectively. The distribution suggests that the operating points of the electric motor after being modified for optimization concentrate more on the high-efficiency region.

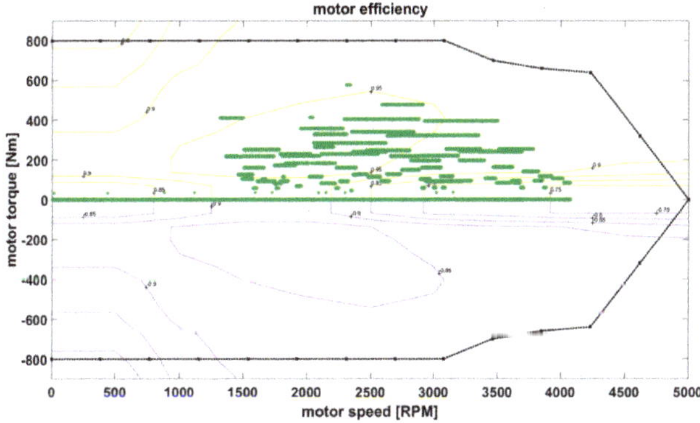

Figure 12. Efficiency points of the HEHV (rule-based) electric motor.

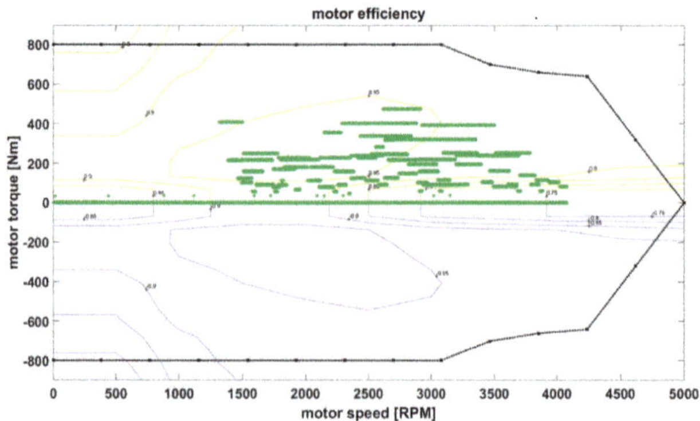

Figure 13. Efficiency points of the HEHV (GA) electric motor.

To understand the motor-use state, the power is compared in this paper, as shown in Figure 14. The usage rate of the electric motor after optimization is less than the original rule-based control strategy, so that better electricity economic performance is reached.

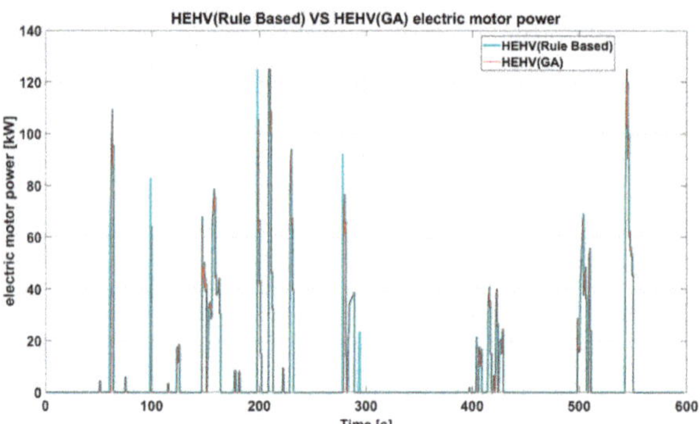

Figure 14. Power comparison between the HEHV (rule-based) and HEHV (GA) electric motors.

The reason why the electric-motor-usage rate after optimization is less can be explained with the help of a comparison of hydraulic motor power, as shown in Figure 15. It is found that the HEHV after optimization uses more hydraulic energy than the original control strategy.

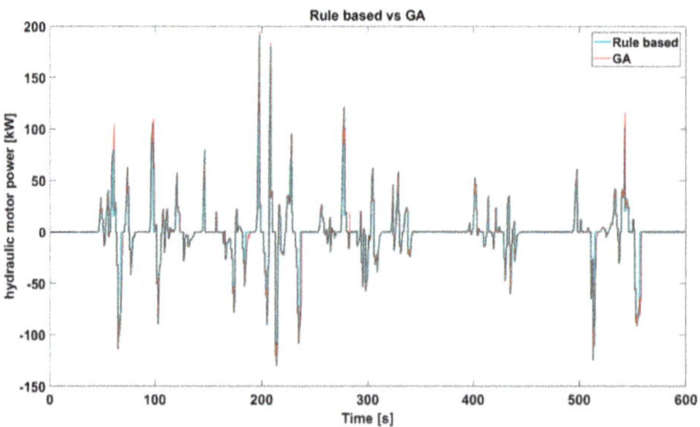

Figure 15. Comparison between HEHV (rule-based) and HEHV (GA) hydraulic motor power.

As suggested by the comparison of operating-point distribution of hydraulic motor/pump of two control strategies (Figure 16) and state of accumulator use (Figure 17), there are more operating points for the hydraulic motor/pump after optimization than for the original strategy, and they tend to be in the high-efficiency zone. The accumulator is applied more completely due to the wider range of applications for the hydraulic motor/pump.

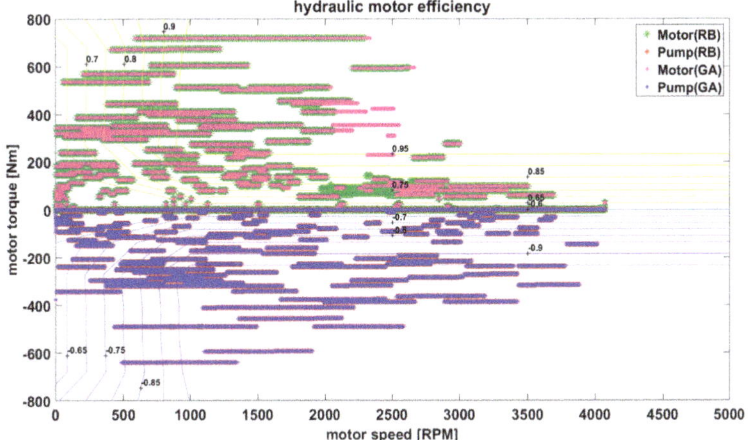

Figure 16. Comparison of operating-point distribution of hydraulic motor/pump (GA vs. rule-based).

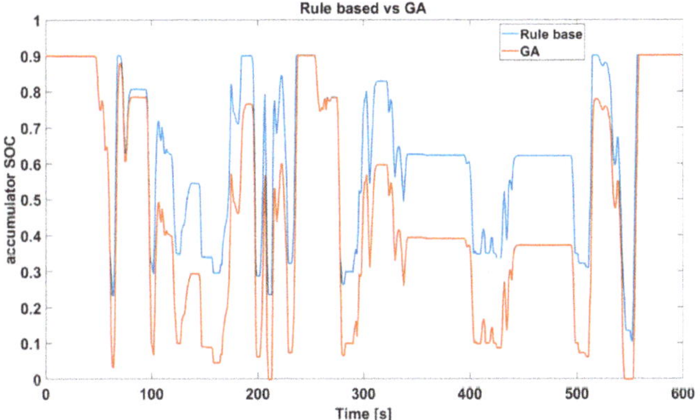

Figure 17. Comparison between HEHV (rule-based) and HEHV (GA) accumulator *SOC*.

The analysis above indicates that the electricity economic performance of the HEHV after optimization is more improved. Figure 18 shows the battery *SOC* comparison of the EV and HEHV (Rule based) and HEHV (GA). It is clear that the HEHV after optimization is more improved than the HEHV with original strategy.

The electricity economic performance of the HEHV (rule-based) and HEHV (GA) is drawn in Table 4. Table 5 shows the percentage improvement of electricity in this study. The HEHV with original rule-based control shows 36.5% improvement over the EV, and the HEHV with modified rule-based control has 43.7% improvement.

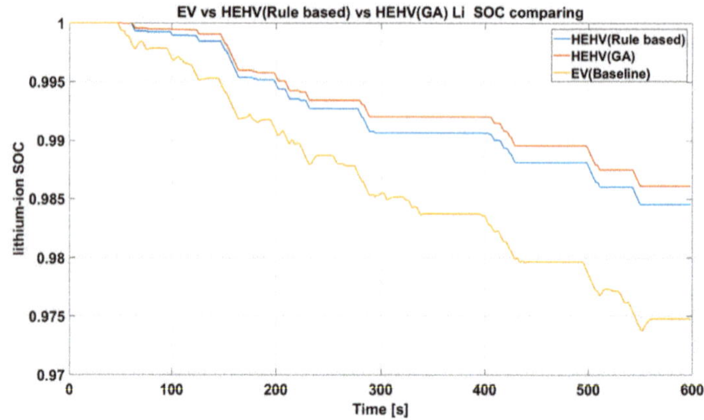

Figure 18. Battery *SOC* comparison between HEHV (Rule based), HEHV (GA), and EV.

Table 4. Comparison between electricity economic performances.

	Power Consumption (kWh)	Electricity Economy (kWh/km)
Hydraulic Electric Hybrid Vehicle, HEHV (Rule-Based)	0.40	0.212
Hydraulic Electric Hybrid Vehicle, HEHV (GA)	0.355	0.188
Percent difference	+11.3%	+11.4%

Table 5. Electricity improvement percentage.

	Percent of Improvement
Electric Vehicle (EV)	—
Hydraulic Electric Hybrid Vehicle, HEHV (Rule-Based)	36.5%
Hydraulic Electric Hybrid Vehicle, HEHV (GA)	43.7%

5. Discussion

This research mainly targeted the medium and large trucks for energy efficiency, and a hydraulic hybrid electric powertrain system was proposed to apply on the medium duty vehicle for energy efficiency. SimuLink simulation models of the EV and HHEV were built to evaluate the efficiency of the HHEV system. Compared to the EV system, the HHEV system had better energy efficiency, but the control algorithm was not optimized. To improve the efficiency, the genetic algorithm was implemented to achieve the optimized energy efficiency. Since GA was a global optimization algorithm which required longer CPU time for calculation and was not suitable for real-time control, the result of design variables of GA was required to apply on the real-time control strategy, rule-based control. In this research, two design variables of GA, thresholds of vehicle acceleration and hydraulic accumulator *SOC*, were optimized. These two optimized variables were applied in the HHEV simulation. From the simulation result, Tables 4 and 5, the rule-based model with GA could further improve the energy efficiency. The simulation results show that the electricity economic performance of the designed hydraulic hybrid vehicle was improved by 36.5% when compared to that of pure electric vehicle. The performance of energy consumption after genetic algorithm optimization was improved by 43.7%.

6. Conclusions

In this study, HEHV energy management strategy was applied, and Matlab/Simulink simulation program was utilized to establish a backward simulation model, to simulate the large-vehicle energy consumption. Movement situation of power components, *SOC* of energy storage components, and overall electricity economic performance of the pure electric vehicle and HEHV were obtained with the NYCC driving cycle. The following results can be achieved after simulation in this study:

- In the performance analysis of the pure electric vehicle, the electricity consumption of the set driving cycle is 0.334 kWh/km, and this is taken as the basic model for future comparison.
- In HEHV analysis (rule-based control), the electricity economic performance after simulation of the set driving cycle is 0.212 kWh/km, which is greatly improved than 0.334 kWh/km of pure electric vehicle, saving 36.5% of electricity. This is mainly because the hydraulic motor/pump in the pumping mode (energy recovery state) is more able to absorb, recover, and store the vehicle kinetic energy than electric motor, and the hydraulic motor/pump also avoids the application of electric motor at low speed.
- In the HEHV optimization analysis (genetic algorithm), 11.3% and 43.7% of electricity can be saved as compared with the HEHV (rule-based control) and pure electric vehicle, respectively.
- Through the HHEV simulation, the genetic algorithm was able to improve the energy efficiency of the HHEV by adjusting the chosen design variables of control strategy.

Author Contributions: Conceptualization, T.-S.L.; investigation, T.-S.L.; methodology, H.-Y.H.; project administration, J.-S.C.; software, J.-S.C.; validation, H.-Y.H. and J.-S.C. All authors have read and agreed to the published version of the manuscript.

Funding: This research received no external funding.

Conflicts of Interest: The authors declare no conflict of interest.

Nomenclature

A_f	front area of the vehicle
Ah	battery capacity (ampere hour)
C_D	aerodynamic resistance coefficient
$D_{P/M}$	maximum hydraulic motor displacement
F_t	vehicle tractive effort
F_{tire}	tyre force
GR	final transmission gear ratio
I_{bat}	battery output current
N	hydraulic motor speed
P	accumulator pressure
P_{bat}	battery power
$Q_{P/M}$	hydraulic pump fluid flow rate
R_a	aerodynamic resistance
R_c	grading resistance
R_{int}	battery internal resistance
R_r	rolling resistance
$R_{r,A}$	front wheel rolling resistance
$R_{r,B}$	rear wheel rolling resistance
R_s	acceleration resistance
S_p	hydraulic pump plate angular position
SOC	state of charge
SOC_{int}	initial value battery state of charge
SOL Li	lithium ion battery *SOC*
T_{drive}	tyre torque
$T_{P/M}$	hydraulic pump torque
V	volume of accumulator container area
V_f	accumulator inlet/outlet fluid

V_{oc}	battery open circuit voltage
V_t	battery terminal voltage
W	vehicle weight
W_b	accumulator boundary movement work
a	vehicle acceleration
g	gravity acceleration
r	tyre radius
v_w	wind speed
η_{fd}	transmission efficiency
$\eta_{tP/M}$	hydraulic pump mechanical efficiency
$\eta_{vP/M}$	hydraulic pump volumetric efficiency
θ	road slope angle
μ_r	rolling resistance coefficient
ρ	air density
ω_{drive}	tyre angular speed
$\Delta P_{P/M}$	hydraulic pump pressure difference at the entry and exit

References

1. Sokar, M. Investigation of Hydraulic Transmissions for Passenger Cars. Ph.D. Dissertation, Aachen University, Aachen, Germany, February 2011.
2. Chen, J.S. Energy efficiency comparison between hydraulic hybrid and hybrid electric vehicles. *Energies* **2015**, *8*, 4697–4723. [CrossRef]
3. Ramakrishnan, R.; Hiremath, S.S.; Singaperumal, M. Theoretical investigations on the effect of system parameters in series hydraulic hybrid system with hydrostatic regenerative braking. *J. Mech. Sci. Technol.* **2012**, *26*, 1321–1331. [CrossRef]
4. Salmasi, F.R. Control strategies for hybrid electric vehicles: Evolution, classification, comparison, and future trends. *IEEE Trans. Veh. Technol.* **2007**, *56*, 2393–2404. [CrossRef]
5. Lu, X.; Chen, Y.; Wang, H. Multi-objective optimization based real-time control for PEV hybrid energy management systems. In Proceedings of the IEEE Applied Power Electronics Conference and Exposition (APEC), San Antonio, TX, USA, 4–8 March 2018; pp. 969–975.
6. Zeng, Y.; Cai, Y.; Kou, G.; Gao, W.; Qin, D. Energy management for plug-in hybrid electric vehicle based on adaptive simplified-ECMS. *Sustainability* **2018**, *10*, 2060. [CrossRef]
7. Wang, X.; He, H.; Sun, F.; Zhang, J. Application study on the dynamic programming algorithm for energy management of plug-in hybrid electric vehicles. *Energies* **2015**, *8*, 3225–3244. [CrossRef]
8. Yu, J.; Liu, N.; Zhang, Y.; Wang, B. Modeling and control strategy simulation of extended-range electric vehicle. In Proceedings of the 2011 International Conference on Transportation, Mechanical, and Electrical Engineering (TMEE), Changchun, China, 16–18 December 2011; pp. 829–832.
9. Gao, J.P.; Zhu, G.M.G.; Strangas, E.G.; Sun, F.C. Equivalent fuel consumption optimal control of a series hybrid electric vehicle. *Proc. Inst. Mech. Eng. Part D J. Automob. Eng.* **2009**, *223*, 1003–1018. [CrossRef]
10. Konev, A.; Lezhnev, L.; Kolmanovsky, I. *Control Strategy Optimization for a Series Hybrid Vehicle*; SAE Paper No. 2006-01-0663; SAE World Congress: Detroit, MI, USA, 2006.
11. Liu, X.; Fan, Q.; Zheng, K.; Duan, J.; Wang, Y. Constant SOC control of a series hybrid electric vehicle with long driving range. In Proceedings of the IEEE International Conference on Information and Automation, Harbin, China, 11–13 August 2010; pp. 1603–1608.
12. Li, S.G.; Sharkh, S.M.; Walsh, F.C.; Zhang, C.N. Energy and battery management of a plug-in series hybrid electric vehicle using fuzzy logic. ieee trans. *Veh. Technol.* **2011**, *60*, 3571–3585. [CrossRef]
13. Ho, P.; Klong, E. *Intelligent Energy Distribution for Series HEVs Using Determined Optimal Driving Patterns via a Genetic Algorithm*; SAE Paper No. 2013-01-0572; Graduate Faculty of North Carolina State University: Raleigh, NC, USA, 2013.
14. Xu, Q.; Luo, X.; Jiang, X.; Zhao, M. Research on double fuzzy control strategy for parallel hybrid electric vehicle based on GA and DP optimization. *IET Electr. Syst. Transp.* **2018**, *8*, 144–151. [CrossRef]

15. Kaur, J.; Saxena, P.; Gaur, P. Genetic algorithm based speed control of hybrid electric vehicle. In Proceedings of the 2013 Sixth International Conference on Contemporary Computing, Noida, India, 8 August 2013; pp. 65–69.
16. Hu, F.; Zhao, Z. Optimization of control parameters in parallel hybrid electric vehicles using a hybrid genetic algorithm. In Proceedings of the 2010 IEEE Vehicle Power and Propulsion Conference, Lille, France, 1–3 September 2010; pp. 1–6.

Article

A Step Up/Down Power-Factor-Correction Converter with Modified Dual Loop Control

Yi-Hung Liao

Department of Electrical Engineering, National Central University, Taoyuan City 32001, Taiwan; yhlmliao@gmail.com; Tel.: +886-3-4227151 (ext. 35153)

Received: 5 November 2019; Accepted: 29 December 2019; Published: 1 January 2020

Abstract: A step up/down AC/DC converter with modified dual loop control is proposed. The step up/down AC/DC converter features the bridgeless characteristic which can reduce bridge-diode conduction losses. Based on the step up/down AC/DC converter, a modified dual loop control scheme is proposed to achieve input current shaping and output voltage regulation. Fewer components are needed compared with the traditional bridge and bridgeless step up/down AC/DC converters. In addition, the intermediate capacitor voltage stress can be reduced. Furthermore, the top and bottom switches still have zero-voltage turn-on function during the negative and positive half-line cycle, respectively. Hence, the thermal stresses can also be reduced and balanced. Simulation and experimental results are provided to verify the validity of the proposed step up/down AC/DC converter and its control scheme.

Keywords: bridgeless; AC/DC converter; power factor correction; zero-voltage switching (ZVS)

1. Introduction

Power factor correction is very popular and necessary for modern power sources in the ac grid. It decreases line current harmonics, line losses, and increases system power capacity due to reducing system reactive power flow [1–3]. Today, boost rectifiers are the most commonly-used circuit structures implemented for power factor correction. However, some consumer electronic devices, portable devices and server power applications [4,5] require lower dc voltage level than the main ac voltage source. The dc output voltage in boost rectifiers is always higher than the peak value of the main input ac voltage. Therefore, in low dc voltage level applications, another dc-dc step-down converter is necessary that follows the boost rectifier to form a two-stage structure as shown in Figure 1. Because of the two-stage structure, power efficiency may degrade and the total number of components in the system is increased. Thus, the efficiency, cost, and volume of the two-stage power conversion system are not a good choice and need to be improved.

Step-down PFC rectifiers, such as buck converters are therefore considered. However, the buck rectifier input current is discontinuous. A dead angle also exists when the line input voltage is lower than the output voltage so that the input current cannot be easily shaped [6–8]. As a result, the step up/down AC-DC topologies are developed including buck-boost, Cuk, and Sepic type rectifiers [9–11]. The buck-boost rectifier also has inherent discontinuous input current like the buck converter, and needs an additional filter to smooth the input current. Although the Sepic rectifier has continuous input current, the output current is still discontinuous and easily causes output voltage ripples.

Bridgeless rectifier topologies are explored in [12,13] to reduce the diode bridge conduction losses and increase the conversion efficiency. The bridgeless PFC boost rectifiers, such as the dual boost rectifier and the totem-pole boost rectifier, have been discussed [14]. Due to the need for lower output voltage applications, the bridgeless Cuk/Sepic rectifiers [15,16] with two dc/dc Cuk/Sepic circuit structures were proposed. The bridgeless Cuk rectifier [16] is shown in Figure 2. However, four diodes are still needed to achieve step up/down output voltage. Other bridgeless Sepic [17] and Cuk [18] power-factor-correction rectifiers were also proposed with reduced number of components and conduction losses. These rectifiers were operated in discontinuous conduction mode without current loop control. A control method for bridgeless Cuk/Sepic power factor correction rectifier operated in continuous conduction mode was also proposed to achieve power decoupling [19]. Although, the bulky electrolytic capacitor can be replaced with a small film capacitor, this control method requires an extra voltage sensor for the intermediate capacitor and the system cost is increased.

Pulsating power buffering technology [8,20,21] has recently expanded, which can reduce the number of components including passive and active ones. Although rectifiers using pulsating power buffering technology have high power density, high conversion efficiency and high reliability, high voltage stress is still present in the switches and diodes [22], which leads to high switching and conduction losses and reduces the rectifier life-span.

This paper proposes a bridgeless Cuk rectifier with modified dual loop control scheme. The voltage stresses in the switches and diodes can be adjusted to low voltage levels by the proposed control scheme, which may reduce the switching and conduction losses and increase the rectifier life-span. The detailed operation principle and switching sequence of the bridgeless Cuk rectifier are explained. Simultaneously, a modified dual loop control scheme is also proposed to achieve input current shaping and output voltage regulation as well as voltage stress reduction.

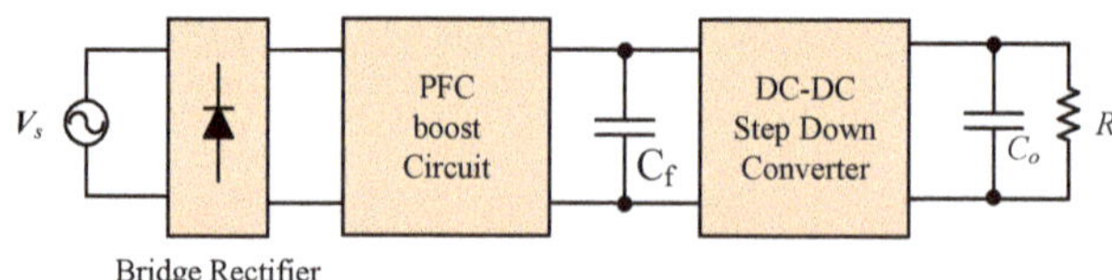

Figure 1. Two-stage AC/DC conversion structure.

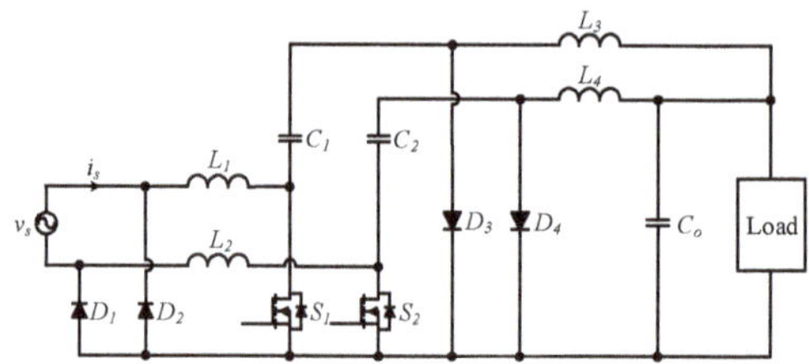

Figure 2. Bridgeless Cuk rectifier [16].

2. Circuit Topology and Switching Sequence

The bridgeless Cuk converter [19] discussed in this paper is shown in Figure 3. The proposed control switching sequence and key waveforms in one switching period during the positive and negative half line cycle are shown in Figure 4. For convenience of discussion the active switches are assumed to be ideal active switches with anti-paralleling body diode. Both the input inductor L_s and output inductor L_o are assumed to be operated in continuous conduction mode. The circuit operation can be divided into three operation states in one switching period T for both positive and negative half-line cycles. The circuit operation principle of the bridgeless Cuk converter during the positive half-line cycle is discussed first, as follows:

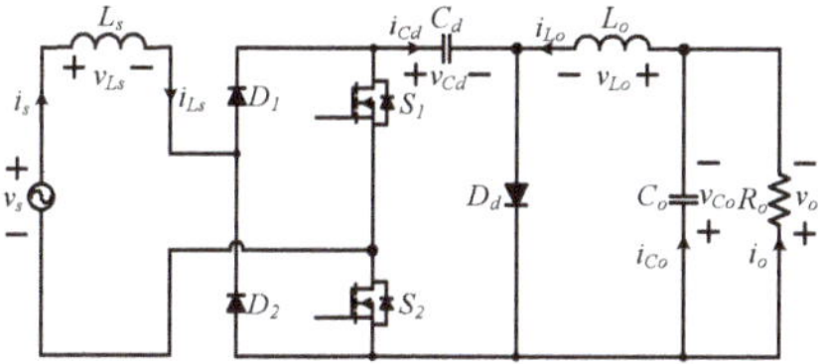

Figure 3. Bridgeless Cuk converter [19].

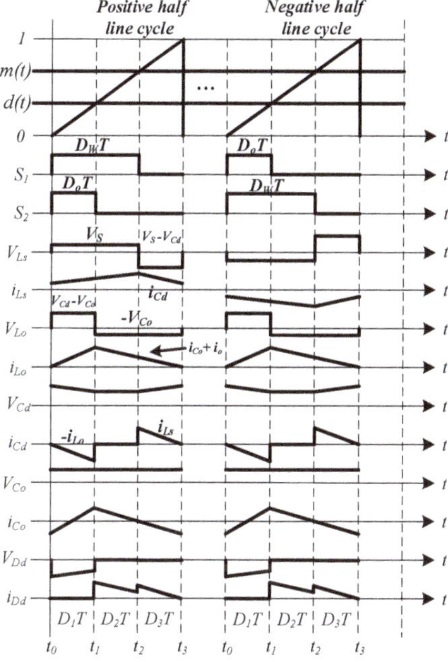

Figure 4. Control switching sequence and key waveforms in one switching period.

(1) State 1 ($t_0 \leq t < t_1$): In this state, as shown in Figure 5, both switches S_1 and S_2 are turned on. The zero-voltage switching of S_2 is obtained due to body diode conducting in switch S_2 in the pre-state, i.e., State 3. The input inductor L_s is magnetized by the input voltage V_s so as to increase the inductor current i_{Ls}. The inductor current i_{Ls} flows through diode D_1 and switch S_1 and goes back to the main ac source. Simultaneously, the intermediate capacitor C_d releases energy to the output inductor L_o and load. The equivalent circuit equations are described as Equations (1)–(4).

$$L_s \frac{di_{Ls}}{dt} = v_s,$$ (1)

$$L_o \frac{di_{Lo}}{dt} = v_{Cd} - v_o,$$ (2)

$$C_d \frac{dv_{Cd}}{dt} = -i_{Lo},$$ (3)

$$C_o \frac{dv_o}{dt} = i_{Lo} - \frac{v_o}{R_o},$$ (4)

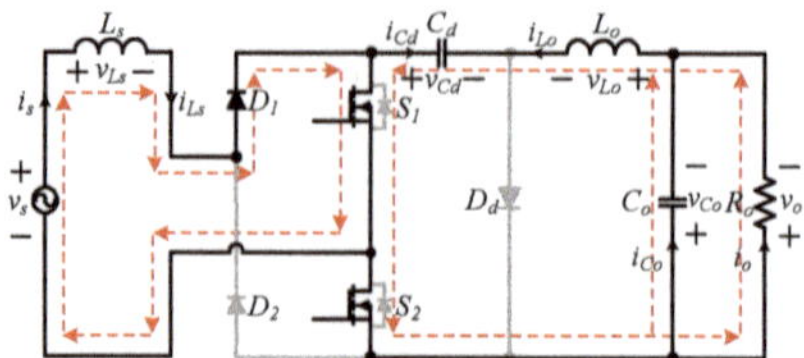

Figure 5. Equivalent circuit of the bridgeless Cuk converter in State 1 during positive half line cycle.

(2) State 2 ($t_1 \leq t < t_2$): In this state, as shown in Figure 6, switch S_1 is turned on and switch S_2 is turned off. Switch current i_{ds1} is increasing. Input inductor L_s is still magnetized by the input voltage V_s so as to increase the inductor current i_{Ls} which still flows through diode D_1 and switch S_1 and then goes back to the main ac source. The voltage of intermediate capacitor C_d remains constant. Simultaneously, the output inductor L_o is demagnetized and releases energy to the load through the diode D_d. The equivalent circuit equations are expressed as Equations (5)–(8).

$$L_s \frac{di_{Ls}}{dt} = v_s, \tag{5}$$

$$L_o \frac{di_{Lo}}{dt} = -v_o, \tag{6}$$

$$C_d \frac{dv_{Cd}}{dt} = 0, \tag{7}$$

$$C_o \frac{dv_o}{dt} = i_{Lo} - \frac{v_o}{R_o}, \tag{8}$$

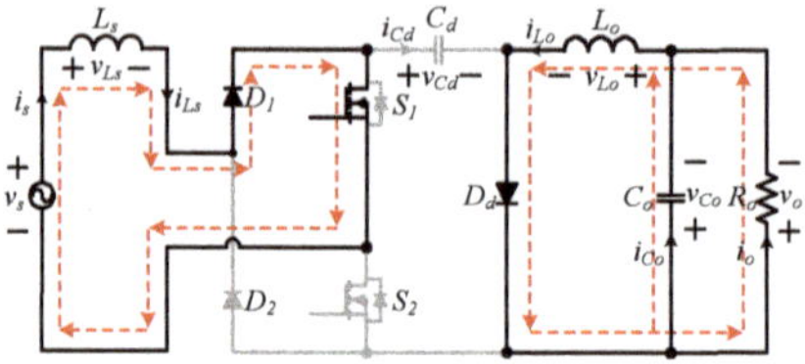

Figure 6. Equivalent circuit of the bridgeless Cuk converter in State 2 during positive half line cycle.

(3) State 3 ($t_2 \leq t < t_3$): In this state, as shown in Figure 7, switch S_1 is turned off and S_2 is also turned off. Input inductor L_s is demagnetized by the voltage $-(V_{cd}-V_s)$ so as to decrease the inductor current i_{Ls} which flows through diodes D_1 and D_d, and the body diode of switch S_2 and goes back to the main ac source. The intermediate capacitor C_d is charged by the input inductor current i_{Ls}. Simultaneously, the output inductor L_o still releases energy to the load through diode D_d. The equivalent circuit equations are given by Equations (9)–(12).

$$L_s \frac{di_{Ls}}{dt} = v_s - v_{Cd}, \tag{9}$$

$$L_o \frac{di_{Lo}}{dt} = -v_o, \tag{10}$$

$$C_d \frac{dv_{Cd}}{dt} = i_{Ls}, \tag{11}$$

$$C_o \frac{dv_o}{dt} = i_{Lo} - \frac{v_o}{R_o}, \tag{12}$$

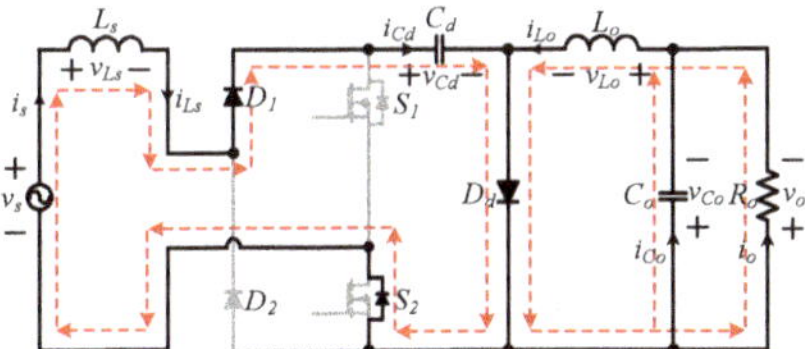

Figure 7. Equivalent circuit of the bridgeless Cuk converter in State 3 during positive half line cycle.

Referring the gate signals shown in Figure 4, while the bridgeless Cuk converter is operated during the negative half line cycle, the circuit operation principle in the proposed control switching sequence can be described as follows:

(1) State 1 ($t_0 \leq t < t_1$): In this state, as shown in Figure 8, both the switches S_1 and S_2 are turned on. The zero-voltage switching of S_1 is obtained due to body diode conducting in switch S_1 in the pre-state, i.e., State 3. The input inductor L_s is magnetized by the input voltage V_s so as to increase the inductor current i_{Ls} in the inverse direction. The inductor current i_{Ls} flows through diode D_2 and switch S_2 and goes back to the main ac source. Simultaneously, the intermediate capacitor C_d releases energy to the output inductor L_o and load. The equivalent circuit equations are described as Equations (13)–(16).

$$L_s \frac{di_{Ls}}{dt} = v_s, \tag{13}$$

$$L_o \frac{di_{Lo}}{dt} = v_{Cd} - v_o, \tag{14}$$

$$C_d \frac{dv_{Cd}}{dt} = -i_{Lo}, \tag{15}$$

$$C_o \frac{dv_o}{dt} = i_{Lo} - \frac{v_o}{R_o}, \tag{16}$$

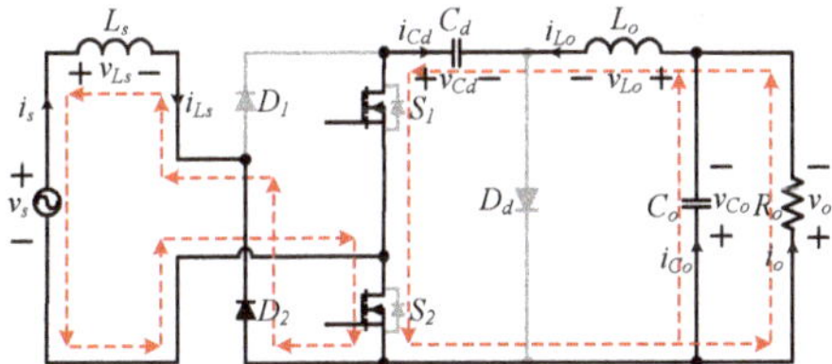

Figure 8. Equivalent bridgeless Cuk converter circuit in State 1 during negative half line cycle.

(2) State 2 ($t_1 \leq t < t_2$): In this state, as shown in Figure 9, switch S_2 is turned on and switch S_1 is turned off. The switch current i_{ds2} is increasing. Input inductor L_s is still magnetized by the input voltage V_s so as to increase the inductor current i_{Ls} in the inverse direction which still flows through diode D_2 and switch S_2 and then goes back to the main ac source. The intermediate capacitor C_d voltage remains constant. Simultaneously, the output inductor L_o is demagnetized and releases energy to the load through diode D_d. The equivalent circuit equations are expressed as Equations (17)–(20).

$$L_s \frac{di_{Ls}}{dt} = v_s, \tag{17}$$

$$L_o \frac{di_{Lo}}{dt} = -v_o, \tag{18}$$

$$C_d \frac{dv_{Cd}}{dt} = 0, \tag{19}$$

$$C_o \frac{dv_o}{dt} = i_{Lo} - \frac{v_o}{R_o}, \tag{20}$$

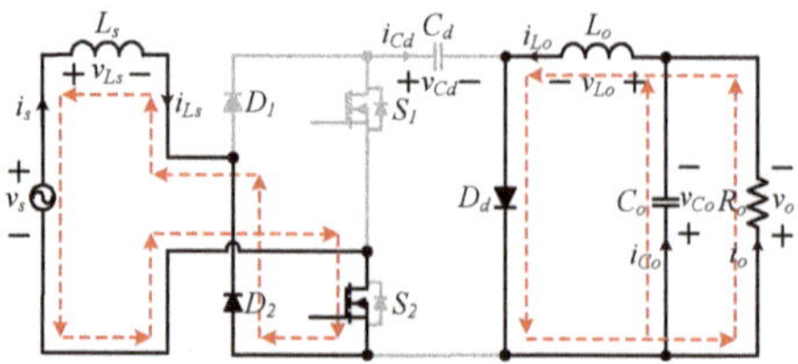

Figure 9. Equivalent circuit of the bridgeless Cuk converter in State 2 during negative half line cycle.

(3) State 3 ($t_2 \leq t < t_3$): In this state, as shown in Figure 10, switch S_2 is turned off and S_1 is also turned off. Input inductor L_s is demagnetized in the inverse direction by the voltage ($V_{cd} + V_s$) so as to decrease the inductor current i_{Ls} which flows through diodes D_2, D_d and the body diode of switch S_1 and goes back to the main ac source. The intermediate capacitor C_d is charged by the input inductor current i_{Ls} in the inverse direction. Simultaneously, the output inductor L_o still releases energy to the load through diode D_d. The equivalent circuit equations are given by Equations (21)–(24).

$$L_s \frac{di_{Ls}}{dt} = v_s + v_{Cd}, \tag{21}$$

$$L_o \frac{di_{Lo}}{dt} = -v_o, \tag{22}$$

$$C_d \frac{dv_{Cd}}{dt} = -i_{Ls}, \tag{23}$$

$$C_o \frac{dv_o}{dt} = i_{Lo} - \frac{v_o}{R_o}, \tag{24}$$

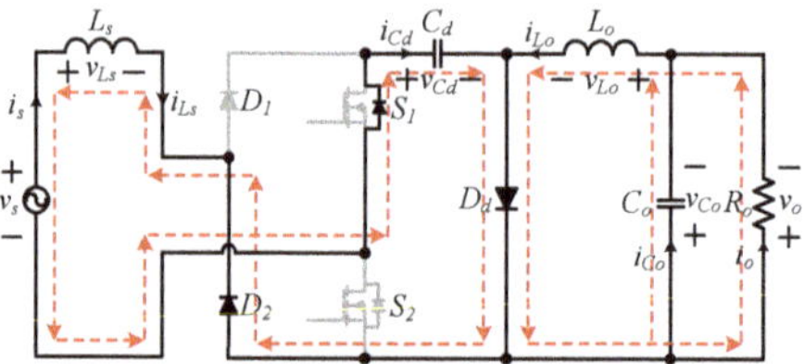

Figure 10. Equivalent circuit of the bridgeless Cuk converter in State 3 during negative half line cycle.

To further reveal the potential merits of the proposed step up/down converter with modified dual loop control, Table 1 is provided to summarize comparisons for the bridge Cuk [11], bridgeless Cuk [16], and the proposed step up/down converter with modified dual loop control. It is worth mentioning that the power levels of the three converters in Table 1 are all at small power levels like the fly-back converter. Although the control methods may be different, the harmonics of the three converters all meet the IEC61000-3-2 Class D standard.

Table 1. Comparisons of step up/down converters.

Topology	Bridge Cuk [11]	Bridgeless Cuk [16]	Proposed Step Up/Down Converter with Control
Control	Dual Loop	Dual Loop	Modified Dual Loop
Switch	1	2	2
Diode	5	4	3
Inductor	2	4	2
Capacitor	2	3	2
Total Number of Components	10	13	9
Voltage Gain	$\frac{v_o}{\|v_s\|} = \frac{D}{(1-D)}$	$\frac{v_o}{\|v_s\|} = \frac{D}{(1-D)}$	$\frac{v_o}{\|v_s\|} = \frac{D_o}{(1-D_w)}$
Voltage stresses of switches	v_o/D	v_o/D	v_o/D_o
D or D_o	one solution	one solution	Multiple solutions
Harmonics	meet the standard	meet the standard	meet the standard

3. Control Scheme and Parameter Design

3.1. Control Scheme

According to the circuit analysis in the previous section, assume the duty ratio $D_W = D_1 + D_2$ and $D_0 = D_1$. While the main ac voltage is operating in the positive half line cycle $v_s > 0$, by utilizing state-space averaged technique and flux balance theory in the input inductor L_s and output inductor L_o, one can obtain the equations

$$v_{Cd} = \frac{v_S}{(1 - D_W)}, \tag{25}$$

$$v_{Cd} = \frac{v_o}{D_o}, \tag{26}$$

Similarly, while the main ac voltage is operating in the negative half line cycle $v_s < 0$, the corresponding symmetrical equations can also be obtained as

$$v_{Cd} = \frac{-v_S}{(1 - D_W)}, \tag{27}$$

$$v_{Cd} = \frac{v_o}{D_o}, \tag{28}$$

Merging Equations (25)–(28) in both the positive and negative half line cycles of the main ac voltage, the voltage gain of the bridgeless Cuk converter is obtained as

$$\frac{v_o}{|v_S|} = \frac{D_o}{(1 - D_W)}, \tag{29}$$

As can be observed from Equation (29), the output voltage is related to the two parameters D_o and D_W. If the input and output voltages are given, infinite different kinds of solutions exist in the Equation (29). However, in the same operation condition for the conventional dual loop control scheme shown in Figure 11, only one solution is obtained, i.e., $D_o = D_W$. Therefore, in order to reduce the voltage stresses of all switches and diodes in the circuit, the conventional dual loop control scheme is not suitable.

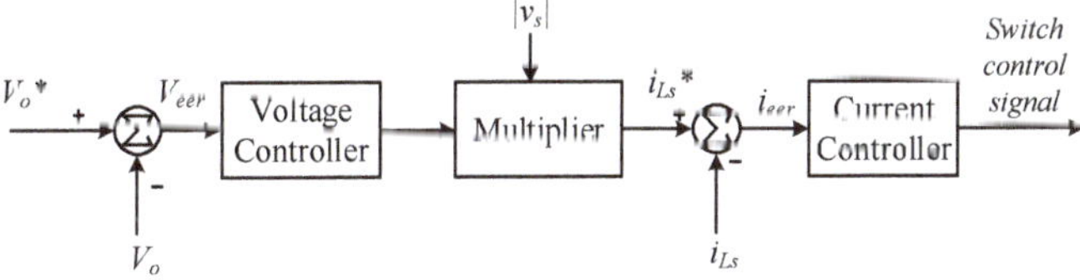

Figure 11. Conventional dual loop control scheme.

A modified dual loop control scheme is proposed. The proposed control scheme for the bridgeless Cuk converter is shown in Figure 12. The actual input current i_{Ls} compared with the current command $i_{Ls}{}^*$ to generate the current error as the input of the current controller and then produce the control signal V_{Dw}. The actual output voltage V_o compared with the output voltage command $V_o{}^*$ generates the voltage error as the voltage controller input. The voltage controller generates the current command amplitude and also the control signal V_{Do}. In the conventional dual loop control scheme, only one control signal is produced to achieve both input current shaping and output voltage regulation. In the proposed control scheme, two control signals V_{Dw} and V_{Do} are produced to control the input current shaping and output voltage regulation. Thus, the intermediate capacitor voltage is not fixed and can be adjusted to fit a better low voltage level. Hence, the intermediate capacitor voltage stress could be reduced and the adopted electrolytic capacitor life span could also be increased. According to the circuit analysis in Section 2, the voltage stresses of active switches S_1 and S_2, diodes D_1, D_2, and D_d are clamped and equal to the intermediate capacitor voltage. The average switching power loss P_s in one switching period caused by transitions can be defined as

$$P_s = 0.5 V_{DS} I_{DS} [t_{c(on)} + t_{c(off)}], \tag{30}$$

where $t_{c(on)}$ and $t_{c(off)}$ are the turn-on and turn-off crossover intervals, respectively. For simplification, the switches are operated in the same turn-on and turn-off crossover intervals and at the same switching frequency f_s. The average switching power loss is then proportional to the voltage across the switch V_{DS} and the entire current I_{DS} which flows through the switch as

$$P_s \propto V_{DS} I_{DS}, \tag{31}$$

According to the above equation, if the intermediate capacitor voltage is adjusted to fit a better low voltage level, the average switching power loss is also reduced. This is also true for the diodes. Therefore, the total losses in semiconductor devices can be reduced and the efficiency can be lifted.

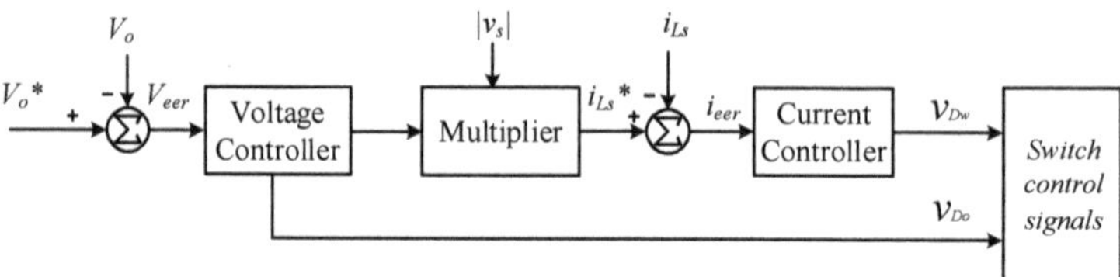

Figure 12. Proposed modified dual loop control scheme.

3.2. Parameter Design

To verify the feasibility of the proposed step up/down AC/DC converter with modified dual loop control, a parameter design for inductor and capacitor is discussed. In order to find the boundary between the continuous and discontinuous modes for input inductor L_s, one can find that the critical value of K_1 at boundary between modes, $K_{crit}(D_w)$, is function of duty cycle D_w and can be expressed as

$$K_1 > K_{crit}(D_w), \text{ where } K_1 = \frac{2L_s}{R_o T_s} \text{ and } K_{crit}(D_w) = \frac{(1 - D_w)^2}{D_w} \tag{32}$$

The critical value $K_{cirt}(D_w)$ is plotted vs. duty cycle D_w in Figure 13. Consider inductor L_s is operated in CCM and the switching frequency is f_s. The maximum input current ripple is less than 25% of the fundamental current. The minimum input inductor L_s value can be derived by the equation

$$L_s \geq \frac{v_{s,\max}}{0.25 \cdot \Delta i_{Ls,BCM}} \cdot \frac{D_W}{f_s}, \tag{33}$$

where $\Delta i_{Ls,BCM}$ is the input current ripple while inductor L_1 is operated in BCM. Consider that inductor L_o is operated in BCM and one can find that the critical value for K_2 at the boundary between modes, $K_{crit}(D_o)$, is function of the duty cycle D_o and can be expressed as

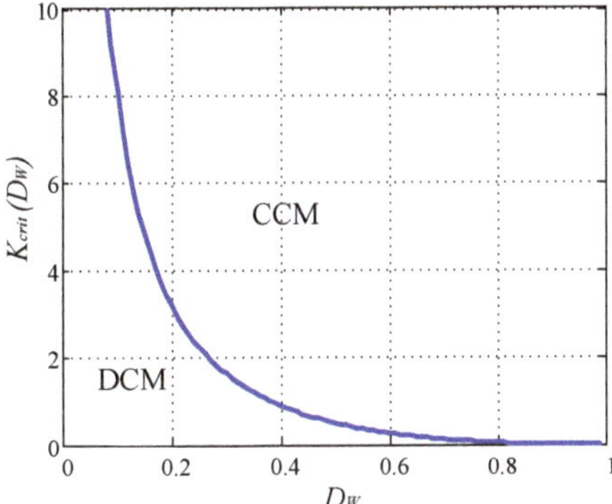

Figure 13. Proposed step up/down AC/DC converter K_{cirt} (D_w) vs. D_w.

$$K_2 > K_{crit}(D_o), \text{ Where } K_2 = \frac{2L_o}{R_o T_s} \text{ and } K_{crit}(D_o) = \frac{1 - D_o}{2} \tag{34}$$

The critical value K_{cirt} (D_o) is plotted vs. duty cycle D_o in Figure 14. Similarly, the minimum value of inductor Lo also can be derived as

$$L_o \geq \frac{v_{Cd,max}}{\Delta i_{Lo,BCM}} \cdot \frac{D_o}{f_s}, \tag{35}$$

where $\Delta i_{Lo,BCM}$ is the output current ripple while inductor L_o is operated in BCM.

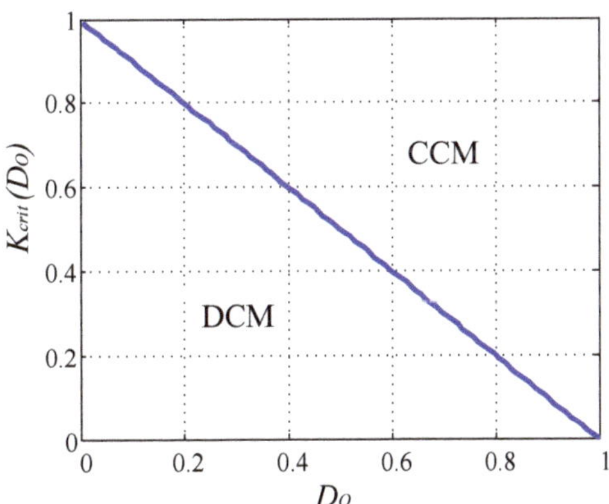

Figure 14. Proposed step up/down AC/DC converter K_{cirt} (D_o) vs. D_o.

Consider the output capacitor and assume the switching ripple is neglected. The output capacitor must be large enough to minimize the output ripple because the output voltage ripple frequency is twice the input line frequency. The output filter capacitor can be determined by

$$C_o = \frac{P_o}{\omega V_o (2\Delta V_o)}, \tag{36}$$

where ΔV_o is the output voltage ripple and ω is the input line angular frequency.

4. Simulation and Experimental Results

To verify the validity of the bridgeless step up/down AC/DC converter, some simulation results are executed and a prototype system is constructed to facilitate the theoretical results as verification. The simulation and experimental parameters are listed in Table 2. The input voltage is the AC grid with 110 V_{rms} and 60 Hz fundamental frequency. The controlled output voltage is 48 V and the load is 48 Ω. The assigned output power rating is 48 W. The simulation results for the input voltage V_s, input current i_s and the corresponding intermediate capacitor voltage V_{cd} are shown in Figure 15. It follows from Figure 15 that the input current shaping can be achieved. Figure 16 shows the switching control signals for switch S_1 and S_2 and the corresponding voltage and current of switch S_2 during the positive half-line cycle. As can be seen from Figure 16, the ZVS turn-on of switch S_2 is obtained during the positive half-line cycle. Similarly, Figure 17 shows the switching control signals for switch S_1 and S_2 and the corresponding voltage and current of switch S_1 during the negative half-line cycle. It also can be seen from Figure 17 that the ZVS turn-on of switch S_1 is obtained during the negative half-line cycle.

Consider that the load is a dynamic load and/or RL load such as a dc motor whose armature winding resistance is Ra = 0.5 Ω, armature winding inductance is La = 0.5 mH, back electromotive force is 47 V. Figure 18 shows the simulation results for the input voltage, input current and the corresponding intermediate capacitor voltage. As can be observed from Figure 18, the output power is about 120 W and the power factor correction is also achieved. Hence, the proposed converter can indeed be operated in the RL load. Consider the intermediate capacitor voltage which can be adjusted using the control signal V_{Do} based on Equations (26) and (28). Figure 19 shows the simulation results for the input voltage and the corresponding input current, and the control signal V_{Do} and the corresponding intermediate capacitor voltage V_{Cd} under the low control signal V_{Do} value. Figure 20 shows the same simulated condition under the high control signal V_{Do} value. It can be seen from Figures 19 and 20 that the lower the control signal V_{Do} value, the higher the intermediate capacitor voltage V_{Cd}. That the duty ratio Do affects the intermediate capacitor voltage level and also the voltage stresses of the switches and diodes in the circuit is very important information. This also implies that the duty ratio Do affects the converter power losses and efficiency. Finally, to facilitate understanding of the proposed step up/down converter with modified dual loop control and as verification, a prototype is constructed with a TMS320F28335 digital signal processor (DSP). The experimental hardware construction block diagram is shown in Figure 21. Figures 22 and 23 show the experimental results for the switching control signals and the corresponding voltage and current of switches S_2 and S_1 during positive and negative half-line cycles, respectively. As can be observed from Figures 22 and 23, the ZVS soft switching of switches S_2 and S_1 were indeed achieved and agreed with the simulation results. The measured harmonic distribution of the input current is shown in Figure 24. One can find that the measured harmonic currents meet the IEC 61000-3-2 Class D harmonic standards.

In order to understand the total harmonic distortion THD_i of the input currents in the three converters listed in Table 1, the PSIM software is adopted to carry out the simulation. The input voltage is 110Vrms, the output voltage is controlled at 48 V and the load is 2 A. The corresponding parameters and simulated results are shown in Table 3. As can be seen from Table 3, the input current THD_i of the bridge Cuk [11] is better than that of the bridgeless Cuk [16] and the proposed Cuk with modified dual loop control scheme. Nevertheless, the parameter value of the bridge Cuk input inductor [11] is larger than those for the other two. Although the bridge Cuk [11] has the smallest input current THD_i, the input inductor may make it appear bulky.

Table 2. Parameters of the bridgeless Cuk converter for simulation and experimentation.

Parameters	Value
Input Inductor L_s	1.5 mH
Output Inductor L_o	50 uH
Intermediate Capacitor C_d	5 uF
Output Capacitor C_o	470 uF
Switching frequency f_s	50 kHz

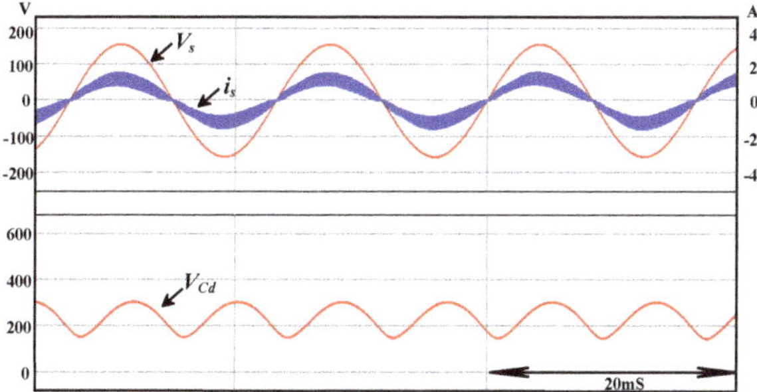

Figure 15. Simulation results for (top) the input voltage V_s, current i_s, and (bottom) corresponding intermediate capacitor voltage.

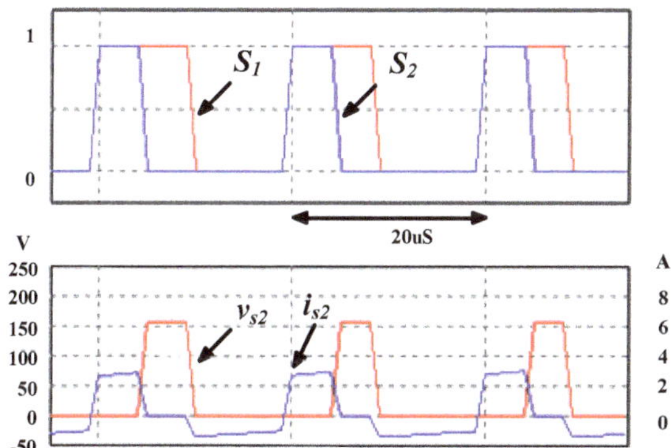

Figure 16. Simulation results for (top) switching control signals and (bottom) corresponding voltage and current of switch S_2 during positive half line cycle.

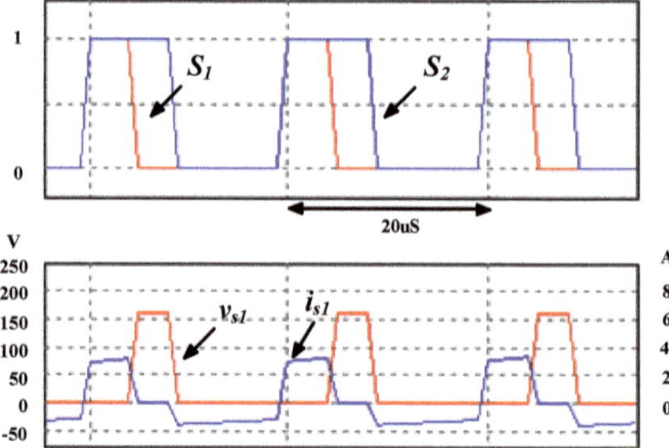

Figure 17. Simulation results for (top) switching control signals and (bottom) corresponding voltage and current of switch S_1 during negative half line cycle.

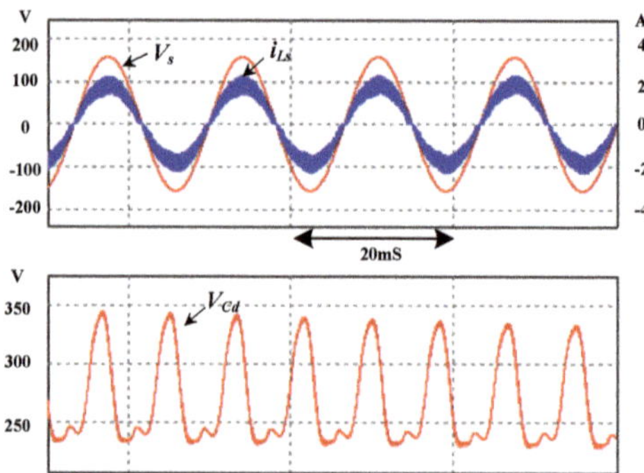

Figure 18. Simulation results for (top) the input voltage V_s, current i_s, and (bottom) corresponding intermediate capacitor voltage while the load is a dc motor.

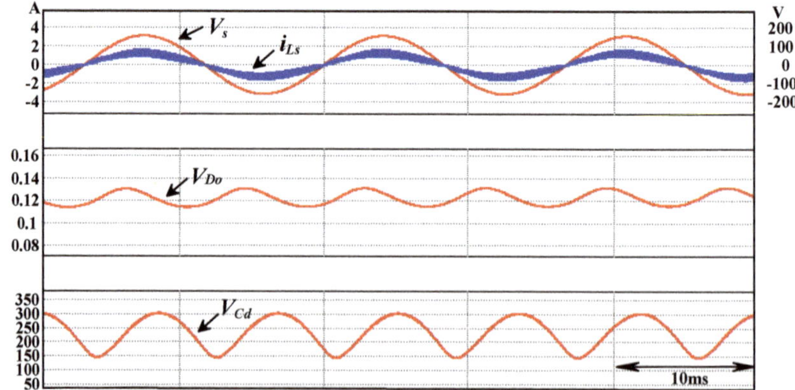

Figure 19. Simulation results for (top) the input voltage V_s, current i_s, (middle) the control signal V_{Do} with low parameter value, and (bottom) corresponding intermediate capacitor voltage.

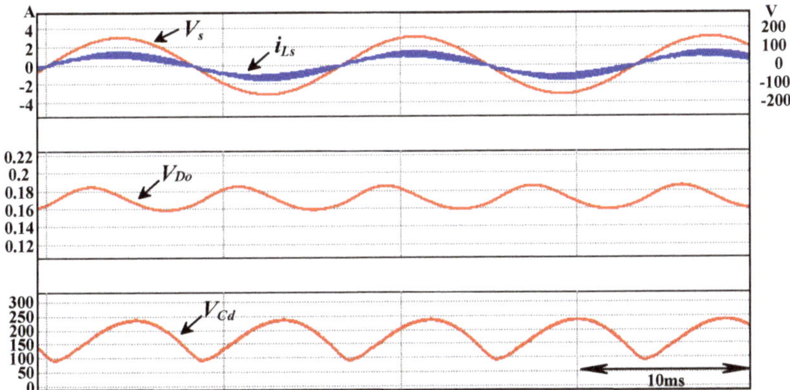

Figure 20. Simulation results for (top) the input voltage V_s, current i_s, (middle) the control signal V_{Do} with high parameter value, and (bottom) corresponding intermediate capacitor voltage.

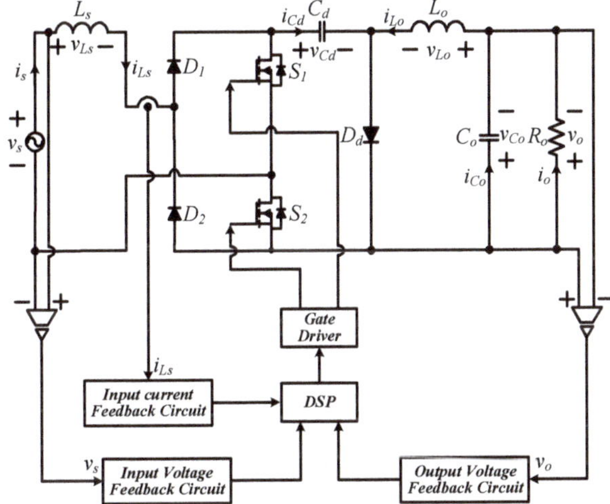

Figure 21. Experimental hardware construction block diagram.

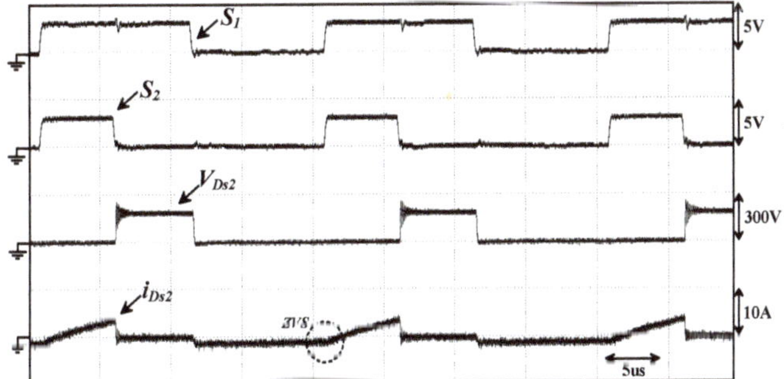

Figure 22. Experimental results for (top) switching control signals S_1, S_2 and (bottom) corresponding voltage and current of switch S_2 during positive half line cycle.

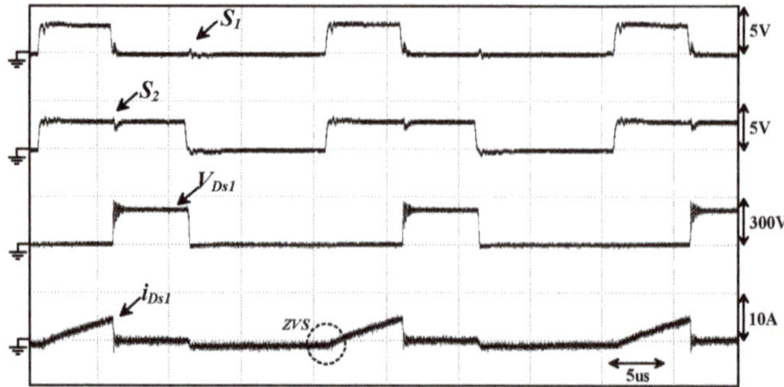

Figure 23. Experimental results for (top) switching control signals S_1, S_2 and (bottom) corresponding voltage and current of switch S_1 during negative half line cycle.

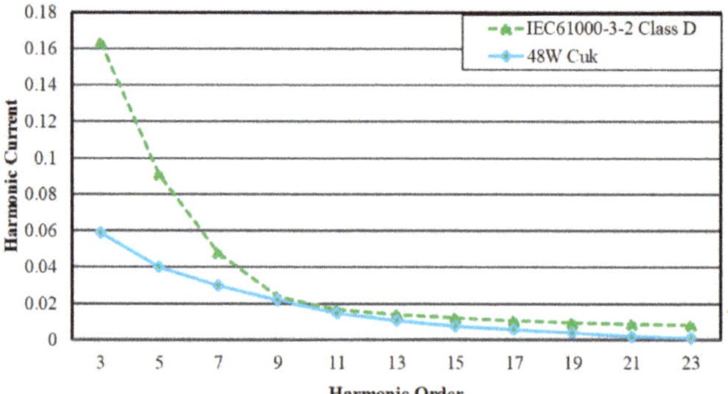

Figure 24. The measured harmonic distribution of the input current compared with IEC61000-3-2 Class D standard.

Table 3. Comparisons of the total harmonic distortion of the step up/down converters.

Parameters	Bridge Cuk [11]	Bridgeless Cuk [16]	Proposed Step Up/Down Converter with Control
Input inductor	6.4 mH	1mH × 2	1.5 mH
Output inductor	206 uH	22uH × 2	50 uH
Intermediate capacitor	0.61 uF	1uF × 2	5 uF
THD$_i$ of Input current	5.6%	15.2%	13.3%

5. Conclusions

This paper presented a bridgeless step up/down converter with modified dual loop control scheme. The proposed system has ZVS soft switching in switches S_1 and S_2 during the negative and positive half-line cycle operation, respectively. Thus, the switching losses can be reduced and the thermal stress can be balanced between switches S_1 and S_2. There are fewer components compared to the bridge Cuk and the bridgeless dual Cuk configuration. Therefore, the size and cost can be reduced. In addition, based on the proposed control scheme, the voltage stresses of the intermediate capacitor, active switches, and diodes can all be reduced. To verify the validity of the proposed step up/down converter, simulation, and experimental results are offered. From simulation and experimental results, the proposed bridgeless step up/down converter can indeed achieve input current shaping and output voltage regulation as well as reduce the switching and conduction losses.

Author Contributions: The author contributed to the theoretical analysis, modeling, simulation, experiment, and manuscript preparation. The author have read and agreed to the published version of the manuscript.

Funding: This research is sponsored by the Ministry of Science and Technology of R.O.C. under grant MOST 108-3116-F-008-001.

Conflicts of Interest: The authors declare no conflict of interest.

References

1. Bodetto, M.; El Aroudi, A.; Cid-Pastor, A.; Calvente, J.; Martinez-Salamero, L. Design of ac–dc PFC high-order converters with regulated output current for low-power applications. *IEEE Trans. Power Electron.* **2016**, *31*, 2012–2025. [CrossRef]
2. Chang, C.H.; Cheng, C.A.; Chang, E.C.; Cheng, H.L.; Yang, B.E. An integrated high-power-factor converter with ZVS transition. *IEEE Trans. Power Electron.* **2016**, *31*, 2362–2371. [CrossRef]
3. Kim, J.S.; Lee, S.H.; Cha, W.J.; Kwon, B.H. High-Efficiency Bridgeless Three-Level Power Factor Correction Rectifier. *IEEE Trans. Ind. Electron.* **2017**, *64*, 1130–1136. [CrossRef]
4. Dusmez, S.; Choudhury, S.; Bhardwaj, M.; Akin, B. A Modified Dual-Output Interleaved PFC Converter Using Single Negative Rail Current Sense for Server Power Systems. *IEEE Trans. Power Electron.* **2014**, *29*, 5116–5123. [CrossRef]
5. Li, Y.C.; Lee, F.C.; Li, Q.; Huang, X.; Liu, Z. A novel AC-to-DC adaptor with ultra-high power density and efficiency. In Proceedings of the 2016 IEEE APEC Conference and Exposition, Long Beach, CA, USA, 20–24 March 2016.
6. Jang, Y.; Jovanović, M.M. Bridgeless buck PFC rectifier. In Proceedings of the 2010 IEEE APEC Conference and Exposition, Palm Springs, CA, USA, 21–25 February 2010.
7. Huber, L.; Gang, L.; Jovanovic, M.M. Design-Oriented Analysis and Performance Evaluation of Buck PFC Front End. *IEEE Trans. Power Electron.* **2010**, *25*, 85–94. [CrossRef]
8. Ohnuma, Y.; Itoh, J.I. A Novel Single-Phase Buck PFC AC-DC Converter with Power Decoupling Capability Using an Active Buffer. *IEEE Trans. Ind. Appl.* **2014**, *50*, 1905–1914. [CrossRef]
9. Firnao Pires, V.; Fernando Silva, J. Half-bridge single-phase buck-boost type AC-DC converter with sliding-mode control of the input source current. *IEE Proc. Electr. Power Appl.* **2000**, *147*, 61–67. [CrossRef]
10. Abdelsalam, I.; Adam, G.P.; Holliday, D.; Williams, B.W. Single-stage, single-phase, ac–dc buck–boost converter for low-voltage applications. *IET Power Electron.* **2014**, *7*, 2496–2505. [CrossRef]
11. Simonetti, D.S.L.; Sebastian, J.; Uceda, J. The discontinuous conduction mode Sepic and Cuk power factor preregulators: Analysis and design. *IEEE Trans. Power Electron.* **1997**, *44*, 630–637. [CrossRef]
12. Mitchell, D.M. AC-DC Converter Having an Improved Power Factor. U.S. Patent 4,412,277, 25 October 1983.
13. Choi, W.Y.; Kwon, J.M.; Kim, E.H.; Lee, J.J.; Kwon, B.H. Bridgeless boost rectifier with low conduction losses and reduced diode reverse-recovery problems. *IEEE Trans. Ind. Electron.* **2007**, *54*, 769–780. [CrossRef]
14. Huber, L.; Jang, Y.; Jovanovic, M.M. Performance evaluation of bridgeless PFC boost rectifiers. *IEEE Trans. Power Electron.* **2008**, *23*, 1381–1390. [CrossRef]
15. Sabzali, A.J.; Ismail, E.H.; Al-Saffar, M.A.; Fardoun, A.A. New Bridgeless DCM Sepic and Cuk PFC Rectifiers with Low Conduction and Switching Losses. *IEEE Trans. Ind. Appl.* **2011**, *47*, 873–881. [CrossRef]
16. Fardoun, A.A.; Ismail, E.H.; Sabzali, A.J.; Al-Saffar, M.A. New Efficient Bridgeless Cuk Rectifiers for PFC Applications. *IEEE Trans. Power Electron.* **2012**, *27*, 3292–3301. [CrossRef]
17. Mahdavi, M.; Farzanehfard, H. Bridgeless SEPIC PFC Rectifier with Reduced Components and Conduction Losses. *IEEE Trans. Ind. Electron.* **2011**, *58*, 4153–4160. [CrossRef]
18. Mahdavi, M.; Farzaneh-Fard, H. Bridgeless CUK power factor correction rectifier with reduced conduction losses. *IET Power Electron.* **2012**, *5*, 1733–1740. [CrossRef]
19. Liu, Y.; Sun, Y.; Su, M. A Control Method for Bridgeless Cuk/Sepic PFC Rectifier to Achieve Power Decoupling. *IEEE Trans. Ind. Electron.* **2017**, *64*, 7272–7276. [CrossRef]
20. Wang, H.; Chung, H.S.H.; Liu, W. Use of a series voltage compensator for reduction of the dc-link capacitance in a capacitor-supported system. *IEEE Trans. Power Electron.* **2014**, *29*, 1163–1175. [CrossRef]

Energies **2020**, *13*, 199

21. Qin, S.; Lei, Y.; Barth, C.; Liu, W.C.; Pilawa-Podgurski, R.C.N. A high power density series-stacked energy buffer for power pulsation decoupling in single-phase converters. *IEEE Trans. Power Electron.* **2017**, *32*, 4905–4924. [CrossRef]

22. Qi, W.; Li, S.; Yuan, H.; Tan, S.; Hui, S. High-Power-Density Single-Phase Three-Level Flying-Capacitor Buck PFC Rectifier. *IEEE Trans. Power Electron.* **2019**, *34*, 10833–10844. [CrossRef]

 *energies*

Article

The Optimal Control of Fuel Consumption for a Heavy-Duty Motorcycle with Three Power Sources Using Hardware-in-the-Loop Simulation

Chien-Hsun Wu * and Yong-Xiang Xu

Department of Vehicle Engineering, National Formosa University, Yunlin 63201, Taiwan;
10658109@gm.nfu.edu.tw
* Correspondence: chwu@nfu.edu.tw

Received: 9 November 2019; Accepted: 18 December 2019; Published: 19 December 2019

Abstract: This study presents a simulation platform for a hybrid electric motorcycle with an engine, a driving motor, and an integrated starter generator (ISG) as three power sources. This platform also consists of the driving cycle, driver, lithium-ion battery, continuously variable transmission (CVT), motorcycle dynamics, and energy management system models. Two Arduino DUE microcontrollers integrated with the required circuit to process analog-to-digital signal conversion for input and output are utilized to carry out a hardware-in-the-loop (HIL) simulation. A driving cycle called worldwide motorcycle test cycle (WMTC) is used for evaluating the performance characteristics and response relationship among subsystems. Control strategies called rule-based control (RBC) and equivalent consumption minimization strategy (ECMS) are simulated and compared with the purely engine-driven operation. The results show that the improvement percentages for equivalent fuel consumption and energy consumption for RBC and ECMS using the pure software simulation were 17.74%/18.50% and 42.77%/44.22% respectively, while those with HIL were 18.16%/18.82% and 42.73%/44.10%, respectively.

Keywords: heavy-duty motorcycle; simulation platform; energy management system; equivalent consumption minimization strategy (ECMS); hardware in-the-loop; signal processing

---

1. Introduction

Hybrid electric vehicles received much attention in the past few decades due to urgent concerns about the emission of carbon and exhaust gases. The reasons are that, on the one hand, the energy efficiency of the entire powertrain can be enhanced so that the emission of carbon can be reduced; on the other hand, the purely electric drive and smoothness of engine operation during city driving can effectively reduce the emission of exhaust gases [1]. The architectures of hybrid electric vehicles (HEVs) can be divided into two types: series and parallel HEVs [2]. The series operation can solve the problems of low efficiency and high fuel consumption caused by cold starting and the preheating process of engines [3]. Furthermore, ultracapacitors can be used as energy storage systems. Their capacities can affect the times of engine starting. The kinetic energy of vehicles can be restored using regenerative braking. Therefore, the comfort and system efficiency of the integrated powertrain can be improved [2]. Furthermore, if the reactively controlled compression ignition (RCCI) engines can be implemented into a series HEV, not only can the feature of good fuel economy be promoted, but the advantages of low emissions of nitrogen oxides (NO_x) and particulate matter (PM) can also be gained [4]. As for the parallel HEVs, the flexibility of utilizing multiple energy storage systems and electric power systems possesses the potential of further reducing fuel consumption, as well as emissions of pollutants. Moreover, compared with conventional vehicles, the energy management strategies can be designed through optimal numerical analysis without sacrificing their drivability [5]. However, the relatively complicated architectures of parallel HEVs result in a higher cost of manufacture and maintenance [6]. Unlike conventional vehicles with internal combustion engines, one of the effective ways to improve fuel economy and emissions of pollutants is utilizing the optimal strategy of energy management [7]. An efficient strategy of energy management for real-time control can be designed through a simulation on an entire power system. This is further modified according to effective engineering experiences. Consequently, the resulting performances of the designed vehicle can be significantly improved especially when dealing with some rather complicated architectures of powertrain [8]. Currently, when facing an urgent impact of ISO 26262, global industries of vehicles, such as factories of original equipment manufacturing (OEM), tier 1 components, vehicular chips, and development tools, started implementing ISO 26262 in the processes of product development or adjusting the software/hardware development tools to meet the requirements of ISO 26262 [9]. To efficiently simplify the extremely high-level and complicated tasks of vehicle dynamic computations, the required core technology will definitely involve low-cost and high-efficiency embedded systems that are integrated with software and hardware and capable of customization of the basic architecture for computing according to the specific functionality [10]. Currently, a real-time simulation coupled with the hardware-in-the-loop (HIL) platform is widely applied in industry [11,12]. A performance evaluation on a hybrid electric system called a molten carbonate fuel cell/micro gas turbine (MCFC-MGT) using an HIL simulation was proven to be an effective means of saving development cost [13]. Furthermore, when performing tests on flexible alternating current transmission systems (FACTSs) under either conventional or newly revised industry specifications, the HIL simulations are regarded as an efficient testing method prior to the actual field tests [14].

With regard to past studies, the main goal to optimize synergy/electricity systems is enhancing the energy efficiency and performance of an entire system for fuel saving coupled with cost reduction. Accordingly, two major factors are the strategy of energy management and design of the synergy/electricity powertrain. For the optimization of the former, especially on HEVs, one of the widely applied control methods is rule-based control (RBC) [15]. Its advantages are easy implementation, highly efficient computation, and fast experimental verification. However, its inherent propensity to evaluate by engineering intuition showed its weakness when dealing with rather complicated and highly nonlinear systems. To overcome this and enhance robustness due to system uncertainty, fuzzy logic control (FLC) was introduced, and it is especially applicable to various types of HEVs [16,17]. Sunddararaj et al. proposed the proportional/integral and fuzzy logic control strategies for direct current DC/DC converters. This hybrid controller further improved the performance of the bi-directional DC/DC converter and the power circuits for voltage gain, filtering, system efficiency, and electricity quality [18]. However, for systems with considerable numbers of control variables requiring a mass of logical rules for effective control, a global search using equivalent consumption minimization strategies (ECMS) is adopted for optimization in a more numerically precise manner [19,20]. Multi-dimensional look-up tables derived from the simulated results based on the ECMS concept can be constructed through computer coding and directly downloaded to the control unit to realize the practical implementation of optimal energy management in the synergy/electricity powertrains. This research can be employed for other types of electric vehicles as long as the characteristics of energy/power sources are built and the objective function (goal) is set. The major advantage of EMCS is to accommodate both the theoretical optimization analysis and the direct usage for the control unit because of the resulting matrix. The drawback is that, compared with the time-independent or predictive control laws (such as dynamics programming), the ECMS optimization cannot handle the system constraints (such as the start-to-end balance for state of charge). However, to deal with the unknown out loads and the complicated system dynamics, it is one of the best candidates for the optimization [21]. Other algorithms are introduced in the paragraph below.

The dynamic programming (DP) algorithm is a commonly used approach for optimization. Basically, the optimal scheme for power distribution can be precisely calculated based on known driving cycles [22,23]. However, traditional DP can only be used to attain the optimal solutions under the designated driving cycles rather than the unpredictable driving patterns in reality. An evolved approach from DP called stochastic dynamic programming (SDP) is capable of dealing with uncertain situations during actual driving and, thus, may further extend the application to real-time control [24]. However, the undesirably high computing load and sophisticated derivation of mathematics limit its application to the practical control of vehicles. Alternatively, the genetic algorithm (GA) is a feasible method to obtain accurate solutions especially for the control systems of vehicles with high nonlinearity [25]. Other control strategies such as bionic optimization algorithms (e.g., particle swarm optimization, bacterial foraging algorithm, etc.) can search for the best solution (minimal equivalent fuel or minimal energy consumption) in real time. The optimal control variable is the power (energy) distribution ratio [26,27]. Nevertheless, the resulting lack of analytical approaches makes control rules hard to amend. Compared with the algorithms mentioned above, predictive control, linear programming, robust control, etc. possess relatively clear expressions of mathematics for advanced modification.

In this paper, the algorithms were implemented into a real vehicular controller for a hybrid electric motorcycle. Therefore, considering the requirements of high computing efficiency and quantitative analysis for control, both RBC and ECMS were adopted for comparison. For ECMS, the cost function was set as the minimum energy consumption of the engine, motor, and integrated starter generator (ISG). Furthermore, the penalty and the relationship between input and output for the optimal search needed to be specified [28,29]. Thereby, a simulation platform for a hybrid electric motorcycle with three power sources, i.e., the engine, motor, and ISG, was constructed. Then, through HIL as the effective testing method, the performance, energy efficiency, and some other related data could be obtained and evaluated.

2. System Configuration and Dynamic Model

The schematic configuration of the hybrid electric motorcycle with three power sources in this study is shown in Figure 1. The basic specifications of the drivetrain, engine, and ISG were mainly based on a commercially available motorcycle called the Honda PCX Hybrid. Moreover, a 48-V driving motor, a lithium-ion battery, a continuously variable transmission (CVT), and a final drive were included. The control strategy was formulated to optimize the power distribution among three power sources based on instant data of minimum equivalent fuel consumption. The main specifications of the test motorcycle are listed in Table 1.

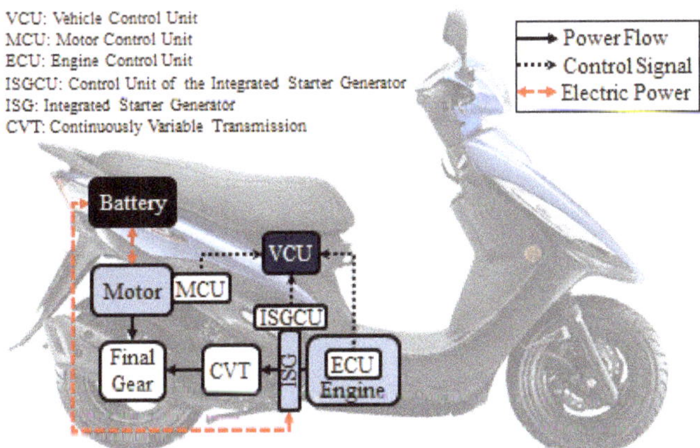

Figure 1. The schematic configuration of the test motorcycle.

Table 1. The main specifications of the test motorcycle. CVT—continuously variable transmission.

	Item		Parameters
Vehicle	Weight	Curb (kg)	≤225
		Gross (kg)	285
Propulsion	Engine	Type	Internal combustion engine
		Peak power (kW)	9 kW/8500 rpm
		Peak torque (Nm)	11.8 Nm/3000 rpm
	Motor	Type	Permanent magnet synchronous motor
		Peak power (kW)	3 kW/2500 rpm
		Peak torque (Nm)	12 Nm/2500 rpm
	Integrated starter generator	Type	Permanent magnet synchronous motor
		Peak power (kW)	1.2 kW/3000 rpm
		Peak torque (Nm)	6 Nm/3000 rpm
	Transmission	Type	CVT
		Gear ratio	2.55–0.81

The test motorcycle was a type of parallel hybrid. The engine was coaxially coupled with the ISG. The ISG was placed between the engine and CVT. The driving motor was mechanically coupled to the final drive gear and the output shaft of the CVT. Finally, the final drive was directly linked to the rear wheel for propulsion. The simulation procedure of the test drive is stated below. The control model tracked the instant driving pattern. This information was imported into the control model of energy management to evaluate it with respect to the instant condition of power distribution. Then, a demanded vehicle speed for the simulation platform was generated and compared with the actual vehicle speed simulated from the vehicle dynamic model. The resulting error of driving speed was transmitted to the driver model where the instant throttle opening and braking torque were calculated based on proportional integral (PI) control. Then, this evaluated torque command was transferred to the energy management system where an appropriate distribution ratio of torque outputs was computed with reference to the instant state of charge (SOC) of the battery and the torque constraints of power source. These evaluated torque commands were sent to the three models of power source. Then, the actual vehicle speed at the next time step could be calculated via the motorcycle dynamics model.

2.1. Driving Cycle Model

The motorcycle density in Taiwan is now the highest in Asia. Even though motorcycles are smaller than cars, their total emissions of pollutants are relatively higher. All the newly produced motorcycles are required to be tested for the emission of pollutants under the specific driving cycle, and they can only be put onto the market when the test is passed. In order to simulate the performance and energy consumption of the test motorcycle when actual driving on roads, the world motorcycle test cycle (WMTC), in compliance with the Stage 6 Gasoline Vehicle Emissions Standards in Taiwan, was utilized, as shown in Figure 2. According to the testing regulations of Taiwan driving cycles in WMTC Class 1, this research focused on engine displacement < 150 cc and maximal speed < 100 km/h, referring to urban driving. The total driving time and maximum driving speed for WMTC were 600 s and 50 km/h, respectively.

2.2. Driver Model

The driver model in the simulation platform was used to simulate the driver's control and operational response for the accelerator and brake. The actual vehicle speed and demanded vehicle speed plus the difference between them calculated from the simulation platform were input into the driver model and converted to the required throttle opening and brake torque as outputs via the PI controller for modulating the vehicle dynamics at the next instant.

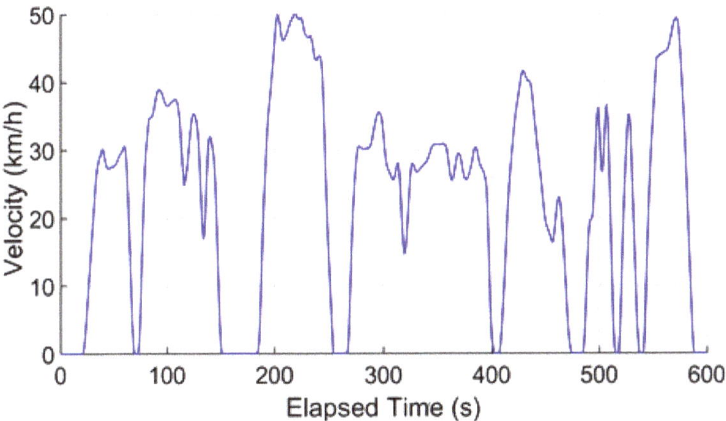

Figure 2. Speed profile of the world motorcycle test cycle (WMTC) driving cycle.

2.3. Engine Model

The specifications of the internal combustion engine model in this study were adopted from the original design data of the Honda PCX Hybrid and, accordingly, the brake specific fuel consumption (BSFC) in terms of engine torque and rotational speed was constructed. Then, the required torque output of the engine plus that of the ISG due to the coaxial coupling between the engine and ISG was sent to the CVT model.

2.4. Motor and ISG Model

The type of motor used for the motor and ISG models was a brushless DC motor. The torque commands of the motor and ISG were sent from the energy management system to the motor and ISG models where the appropriate torque for propulsion was evaluated by considering the maximum physical constraint for protection. The corresponding efficiencies of the motor and ISG can be calculated through two-dimensional (2D) look-up tables in terms of input torque and rotational speed as shown below.

$$\eta_m(t) = f(T_m(t),\, N_m(t)),\tag{1}$$

$$\eta_{isg}(t) = f\!\left(T_{isg}(t),\, N_{isg}(t)\right),\tag{2}$$

where η_m is the efficiency of the motor, T_m is the torque of the motor, N_m is the rotational speed of the motor, η_{isg} is the efficiency of the ISG, T_{isg} is the torque of the ISG, and N_{isg} is the rotational speed of the ISG.

2.5. Lithium-Ion Battery Model

A lithium-ion battery was used to provide the electric energy to the driving motor and ISG. The relationships between internal resistance and voltage were obtained from the experimental data, allowing the SOC of the battery to be estimated. Since the precision of estimating the SOC affects the vehicle dynamics and strategy of energy management, an internal resistance model was adopted. Accordingly, the equivalent internal resistance of battery was expressed in terms of *SOC* and temperature, as shown below.

$$R_b = R_b(SOC, T_b),\tag{3}$$

where *SOC* is in a range of 0–1 (0%–100%). The formula to calculate *SOC* is described as

$$SOC = \frac{SOC_{int} \times AH - \int \frac{I_b}{3600} dt}{SOC_{int} \times AH},$$ (4)

where SOC_{int} is the initial state of charge of the battery, AH is the nominal capacity of the battery, I_b is the discharging current of the battery, I_b is the current flowing to the motor, and ISG is calculated based on the following expression:

$$I_b = \frac{V_{OC} - \sqrt{V_{OC}^2 - 4 \times P_b \times R_b}}{2 \times R_b},$$ (5)

where V_{OC} is the open-circuit voltage of the battery that is a function of SOC and T_b, i.e., $V_{OC} = V_{OC}(SOC, T_b)$, and determined by a look-up table from experimental data; P_b is the power of the motor or ISG. Then, the voltage of the battery under charge/discharge loading, V_L, can be calculated as follows:

$$V_L = V_{OC} - I_b \times R_b.$$ (6)

2.6. CVT Model

The CVT model was used to calculate the required torque and rotational speed of the engine coupled with the ISG via the CVT transmission system. The output torque and rotational speed of the CVT were related to those of the wheel, T_w and N_e, through the final drive with a final speed reduction ratio. According to the original data of the Honda PCX Hybrid, two one-dimensional (1D) lookup tables of the gear ratio and gear efficiency versus the engine speed were constructed, as shown in Figure 3.

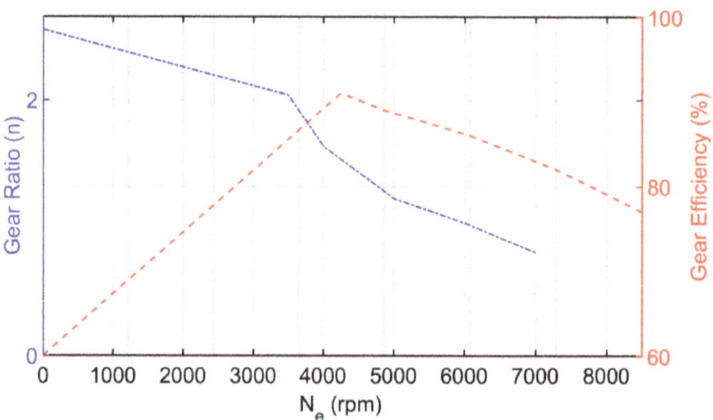

Figure 3. Gear ratio and gear efficiency vs. time for continuously variable transmission (CVT).

2.7. Motorcycle Dynamics Model

Considering the various types of resistance in driving, the wheel driving force is mainly the sum of air resistance, rolling resistance, grade resistance, braking force, and inertia resistance. Based on this mathematics model of dynamics, the rotational speed of wheel and vehicle speed at every instant could be calculated, serving as feedback to the CVT model and driver model for correcting the instantaneous motion of the motorcycle. Accordingly, the acceleration of the motorcycle in terms of the various forces and resistances mentioned above can be written as

$$m_v \frac{dV_v}{dt} = \frac{T_f \eta_f}{R_w} - \frac{1}{2} \rho_a C_d A_f V_v^2 - \mu m_v g \cos(\theta) - m_v g \sin(\theta) - F_{brk},$$ (7)

where m_v is the gross vehicle mass, i.e., the sum of vehicle mass and driver mass, T_f is the torque of the drivetrain, R_w is the tire radius, η_f is the total efficiency of the drivetrain, μ is the coefficient of rolling resistance, g is the acceleration of gravity, C_d is the coefficient of air drag, A_f is the frontal area of motorcycle, ρ_a is air density, F_{brk} is the braking force, V_v is the vehicle speed, and θ is the inclined angle.

2.8. HIL Architecture

In this study, the HIL system of HEV consisted of two real-time Arduino DUE controllers, as shown in Figure 4. One served as the embedded system of the vehicle called the energy management system, and the other acted as the vehicle simulation platform comprising the driving cycle, driver, motor, ISG, engine, lithium-ion battery, CVT, and motorcycle dynamics models. A transit circuit of UART to USB was required for the computer to communicate with the Arduino controllers through the interface pin locations named UART, SPI, andI2C. Furthermore, the Simulink Support Package for Arduino Hardware in Matlab/Simulink®(2017a, MathWorks, Natick, MA, USA) was utilized to connect with the HIL system so that a real-time simulation model could be built.

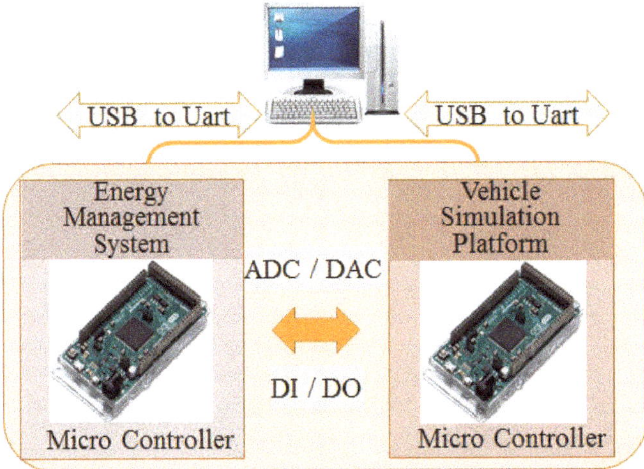

Figure 4. Hardware-in-the-loop (HIL) architecture for heavy-duty motorcycle.

The schematic diagrams of the input/output interface between two real-time controllers and the software structure for the energy management system controller are illustrated in Figures 5 and 6, respectively. To manage the insufficient analog outputs of the microcontroller, the digital outputs were combined with analog signals as shown in Figures 5 and 6. By using the signal scale transform, the resolution of the analog signal was maximized. Furthermore, for the digital-to-analog transformation, values 0–4095 represent 0.56–2.76 V of output voltage. Therefore, the initial value needed to be compensated for the signal inverse as best as possible. The simulation process is briefly described below. Initially, the information of actual and demanded speeds of the vehicle were taken from the driving cycle model and transmitted to the driver model. Then, the demanded torque of vehicle, T_{dem}, was calculated based on PI control. In addition, the SOC was evaluated via the lithium-ion battery model, and the rotational speed of the engine, as well as the ISG, N_e, was determined through the CVT model. These three variables, i.e., T_{dem}, SOC, and N_e, were converted to voltage signals as shown in the input interface of Figure 5 and then directed to the receiving port of the controller of the energy management system as shown in Figure 6. Based on these input data, the demanded torques of the engine, motor, and ISG, as well as the switching mode, were evaluated via the energy management system model and transmitted through the transmitting port shown in Figure 6 back to the output interface of Figure 5. During the simulation, the processed variables were constantly switched between digital values and voltage signals back and forth.

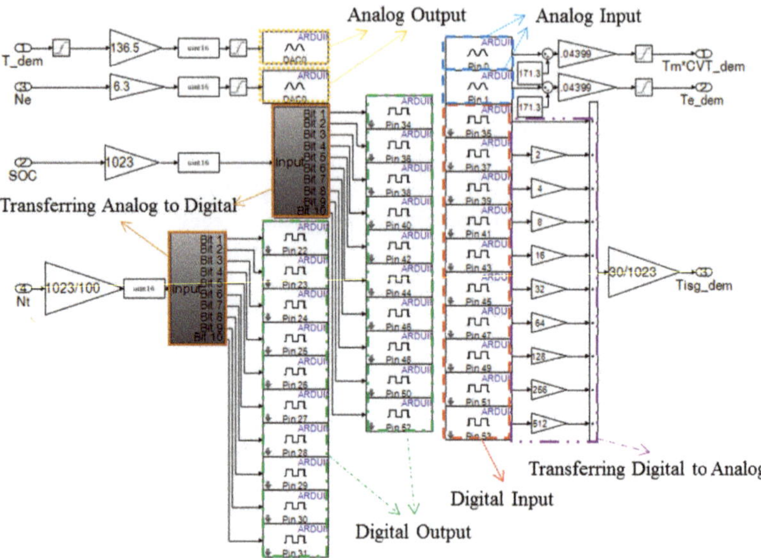

Figure 5. Schematic diagram of the input/output interface between two real-time controllers.

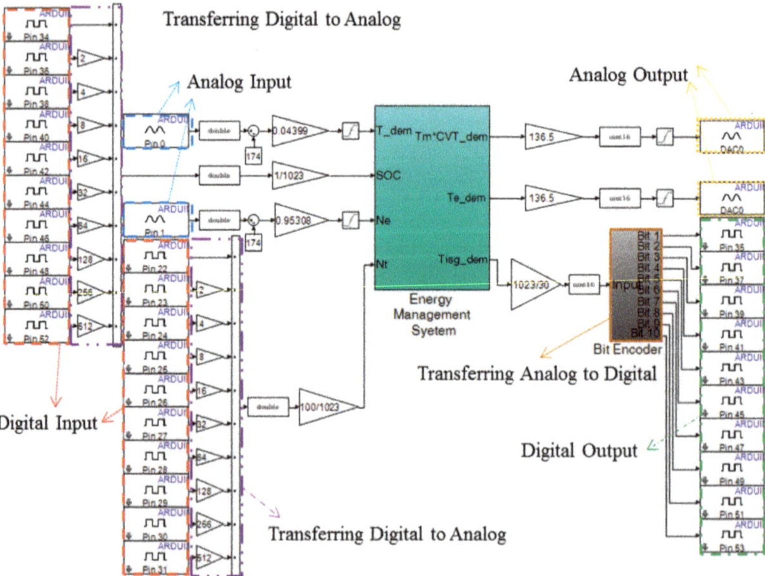

Figure 6. Schematic diagram of the software structure for the energy management system controller.

3. Energy Management System

RBC and ECMS were used to develop the control strategies for the energy management system capable of optimization in favor of fuel economy. Furthermore, the physical characteristics of the related vehicle dynamics were built into mathematical models. By combining these models and strategies, various simulation cases could be performed. In consequence, the favored results of maximum energy efficiency accompanied by the optimized strategy could be attained.

3.1. RBC Method

In this study, the RBC method was utilized to establish the decision-making algorithms for driving torque distribution of three power sources for a hybrid electric motorcycle. Initially, the demanded torque of the vehicle T_{dem} was calculated from the energy management system through PI control, while SOC was evaluated in the lithium-ion battery model. Furthermore, N_e, which is equal to the rotational speed of ISG due to their coaxial coupling, was obtained, while N_m was determined via the CVT model based on a parallel configuration. Then, based on these three inputs of rotational speed, an appropriate set of driving torque distribution of three power sources was calculated through three 1D lookup tables along with the related power limitations and physical constraints. Accordingly, the demanded torques of the motor, engine, and ISG ($T_{mCVT,dem}$, $T_{e,dem}$, and $T_{isg,dem}$), as well as the operation mode (Mode), were determined based on the control strategy of the energy management system.

The RBC method is briefly described below. Three driving modes were formed by two torque constraints as shown in the performance map of the engine in Figure 7. The blue line and black line in Figure 7 indicate the torque constraints of 2 kW and 6 kW, defined as "low power" and "high power", respectively. Then, the lower-left region of the blue line is designated as the purely electric drive mode, and the region between the blue and black lines is defined as the purely engine-driven mode, while the upper-right area of the black line is set as the fully hybrid drive mode. The corresponding operational conditions are described in Table 2. The RBC strategy was based on the engine characteristics. The efficient BSFC area was chosen to be used to modify the engine output. A lower BSFC value would result in a higher engine efficiency. Therefore, from Figure 7, the low-load area was avoided to raise the average engine efficiency.

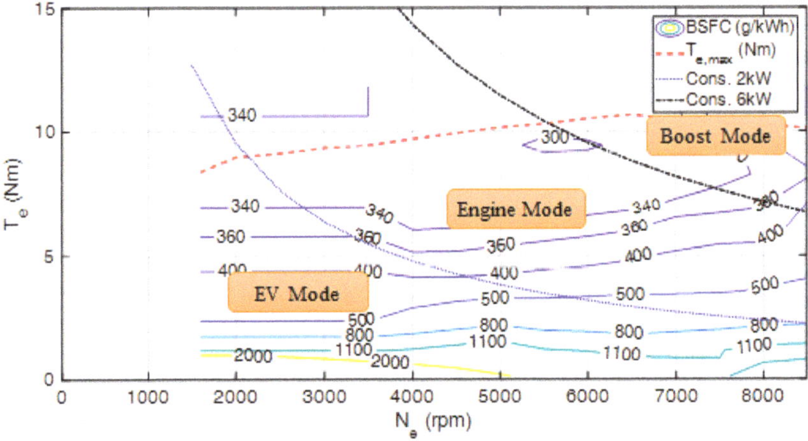

Figure 7. Constraint diagram for rule-based control (RBC) strategy.

When the demanded torque of the driver (T_dem) is zero, the motorcycle is operated in the system ready mode where the demanded torques of the engine, motor, and ISG are all set to be zero. When the demanded torque of driver is greater than 0.2 Nm, the operation is switched to the purely electric drive mode, i.e., the EV mode. In this mode, if T_dem exceeds the limit of maximum engine torque (T_{e,max}) indicated by the red dotted line in Figure 7, the motor and ISG simultaneously deliver driving torque in a ratio of 7:3; otherwise, the ratio is 8:2. When T_dem lies between the low-power and high-power lines, or the instant SOC is too low to sustain the supply of electricity for the motor and ISG, the operation is then changed to purely engine-driven mode, i.e., engine mode. In this mode, if T_dem is greater than $T_{e,max}$, the deficiency of engine torque is supplemented with the assistance of the motor. The output torques of the engine and motor are in a ratio of 7:3. Otherwise, the demanded torque is totally supported by the engine. Now, if T_dem is beyond the high-power limit, the operation enters into boost mode. In this mode, the genset mode is activated at SOC < 0.2, while the safety mode is on if both T_dem = 0 and SOC = 0.

Table 2. Modes and conditions in rule-based control (RBC). SOC—state of charge.

Mode	Condition
System Ready	T_dem = 0
EV	T_dem > 0.2 & T_dem < low power
Engine	T_dem ≥ low power & T_dem < high power‖SOC < 0.2
Genset	SOC < 0.2
Boost	T_dem ≥ high power
Safety	T_dem = 0 && SOC = 0

3.2. ECMS Method

The power ratio (PR) was determined based on the ECMS that converts the variation of SOC into an equivalent expression of fuel consumption. Therefore, the algorithm flow was constructed as follows:

1. By constructing the performance models of the powertrain, defining the corresponding objective function, and planning a global search for optimization, the minimum solutions of equivalent fuel consumption with the optimal results of the power ratio of engine (α) and power ratio of motor (β) were then obtained. Then, the multi-dimensional table was constructed and implemented into the controller of energy management system for optimization. The power ratios of the three power sources are expressed below.

$$T_e = \alpha \times T_d, \tag{8}$$

$$T_{m,CVT} = \beta \times T_d \times CVT, \tag{9}$$

$$T_{isg} = (1 - \alpha - \beta CVT) \times T_d, \tag{10}$$

$$T_e + T_{isg} + T_{m,CVT} = \alpha T_d + \beta CVT T_d + T_d - \alpha T_d - \beta CVT T_d, \tag{11}$$

$$T_d = T_e + T_{isg} + T_{m,CVT}, \tag{12}$$

where T_d is the total demanded torque, T_e is the demanded torque of the engine, T_{isg} is the demanded torque of the ISG, and $T_{m,CVT}$ is the demanded torque multiplied through the CVT.

2. The test data of the engine from the Honda PCX Hybrid motorcycle, ISG, and 48-V motor coupled with physical constraints were uploaded into the program in the form of 1D or 2D look-up tables. The tables of test data included the open-circuit voltage of battery vs. the SOC, the charge/discharge efficiency of the battery vs. the SOC and current, the charge/discharge efficiency of the motor vs. the torque and rotational speed, the BSFC of the engine vs. torque and rotational speed, and the efficiency of the ISG vs. torque and rotational speed; those of the physical constraint contained the maximum output torque of the engine vs. rotational speed, the maximum output torque of the motor vs. rotational speed, the maximum output current of the motor vs. torque and rotational speed, the maximum output torque of the ISG vs. rotational speed, the maximum output current of the ISG vs. torque and rotational speed, etc. All the quantized data mentioned above were loaded into the program so as to proceed with the optimization algorithm.

3. The global search of optimal solutions based on ECMS was executed by six for loops with respect to the discretized SOC, demanded torques, rotational speeds of the engine and motor, etc. The various operating modes were judged through conditional statements of "if–then–else" in

the program. Then, required variables such as the BSFC of the engine, efficiencies of the motor and ISG, demanded torques of the engine and motor, etc. could be calculated. After that, the results of equivalent fuel consumption based on the ECMS under various conditions could be obtained and stored in a four-dimensional (4D) matrix named mf_total_boost. Therefore, the minimum equivalent fuel consumptions under various engine torques with respect to the specific SOC, rotational speed, and demanded torque could be attained and, accordingly, the optimal power ratios could be derived.

For the purpose of simplifying the algorithm of the control model so as to satisfy the heavy loads on real-time calculation for logical control in the embedded systems, the analytical procedures for determining the optimal power ratios were constructed using some key parameters among the models, such as the BSFC, efficiencies of the motor and ISG, maximum torque constraints of the three power sources, and battery parameters. Then, the optimal solutions of the power ratios with respect to the discretized input parameters, i.e., SOC, demanded torques, and rotational speeds of the engine and motor, were obtained using the global search for the minimum objective function based on the ECMS algorithm, as shown below.

$$J = \left(\dot{m}_e + f(SOC)\dot{m}_m + f(SOC)\dot{m}_{isg} \right) + \gamma, \tag{13}$$

where J is the optimal objective function, $\dot{m}_e$ is the actual consumption rate of fuel mass for the engine, $f(SOC)$ is the factor as a function of the SOC, $\dot{m}_m$ is the equivalent consumption of fuel mass for the motor, $\dot{m}_{isg}$ is the equivalent consumption of fuel mass for the ISG, and γ is the mathematical term indicating the relationship between SOC and the penalty value. Moreover, $\dot{m}_m$ can be written as

$$\dot{m}_m = \frac{\overline{BSFC}}{\eta_m \times \eta_b} \times P_m, \tag{14}$$

where $\overline{BSFC}$ is the average value of $BSFC$, η_b is the efficiency of the battery, and P_m is the power of the motor. Similarly, $\dot{m}_{isg}$ is described below.

$$\dot{m}_{isg} = \frac{\overline{BSFC}}{\eta_{isg} \times \eta_b} \times P_{isg}, \tag{15}$$

where P_{isg} is the power of the ISG.

By substituting Equations (14) and (15) into Equation (13), J becomes

$$J = \left[\dot{m}_e + f(SOC)\frac{\overline{BSFC}}{\eta_m \times \eta_b} \times P_m + f(SOC)\frac{\overline{BSFC}}{\eta_{isg} \times \eta_b} \times P_{isg} \right] + \gamma. \tag{16}$$

When a searching point exceeds the specific physical constraints, a penalty value with an extremely large number is assigned to γ; otherwise, a normal value of zero is replaced instead, as seen below.

$$\gamma = 10^6 \text{ (Penalty)}; \gamma = 0 \text{ (Normal)}. \tag{17}$$

The schematic diagram of the global search for the optimal energy management is shown in Figure 8. The main target was to search for optimal operating points with respect to the minimum J by computing the fuel consumption of the engine and total consumption of electric energy, along with the equivalent fuel consumption of the motor and ISG under the discretized variables of demanded torque, SOC, rotational speeds of the engine and motor, and power ratios of the engine and motor. A final optimal 4D table of distribution could be obtained by way of performing double interpolation for minimization twice, as expressed below.

$$J^*(a, b, c, d) = min[J(a, b, c, d, e, f)]. \tag{18}$$

This optimized 4D parametric table was integrated with operation modes to form a multi-dimensional table for optimal control and supervising the fuel economy of the entire system. This considerably reduced the loading for real-time computation and desirably fulfilled the requirement for real-time control of the embedded systems.

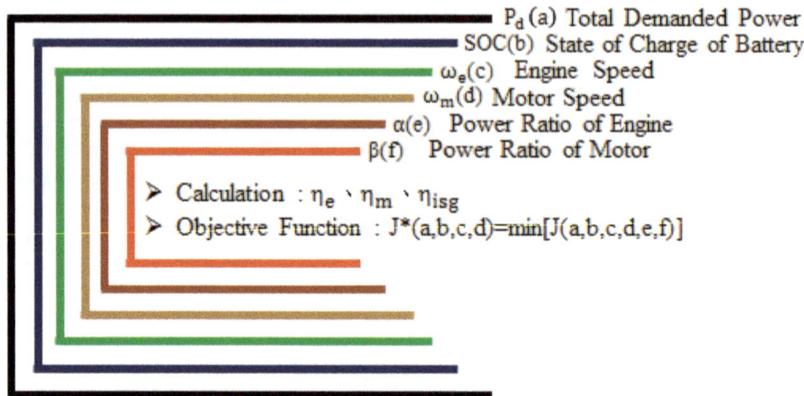

Figure 8. Schematic diagram of software configuration of the global search for the optimal energy management.

4. Simulation and Experimental Results

The simulation results for the output efficiency of the entire vehicle were obtained using an efficient algorithm consisting of an optimized 4D parametric table built into the energy management system as an embedded system. The parameter values are listed in Table 3.

Table 3. Parameter values of targeted vehicle model.

Parameter	Value
Gross vehicle mass, m_v	225 kg
Tire radius, R_w	0.262 m
Air density, ρ_a	1.2 kg/m^3
Frontal area of motorcycle, A_f	0.5 m^2
Coefficient of air drag, C_d	0.7
Coefficient of rolling resistance, μ	0.01
Inclined angle, θ	0 %
Acceleration of gravity, g	9.81 m/s^2
Total efficiency of the drivetrain, η_f	95%
Initial state of charge of the battery, SOC_{int}	80%
Average value of BSFC, $\overline{BSFC}$	372 g/kWh

4.1. HIL Simulation Results of RBC

In this section describing the simulation study, RBC was applied to the HIL system to obtain the results of torque distribution based on the required speed indicated by the WMTC. The simulated results are shown in Figure 9. In Figure 9a, the demanded speeds of the WMTC and the simulated speeds are expressed by a blue solid line and red dotted line, respectively. It can be found that the actual speeds of the hybrid electric motorcycle calculated using the proposed HIL system matched the demanded speeds with satisfactory tolerance when compared with those from the pure software simulation. Furthermore, it can be observed from Figure 9b that, whenever the vehicle started off, the simulated torques increased abnormally high, generating undesired spikes shown in the graph. This was because of the inaccuracy due to signal processing and the delay problem of data transmission between the two Arduino DUE microcontrollers, causing the torques in the HIL environment to exceed those of the software simulation.

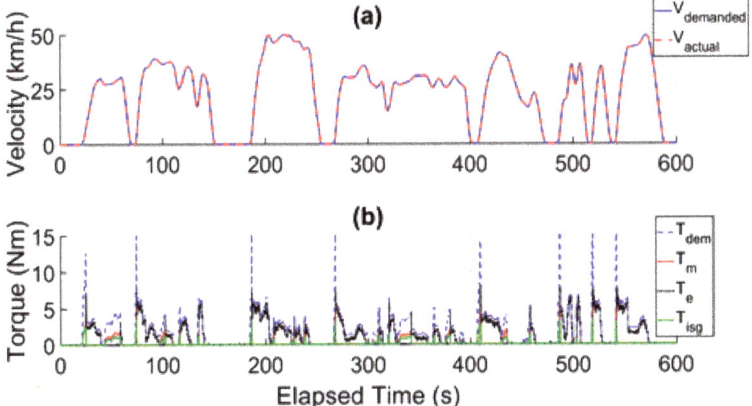

Figure 9. (a) Vehicle speed of driving cycle vs. time for the RBC + HIL; (b) torque distribution vs. time for RBC + HIL.

4.2. HIL Simulation Results of ECMS

ECMS was alternatively applied to the HIL system to simulate the torque distribution based on the demanded speed indicated by the WMTC. The calculated outcomes are shown in Figure 10. Similarly, from Figure 10a, the calculated speeds of the hybrid electric motorcycle using the proposed HIL system were very close to the demanded speeds with satisfactory tolerance when compared with those from the pure software environment. In addition, Figure 10b again shows unduly high spikes of calculated torque mainly due to the inaccuracy from signal processing, transmission delay of data transmission, and AD/DA conversion between the two microcontrollers.

4.3. Energy Consumption Results

This study focused on the comparative evaluation of equivalent total fuel consumption and total energy consumption for various control methods, as well as the pure software simulation vs. the HIL simulation. The results of equivalent fuel consumption and energy consumption under the pure engine mode, RBC mode, and ECMS mode based on the software simulation are listed in Table 4. They reveal that the resulting consumptions were in descending order, and those under ECMS control were the lowest. Compared with the pure engine mode, the improvement percentages of the equivalent fuel consumption and energy consumption for ECMS were 42.77% and 44.22%, respectively.

Table 4. Comparison of equivalent fuel consumption and energy consumption to pure software simulation.

Item	Equivalent Fuel Consumption (g)	Energy Consumption (kJ)	Improvement Compared to Baseline Case (%)
Pure Engine	63.2009	2722.5	-/-
RBC	51.9951	2218.6	17.74/18.50
ECMS	36.1698	1518.5	42.77/44.22

The outcomes of equivalent fuel consumption and energy consumption under the pure engine mode, RBC mode, and ECMS mode using the HIL simulation are described in Table 5. Compared with Table 4, even though the HIL simulation was affected by the accuracy of data processing, transmission delay, and AD/DA conversion, it was found that its results were quite close to those of the pure software simulation without a significant difference. Likewise, the calculated consumptions were in descending order, and ECMS offered the best results. The improvement percentages of equivalent fuel consumption and energy consumption for ECMS with respect to the pure engine mode were 42.73% and 44.10%, respectively. The reason why the results of Tables 4 and 5 were close is because the HIL scheme could successfully simulate the offline simulation by using two microcontrollers in real time. Moreover, it proves that the performance of analog-to-digital or digital-to-analog technology is good for signal transformation. With the same sampling time, the results demonstrate that the ECMS could be implemented with real microcontrollers.

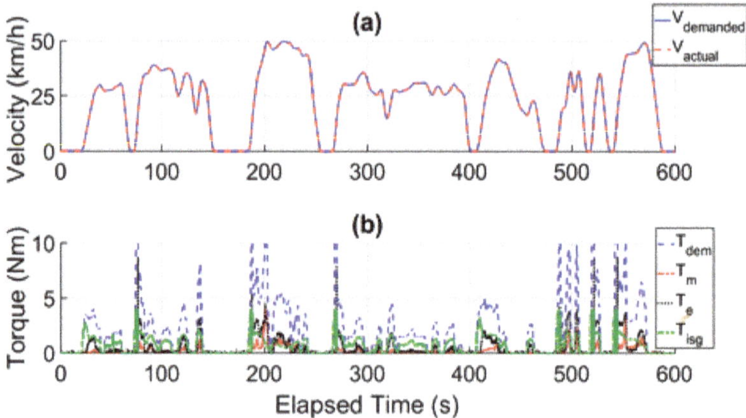

Figure 10. (a) Vehicle speed of driving cycle vs. time for equivalent consumption minimization strategy (ECMS) + HIL; (b) torque distribution vs. time for ECMS + HIL.

Table 5. Comparison of equivalent fuel consumption and energy consumption to HIL simulation.

Item	Equivalent Fuel Consumption (g)	Energy Consumption (kJ)	Improvement Compared to Baseline Case (%)
Pure Engine	63.8562	2750.3	-/-
RBC	52.2597	2232.6	18.16/18.82
ECMS	36.5698	1537.3	42.73/44.10

5. Conclusions

In this study, a simulation platform for a hybrid electric motorcycle with three power sources of engine, driving motor, and ISG was proposed. This platform comprised the driving cycle, driver, lithium-ion battery, continuously variable transmission (CVT), motorcycle dynamics, and energy management system models. Furthermore, a hardware-in-the-loop simulation was performed using TWO Arduino DUE microcontrollers coupled with the required circuit to process analog-to-digital signal conversion for the input and output. The WMTC driving cycle was used for evaluating the performance characteristics and response relationship among subsystems. Control strategies used for optimization were the pure engine, RBC, and ECMS. As a result, the improvement percentages of equivalent fuel consumption and energy consumption for RBC and ECMS compared with the pure engine mode using the pure software simulation were 17.74%/18.50% and 42.77%/44.22%, respectively, while those with the HIL were 18.16%/18.82% and 42.73%/44.10%, respectively.

ECMS is a global optimal search method that searches for the optimal solution based on the system parameters and physical characteristics. The ECMS optimal results can be calculated offline in the form of a multi-dimensional matrix which is downloaded to the microcontroller of the HIL environment. Due to the significant reduction in computational load of the matrix compared to other optimal algorithms, the ECMS has strong advantages in terms of computational loading and computational time. This research successfully transformed the analog-to-digital or digital-to-analog signals without signal distortion as best as possible. At the same sampling time, the results showed that the ECMS can be practically applied with microcontrollers. Compared to the previous HIL system with a more complicated structure and expensive resources, this novel structure can be directly employed on the three-power-source hybrid motorcycles. In future development, the optimization of the microcontroller on the vehicle simulation platform is expected to enhance the accuracy and stability of signal processing so as to realize practical implementation into actual vehicles, greatly contributing to energy saving and environmental protection.

Author Contributions: Conceptualization, C.-H.W.; methodology, C.-H.W.; software, Y.-X.X. and C.-H.W.; validation, Y.-X.X.; formal analysis, C.-H.W. and Y.-X.X.; investigation, C.-H.W.; data curation, Y.-X.X.; writing—original draft preparation, C.-H.W.; writing—review and editing, C.-H.W.; visualization, C.-H.W.; supervision, C.-H.W.; project administration, C.-H.W.; funding acquisition, C.-H.W. All authors have read and agreed to the published version of the manuscript.

Funding: The authors would like to thank the Ministry of Science and Technology of the Republic of China, Taiwan, for financial support of this research under Contract No. MOST 107-2221-E-150-036 and 108-2221-E-150-012; the authors also thank the Industrial Technology Research Institute (ITRI) project for advanced vehicle control.

Conflicts of Interest: The authors declare no conflicts of interest.

References

1. Capata, R. Urban and extra-urban hybrid vehicles: A technological review. *Energies* **2018**, *11*, 2924. [CrossRef]
2. Passalacqua, M.; Carpita, M.; Gavin, S.; Marchesoni, M.; Repetto, M.; Vaccaro, L.; Wasterlain, S. Supercapacitor storage sizing analysis for a series hybrid vehicle. *Energies* **2019**, *12*, 1759. [CrossRef]
3. Wang, D.; Song, C.X.; Shao, Y.L.; Song, S.X.; Peng, S.L.; Xiao, F. Optimal control strategy for series hybrid electric vehicles in the warm-up process. *Energies* **2018**, *11*, 1091. [CrossRef]
4. Solouk, A.; Shahbakhti, M. Energy optimization and fuel economy investigation of a series hybrid electric vehicle integrated with diesel/RCCI engines. *Energies* **2016**, *9*, 1020. [CrossRef]
5. Zou, Y.; Liu, T.; Sun, F.C.; Peng, H. Comparative study of dynamic programming and pontryagin's minimum principle on energy management for a parallel hybrid electric vehicle. *Energies* **2013**, *6*, 2305–2318.
6. Chen, J.S. Energy efficiency comparison between hydraulic hybrid and hybrid electric vehicles. *Energies* **2015**, *8*, 4697–4723. [CrossRef]
7. Fan, J.X.; Zhang, J.Y.; Shen, T.L. Map-based power-split strategy design with predictive performance optimization for parallel hybrid electric vehicles. *Energies* **2015**, *8*, 9946–9968. [CrossRef]
8. Qin, F.Y.; Xu, G.Q.; Hu, Y.; Xu, K.; Li, W.M. Stochastic optimal control of parallel hybrid electric vehicles. Energies. *Energies* **2017**, *10*, 214. [CrossRef]
9. *Road Vehicles-Functional Safety-Part 1~10*; ISO/DIS 26262-1~10; International Organization for Standardization: Geneva, Switzerland, 2009.
10. Sayadi, H.; Makrani, H.M.; Randive, O.; PD, S.M.; Rafatirad, S.; Homayoun, H. Customized machine learning-based hardware-assisted malware detection in embedded device. In Proceedings of the 17th IEEE International Conference on Trust, Security and Privacy in Computing and Communications (IEEE TrustCom-18), New York, NY, USA, 1–3 August 2018.
11. Maclay, D. Simulation gets in the loop. *IEE Rev.* **1997**, *43*, 109–112. [CrossRef]
12. Sung, C.O. Evaluation of motor characteristics for hybrid electric vehicles using hardware-in-the-loop concept. *IEEE Trans. Veh. Technol.* **2005**, *54*, 817–824.
13. Yang, C.; Deng, K.J.; He, H.X.; Wu, H.C.; Yao, K.; Fan, Y.Z. Real-time interface model investigation for MCFC-MGT HILS hybrid power system. *Energies* **2019**, *12*, 2192. [CrossRef]
14. Kotsampopoulos, P.; Georgilakis, P.; Lagos, D.T.; Kleftakis, V.; Hatziargyriou, N. FACTS providing grid services: Applications and testing. *Energies* **2019**, *12*, 2554. [CrossRef]
15. Sheu, K.B. Simulation for the analysis of a hybrid electric scooter powertrain. *Appl. Energy* **2000**, *85*, 589–606. [CrossRef]
16. Tzeng, S.C.; Huang, K.D.; Chen, C.C. Optimization of the dual energy-integration mechanism in a parallel-type hybrid vehicle. *Appl. Energy* **2005**, *80*, 225–245. [CrossRef]

17. Erdinc, O.; Vural, B.; Uzunoglu, M. A wavelet-fuzzy logic based energy management strategy for a fuel cell/battery/ultra-capacitor hybrid vehicular power system. *J. Power Sources* **2009**, *194*, 369–380. [CrossRef]

18. Sunddararaj, S.P.; Rangarajan, S.S.; Gopalan, S. Neoteric fuzzy control stratagem and design of chopper fed multilevel inverter for enhanced voltage output involving plug-in electric vehicle (PEV) applications. *Electronics* **2019**, *8*, 1092. [CrossRef]

19. Katrašnik, T. Analytical method to evaluate fuel consumption of hybrid electric vehicles at balanced energy content of the electric storage devices. *Appl. Energy* **2010**, *87*, 3330–3339. [CrossRef]

20. Liu, X.X.; Qin, D.T.; Wang, S.Q. Minimum energy management strategy of equivalent fuel consumption of hybrid electric vehicle based on improved global optimization equivalent factor. *Energies* **2019**, *12*, 2076. [CrossRef]

21. Hung, Y.H.; Tung, Y.M.; Li, H.W. A real-time model of an automotive air propulsion system. *Appl. Energy* **2014**, *129*, 287–298. [CrossRef]

22. Chen, Z.; Mi, C.C.; Xu, J.; Gong, X. Energy management for a power-split plug-in hybrid electric vehicle based on dynamic programming and neural networks. *IEEE Trans. Veh. Technol.* **2014**, *54*, 1567–1580. [CrossRef]

23. Škugor, B.; Petrić, J. Optimization of control variables and design of management strategy for hybrid hydraulic vehicle. *Energies* **2018**, *11*, 2838. [CrossRef]

24. Wang, Y.Y.; Jiao, X.H.; Sun, Z.T.; Li, P. Energy management strategy in consideration of battery health for PHEV via stochastic control and particle swarm optimization algorithm. *Energies* **2017**, *10*, 1894. [CrossRef]

25. Guerrero, J.I.; Personal, E.; García, A.; Parejo, A.; Pérez, F.; León, C. Distributed charging prioritization methodology based on evolutionary computation and virtual power plants to integrate electric vehicle Fleets on smart grids. *Energies* **2019**, *12*, 2402. [CrossRef]

26. Chen, S.Y.; Wu, C.H.; Hung, Y.H.; Chung, C.T. Optimal strategies of energy management integrated with transmission control for a hybrid electric vehicle using dynamic particle swarm optimization. *Energy* **2018**, *160*, 154–170. [CrossRef]

27. Yuan, G.H.; Yang, W.X. Study on optimization of economic dispatching of electric power system based on hybrid intelligent algorithms (PSO and AFSA). *Energy* **2019**, *183*, 926–935. [CrossRef]

28. Hung, Y.H.; Wu, C.H. An integrated optimization approach for a hybrid energy system in electric vehicles. *Appl. Energy* **2012**, *98*, 479–490. [CrossRef]

29. Hung, Y.H.; Wu, C.H. A combined optimal sizing and energy management approach for hybrid in-wheel motors of EVs. *Appl. Energy* **2015**, *139*, 260–271. [CrossRef]

Article

Optimal Configuration with Capacity Analysis of a Hybrid Renewable Energy and Storage System for an Island Application

Chih-Ta Tsai, Teketay Mulu Beza, Wei-Bin Wu and Cheng-Chien Kuo *

Department of Electrical Engineering, National Taiwan University of Science and Technology 43, Sec. 4, Keelung Rd., Taipei 10607, Taiwan; marcotsai3@gmail.com (C.-T.T.); mteke24@gmail.com (T.M.B.); andytouchwork@gmail.com (W.-B.W.)
* Correspondence: cckuo@mail.ntust.edu.tw; Tel.: +886-920881490; Fax: +886-227376688

Received: 26 October 2019; Accepted: 16 December 2019; Published: 18 December 2019

Abstract: The Philippines consists of 7100 islands, many of which still use fossil fuel diesel generators as the main source of electricity. This supply can be complemented by the use of renewable energy sources. This study uses a Philippine offshore island to optimize the capacity configuration of a hybrid energy system (HES). A thorough investigation was performed to understand the operating status of existing diesel generator sets, load power consumption, and collect the statistics of meteorological data and economic data. Using the Hybrid Optimization Models for Energy Resources (HOMER) software we simulate and analyze the techno-economics of different power supply systems containing stand-alone diesel systems, photovoltaic (PV)-diesel HES, wind-diesel HES, PV-wind-diesel HES, PV-diesel-storage HES, wind-diesel-storage HES, PV-wind-diesel-storage HES. In addition to the lowest cost of energy (COE), capital cost, fuel saving and occupied area, the study also uses entropy weight and the Technique for Order Preference by Similarity to an Ideal Solution (TOPSIS) method to evaluate the optimal capacity configuration. The proposed method can also be applied to design hybrid renewable energy systems for other off-grid areas.

Keywords: renewable energy; hybrid energy systems; cost of energy; energy storage; distributed generation (DG); sensitivity analysis

---

1. Introduction

To achieve good economic life and growth, off-grid communities require an affordable and reliable energy supply. Besides, the growing concern about climate change and environmental pollution, especially since fossil fuels are the main source of energy on Earth, has pushed the power generation systems towards the use of renewable energy [1,2]. In addition to high transportation and fuel cost, energy delivered to isolated areas frequently used fossil fuel-based generators which threaten the anthropogenic and natural ecosystems [3–5]. The main factors pushing increased energy access are regular power interruptions, limited power grid accessibility, availability of renewable resources, increasing concern about the need to decrease fossil fuel dependency and high oil prices, and also the significance of escalating energy access to development and availability of climate- related financial support for low-carbon and climate-resilient development that prioritizes renewables [6–8].

Nowadays, the main challenges facing hybrid systems is to design an energy management strategy to meet the demand for loads, despite the intermittent nature of renewables [9,10], in addition to cost savings and total versatility and multi-faceted goals that can be accomplished [11–13].

PV and wind turbine (WT) have been considered the most promising renewable energy options for off-grid areas or islands to fulfill the energy demand [14–16]. Even stand-alone wind and solar energy may fulfill the low load requirements, while these systems need a significant energy storage for higher loads, resulting in high COE [17–19]. The other option to alleviate this problem is autonomous hybrid renewable energy systems (HRES) which combine two or more energy resources, to fulfill higher energy requirements of off-grid areas and resolve the inherent problem of single renewable energy (RE) resource [1,20–24]. Furthermore, hybridization of energy sources increases the reliability of the system as the shortcomings of any component are compensated for by the selection of other but appropriate components and their sizing is essential during design of such systems [25–27]. Some of the hybrid energy systems with different storage technologies and performance measure criteria found in literatures are presented in Table 1 [12,28,29].

Table 1. Hybrid power systems with various storage technologies.

Hybrid Energy Systems	Storage	Operating Strategy	Grid Connection	Methodology	Performance Measures
PV/Diesel/Batt [12]	LA, Li-ion	LF, CC, CD	Off-grid	HOMER	NPC, COE, EE, RF, DF, CO_2 emissions
PV/Diesel/Batt [13]	LA	LF	Off-grid/grid connected	HOMER	NPC, COE, RF, CO_2 emissions
PV/Diesel/Batt [17]	LA	LF	Off-grid	HOMER	NPC, COE, EE, CO_2 emissions
PV/Diesel/Batt [18]	Li-ion	CC	Off-grid	MATLAB	NPC
PV/Wind/Biogas/Diesel/Batt [19]	LA	-	Off-grid	HOMER	NPC, COE, RF, O&M cost
PV/Wind/Diesel/Batt [28]	CELLCUBE	LF	Off-grid/grid connected	HOMER	NPC, COE, RF, CO_2 emissions
PV/Wind/Biogas/Batt [29]	LA	LF, CC	Off-grid	HOMER	NPC, COE, RF, CO_2 emissions

Key: **CC**: Cyclic Charging; **CD**: Combined Dispatch; **DF**: Duty Factor; **EE**: Excess Energy; **COE**: Cost of Energy; **HOGA**: Hybrid Optimization by Genetic Algorithms; **LA**: Lead Acid; **Li-ion**: Lithium-ion; **RF**: Renewable Friction.

In most of the literature on hybridizing energy systems, cost reduction and CO_2 emission reduction are the focal points of the research work. However, the number of scenarios considered to pinpoint the best trade-off configuration among system reliability, cost of energy and environmental sustainability; the magnitude of excess electricity and the way it is managed; the percentage RF, still needs further investigation, depending on where the HES is going to be installed.

The Batanes Islands are the northernmost provinces of the Philippines. They are composed of 10 islands including Batan, Sabtang, Itbayat and so on. They have a typical volcanic island hilly terrain and a rich natural ecology. The largest island is Batan Island with a longitude of 120.968° and a latitude of 20.445°, as shown in Figure 1, about 161 km from Luzon, Philippines, and only about 190 km from Taiwan [30]. The fuel and other large resources needed on the islands are transported by ship from Luzon. At present, the population of Batan Island is about 12,000 and the main economic activities are sightseeing, agriculture and fishery.

Figure 1. Geographic location of Batan Island.

The power on Batan Island is offered 24 h a day by the Basco diesel power plant of the Philippine National Power Corporation (NPC), as shown in Figure 2. A total of five generators are mainly usied, of which four generators are 600 kW and the one is 450 kW. Because of the problems such as the difficulty of diesel transportation, high pollution and high power generation cost, the NPC is actively seeking to use renewable energy to solve these issues. With a subsidy of the Asian Productivity Organization (APO), the APO Center of Excellence on Green Productivity (APO COE GP) and NPC have conducted renewable energy research, and analyzed hybrid renewable energy systems based on the load demand and climatic conditions on Batan. The proposed capacity allocation scheme in different power supply systems can be as a reference of investment construction for NPC or private power plants.

Figure 2. The Basco diesel power plant.

The rest of this paper is arranged in the following way: In Section 2, the materials and methods followed are discussed. Hybrid energy system descriptions are discussed in Section 3. In Sections 4 and 5 component cost and financial assumptions, results and discussions are described, respectively. Conclusions are discussed in Section 6.

2. Materials and Methods

The study cases in the research contain seven kinds of power systems. Using entropy weight and TOPSIS method to analyze the techno-economic and evaluate optimal capacity configuration for hybrid energy system (HES). The discussions are as follows:

Case 1: Stand-alone diesel system
Case 2: PV-diesel HES
Case 3: Wind-diesel HES
Case 4: PV-wind-diesel HES
Case 5: PV-diesel-storage HES
Case 6: Wind-diesel-storage HES
Case 7: PV-wind-diesel-storage HES

The required information for our simulation analysis include the status of the diesel generators, the load power consumption, the statistics of the oil prices, interest rates and inflation rates in the Philippines in recent years. In addition, the weather data on the island can be obtained on the NASA website, and e market surveys for equipment costs are also needed.

- Based on the load conditions and the specifications of existing diesel generator, the electrical and economic results of the stand-alone diesel system (Case 1) can be analyzed by HOMER (Pro.3.7.6.0, HOMER Energy, Boulder, CO, USA)) and the result is considered as a reference for discussing different hybrid energy systems.

- According to the load conditions and climatic conditions, the electrical and the economic conditions in different capacity configuration are analyzed. The simulation of HES model without energy storage that is, PV-diesel HES (Case 2), wind-diesel HES (Case 3), PV-wind-diesel HES (Case 4) are analyzed. The optimal capacity configuration in the lowest COE for each case needs to be calculated.
- The HES with energy storage analyzed include PV-diesel-storage HES (Case 5), wind-diesel-storage HES (Case 6), PV-wind-diesel-storage HES (Case 7). The RF is in the range of 25–50% and each step is 5%. Finding a capacity configuration with an excess electricity fraction below 5% and a minimum COE for each case are the main points to be considered.
- From Cases 5–7, we select the one has a relatively low COE and use the entropy weight and TOPSIS method to analyze the optimal capacity configuration when the RF is in 25–50%, considering the capital cost, COE, fuel saving, and occupied area.
- The sensitivity analysis of parameters such as global horizontal irradiation data (GHI), wind speed, diesel fuel price, and load consumption. The parameters are used to discuss the economy and electricity of the system.

2.1. Simulation Software Description

The HOMER software was developed by the U.S. National Renewable Energy Laboratory (NREL). The software is helpful in analyzing the power system's electricity and economy to model the optimal power grid. Furthermore, the user can define the input parameters and the constraints as a reference for modeling profitable power system [31]. The architecture diagram of the simulation analysis procedure of the HOMER software is as shown in Figure 3.

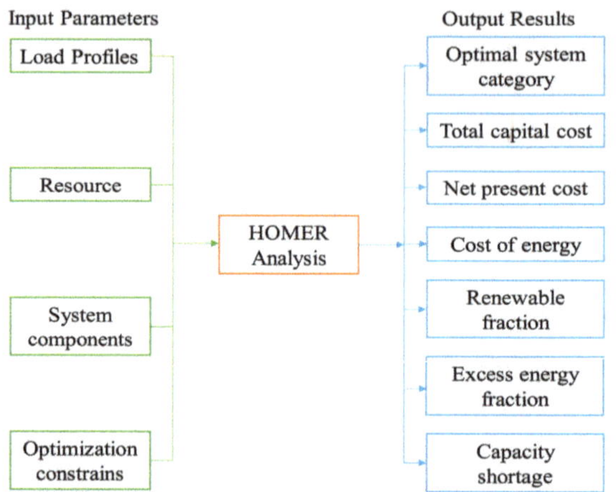

Figure 3. HOMER software simulation analysis process architecture diagram.

2.2. Global Horizontal Irradiation and Wind Speed

Entering the latitude and longitude of Batan into the HOMER software, NASA's global horizontal irradiation data (GHI) and wind speed statistics can be obtained through the Internet connection. The obtained information contains the annual GHI is 1876 kWh/m^2/yr, and the daily average GHI is about 5.14 kWh/m^2/d. The sunshine is abundant and suitable for the development of solar photovoltaic applications. As shown in Figure 4, the varies for the daily average GHI of each month from 3.31 kWh/m^2/d to 6.41 kWh/m^2/d, and the highest and lowest average GHI are in April and December, respectively.

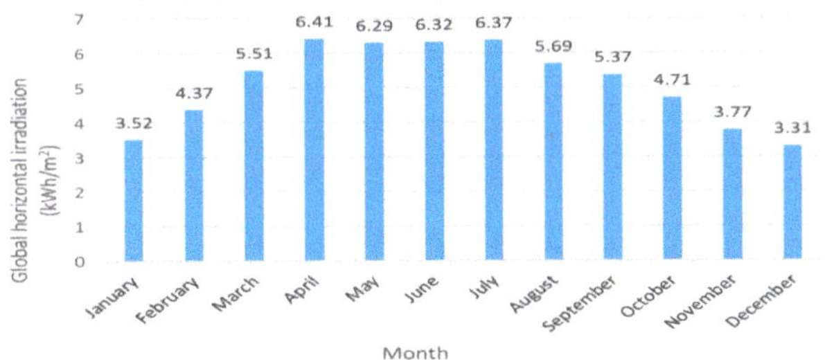

Figure 4. Global horizontal irradiation-monthly data.

According to the NASA anemometer data measured 50 m above the surface of the Earth, the annual average wind speed is 7.22 m/s. As shown in Figure 5, the variation of average wind speed is from 4.95 m/s to 10.04 m/s in each month. The highest and lowest average wind speeds are in December and May, respectively. The monthly trend of the whole year is opposite to that of GHI. The solar energy and wind energy can be supplied power in turn and effectively use natural resources.

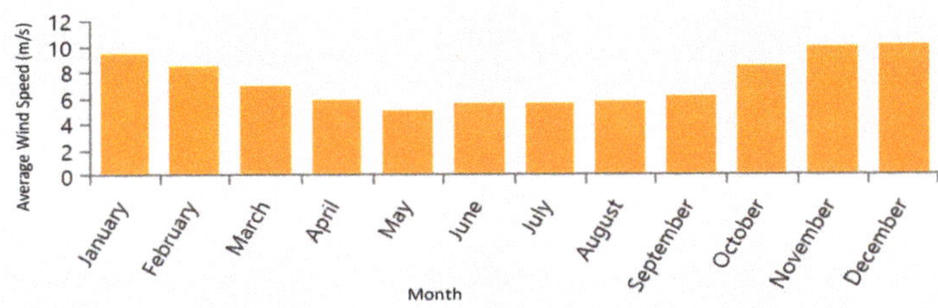

Figure 5. Wind speed monthly data.

2.3. Load Profile

Through the data collection of the frequency distribution of the load power consumption, the average of daily electricity consumption is 16,974 kWh/d, July is the month with higher electricity consumption, and January is the lower month for electricity consumption. The load profile for each month is as shown in Figure 6.

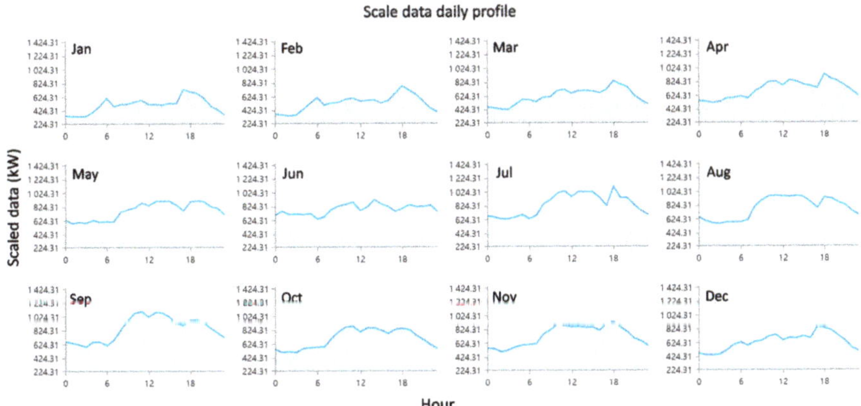

Figure 6. Daily load profile for each 12 months.

Figure 7 shows load power consumption frequency distribution. In the sampling period of hourly, the frequency of in 550 kW–599 kW and 600 kW–649 kW is more than 10%, the total is 20.35%; the frequency exceeding 1000 kW is only 7.02%, and the maximum 1442 kW.

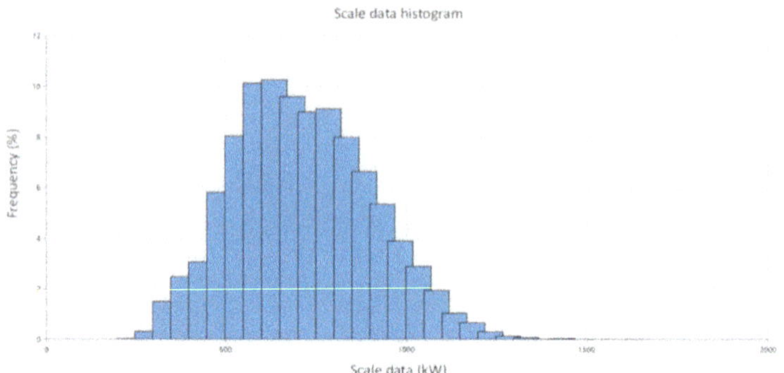

Figure 7. Load power consumption frequency distribution.

2.4. Diesel Price Data

As shown in Figure 8, the oil price information regularly announced by the Philippines' Department of Energy in recent years, the Philippines is affected by fluctuations in international oil prices. From October 2015 to February 2019, domestic oil prices have risen sharply. The retail common price has reached 49.15 PHP/L in 15 October 2018 [32], converted to $ is 0.91 $/L (1 & = 52.26 PHP). The lowest oil price during the statistical period is 20.5 PHP/L, the maximum is 49.15 PHP/L, and average is 33.335 PHP/L. In the offshore islands, the transportation cost must be added to the oil price. Therefore, the actual oil price is according to the geographic location. In more remote areas, the oil price will be affected by the transportation cost. The actual oil price may be 1–1.5 times the estimated oil price [33], in this paper, the average price of the statistical period is 1.25 times the oil price (0.8 $/L), which is used as the basic parameter of the simulation.

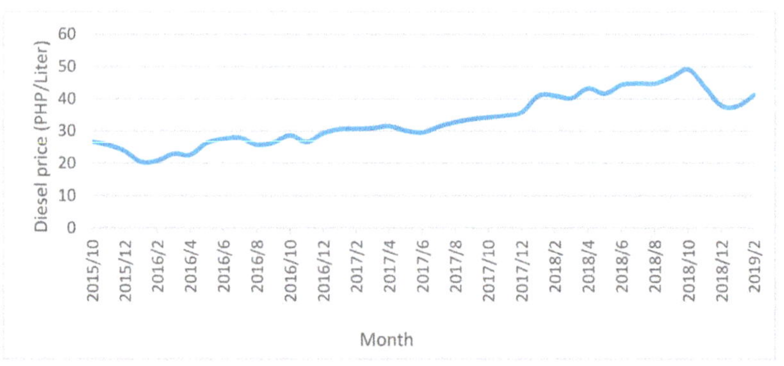

Figure 8. Diesel fuel price data.

2.5. Assignment Indexes

The evaluation indicators used in this paper include economic indicators (Sections 2.5.1–2.5.4), electrical indicators (Sections 2.5.5 and 2.5.6), and occupied area (Section 2.5.7). The calculation formulas and descriptions are as follows [3,34]:

2.5.1. Annual Real Interest Rate

Annual real interest rate (i) is used to adjust between one-time cost and annualized costs. HOMER uses annual real interest rate to compute discount factor and to determine annualized costs from present costs:

$$i = \frac{i' - f}{1 + f} \tag{1}$$

where i is annual real interest rate (%); i' is nominal interest rate (bank board rate) (%); f is expected inflation rate (%).

2.5.2. Net Present Cost

The total of NPC value denotes the cost of system life cycle in HOMER. Equation (2) displays the amount of cash flow of t-year over the factor and initial capital cost. Costs include capital cost, operation cost, replacement cost, maintenance cost, fuel cost and etc. The income includes the selling of electricity and the residual price after the life cycle:

$$i = \frac{i'-f}{1+f} \text{NPC} = CF_0 + \left\{ \frac{CF_1}{(1+i)^1} + \frac{CF_2}{(1+i)^2} + \frac{CF_3}{(1+i)^3} + \cdots + \frac{CF_N}{(1+i)^N} \right\}$$
$$= CF_0 + \sum_{t=1}^{N} \frac{CF_t}{(1+i)^t} \tag{2}$$

where CF_t is cash flow of t-year (according to HOMER software: expenditure is positive and income is negative.) ($), i is the annual real interest rate (%), N is project life time (yr), t is number of year (yr), CF_0 is initial capital cost ($).

2.5.3. Capital Recovery Factor (CRF)

CRF is ratio used to determine present value of the annuity during the project lifetime:

$$CRF(i, t) = \frac{i(1 + i)^t}{(1 + i)^t - 1} \tag{3}$$

where t is the number of years, i is the annual real interest rate (%).

2.5.4. Cost of Energy

HOMER defines levelized COE as average cost per kWh of useful electric energy generated by the system and computed by dividing total annualized cost (TAC) by total annualized useful electric energy generation. The unit of COE is $/kWh. TAC is annualized value of NPC, and its unit is $/yr. The relation is as follows:

$$TAC = \text{NPC} \times CRF(i, N) \tag{4}$$

$$COE = \frac{TAC}{E_{prim}} \tag{5}$$

where E_{prim} is annualized primary served load (kWh/yr), N is project life time (yr).

2.5.5. Renewable Fraction (RF)

RF is the fraction of energy delivered to the load that originated from renewable power sources. The equation is as follows:

$$RF = \left(1 - \frac{E_{non-ren} + H_{non-ren}}{E_{served} + H_{served}} \right) \times 100\% \tag{6}$$

where $E_{non\text{-}ren}$ is the nonrenewable electrical production (kWh/yr), $H_{non\text{-}ren}$ is the nonrenewable thermal production (kWh/yr), E_{served} is the total electrical load served (kWh/yr), H_{served} is the total thermal load served (kWh/yr).

2.5.6. Excess Electricity Fraction

The excess electricity fraction is the ratio of total excess electricity to total electrical production. The equation is as follows:

$$\text{excess electricity fraction} = \frac{E_{excess}}{E_{prod}} \times 100\% \tag{7}$$

where E_{excess} is the total excess electricity (kWh/yr), E_{prod} is the total electrical production (kWh/yr).

2.5.7. Occupied Area

In the calculation of the occupied area, the main considerations are PV array, wind turbine and energy storage [3]. The equation is as follows:

$$\text{Occupied area} = (A_{PV} \times N_{PV}) + (A_{WT} \times N_{WT}) + (A_{Bat} \times N_{Bat}) \tag{8}$$

where A_{PV} is the PV array area per kWp (m^2), N_{PV} is the capacity of PV system (kWp), A_{WT} is the occupied area of one wind turbine (m^2), N_{WT} is the number of wind turbines, A_{Bat} is occupied area of one energy storage rack (m^2), N_{Bat} is the number of energy storage racks.

2.6. Entropy Weight and TOPSIS Method

Entropy weight method is derived from Shannon entropy, that was suggested by Shannon for the numerical measurement of uncertainty in information systems [35]. The weighting factor of this evaluation method depends entirely on the value of the indicator, not on the subjective evaluation. Therefore, it has been considered as an objective method of calculating weights and has been widely applied [3].

TOPSIS is a method for relatively evaluation in limited alternative. This method is evaluated by computing relative closeness of the alternative and the ideal solution, and sorting according to the relative closeness. The higher closeness means the solution is close to the positive ideal solution, that is, the farther away from negative ideal solution, which is the best solution [36].

This study combines these two methods, first using the entropy weight method to calculate the objective weight value of each indicator, and using the TOPSIS method to calculate the relative closeness with the positive ideal solution for the weighted alternative. The calculated results are sorted so that the best solution can be found [37]. The main calculation steps and formulas in this paper are as follows:

Step 1: Initialize matrix

Construct m alternatives, the initial data matrix V of n index, and the eigenvalue of the jth index of the ith alternatives is expressed as v_{ij}, so matrix V can be defined as follows:

$$V = \left(v_{ij}\right)_{m \times n} = \begin{bmatrix} v_{11} & \cdots & v_{1n} \\ \vdots & \ddots & \vdots \\ v_{m1} & \cdots & v_{mn} \end{bmatrix} \tag{9}$$

Step 2: Normalize indexes

The min-max normalization formula is used to normalize the index. If the index is positive, it is used for the benefit type. The higher value is the better. The formula is as follows:

$$r_{ij} = \frac{v_{ij} - \min_j(v_{ij})}{\max_j(v_{ij}) - \min_j(v_{ij})} \tag{10}$$

If the index is negative, it is used for cost type. The lower the value the better. The formula is:

$$r_{ij} = \frac{\max_j(v_{ij}) - v_{ij}}{\max_j(v_{ij}) - \min_j(v_{ij})} \tag{11}$$

The matrix after normalizing can be defined as:

$$R = \left(r_{ij}\right)_{m \times n} \tag{12}$$

Step 3: Calculate proportion and information entropy for each index

Equation (13) is used to calculate the proportion of each j index in each i alternative:

$$P_{ij} = \frac{r_{ij}}{\sum_{i=1}^{m} r_{ij}} \;,\; j = 1, 2 \ldots, n \tag{13}$$

Then calculate the j-index information entropy e_j by using Equation (14), and use the value of e_j to judge the discrete degree of the index. When e_j is smaller, the discrete degree is greater, and the weight as well:

$$e_j = -K \times \sum_{i=1}^{m} P_{ij} \cdot \ln P_{ij} \;,\; j = 1, 2 \ldots, n \tag{14}$$

where the constant $K = \frac{1}{\ln_m}$

Step 4: Calculate weight for each index

Equation (15) is used to calculate the weight ω_j of the jth index:

$$\omega_j = \frac{1 - e_j}{n - \sum_{j=1}^{n} e_j} \;,\; j = 1, 2 \ldots, n \tag{15}$$

where $\sum_{j}^{n} \omega_j = 1$ and $0 \leq \omega_j \leq 1$. The larger ω_j means the higher importance of the objective response of the j indicator.

Step 5: Calculate the weight of normalized matrix R

Multiply the j index corresponding to each i alternatives by ω_j. The matrix Y can be calculated by Equation (16):

$$Y = \left(r_{ij}\right)_{m \times n} \times \omega_j = \begin{bmatrix} r_{11} \cdot \omega_1 & \cdots & r_{1n} \cdot \omega_n \\ \vdots & \ddots & \vdots \\ r_{m1} \cdot \omega_1 & \cdots & r_{mn} \cdot \omega_n \end{bmatrix} \tag{16}$$

Step 6: Compute distance between each i alternatives and positive ideal solution D_i^+ and negative ideal solution D_i^- .

The Euclidean Distance can be calculated as follows:

$$D_i^+ = \sqrt{\sum_{j=1}^{n} \left(y_{ij} - y_j^+\right)^2} \;,\; i = 1, 2 \ldots, m \tag{17}$$

$$D_i^- = \sqrt{\sum_{j=1}^{n}\left(y_{ij} - y_j^-\right)^2} \ , i = 1, 2\ldots, m \tag{18}$$

where $y_j^+ = \left\{ \max_{1\le j\le n} y_{ij} \middle| i = 1, 2\cdots m \right\}$ and $y_j^- = \left\{ \min_{1\le j\le n} y_{ij} \middle| i = 1, 2\cdots m \right\}$.

Step 7: Calculate relative closeness for each alternative

Through the calculation of the relative closeness (C_i), it can be judged that alternative is close to ideal solution. If the C_i is close to 1, the ith alternative is close to the ideal solution:

$$C_i = \frac{D_i^-}{D_i^+ + D_i^-} \ , i = 1, 2\ldots, m \tag{19}$$

Step 8: Sorting the C_i calculated by each i alternative, the alternative has the maximum C_i is the optimization alternative in this study.

3. Hybrid Energy System Description

3.1. Hybrid Energy System Schematic Diagram

The schematic diagram of the hybrid energy system shown in Figure 9 consists of diesel generators, PV system (PV$_S$), wind-power generation system (WG$_S$), storage system, power conversion system (PC$_S$) and load. PV$_S$ and WG$_S$ are operated in parallel with diesel generators and AC coupled. Electricity generated by PV$_S$ and WG$_S$ can be delivered to AC load to decline the output of diesel generator and to minimize fuel consumption [38,39].

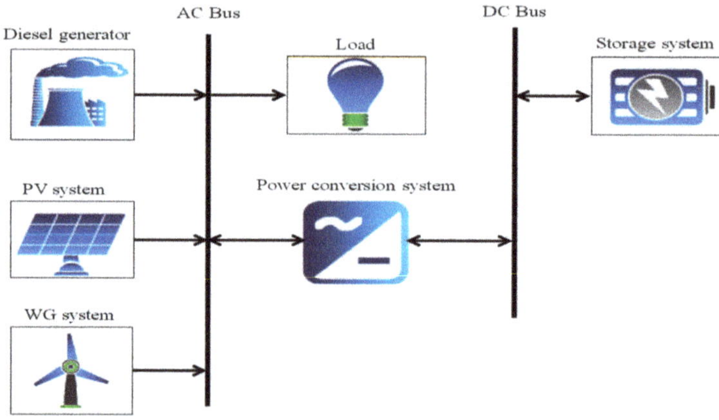

Figure 9. Schematic diagram of the hybrid energy system.

3.2. PV System

The way of electricity generation by PV module is through the conversion of solar energy into DC power. Identification, adoption and utilization of dependable interconnection technology to assembly crystalline silicon solar cells in PV module are very important to guarantee the device functions persistently up to 20 years of design life span [33]. For this study, the PV module selected is the GTEC-G6S6A model mono-crystalline silicon solar cell type which specifications are presented in Table A1 of Appendix A. South installation direction, 10° incline, 5.42 m² per kWp, 25 years of serving life and 80% derating factor are considered. Equation (20) is used by HOMER to calculate the power output of PV system [40,41]:

$$P_{pv} = P_{pv,STC} f_{pv} \frac{G_T}{G_{T,STC}} [1 + K_P(T_C - T_{STC})] \tag{20}$$

where $P_{pv,STC}$ is PV system rated power (kWp), f_{pv} is PV derating factor (%), G_T is solar irradiance on the surface of the PV module (kW/m²), $G_{T,STC}$ is standard solar irradiance (1 kW/m²), K_P is the PV module temperature coefficient (−0.4003%/°C), T_C is temperature of the PV module (°C), T_{STC} is temperature of PV module under the standard test conditions (25 °C).

3.3. WG System

The WGs converts wind kinetic energy into electricity using wind turbine blades. The architecture of WG system in the study is AC coupled hybrid. The wind turbine used for simulation is the Bergey XL 10 Wind Turbine complete set [42] with AC output, rated power 10 kW at 12 m/s, minimum wind speed 2.5 m/s and swept area 38.5 m². The hub height is 30m, power curve of wind turbine is displayed in Figure 10, and the lifetime is set to 20 years [43].

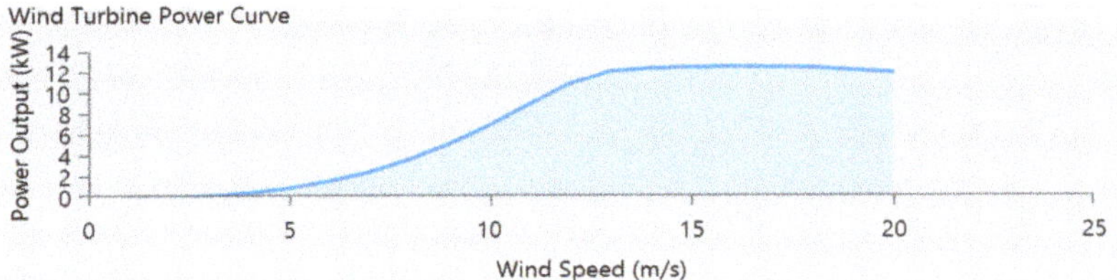

Figure 10. Wind turbine power curve.

The calculation of the power law profile in HOMER is based on Equation (21) to estimate the wind speed of different hub heights [44]:

$$U_{hub} = U_{ref} \times \left(\frac{H_{hub}}{H_{ref}}\right)^{\alpha} \tag{21}$$

where U_{ref} is the reference wind speed (m/s), the value is provided by NASA measures at a height of 50 m from the surface. H_{hub} is the hub height (m). H_{ref} is the measured height of the reference wind speed (m). α is surface roughness, the value is set to 0.14 [44,45].

3.4. Storage System

A storage system will be used to regulate the power and minimize the influence the quality of energy on intermittent renewable energy sources in the proposed HES. It accumulates and transfers energy from PVs, WGs and diesel to a rechargeable Li-ion thin film during excess electricity generation. In this paper a high energy density Samsung SDI M2-R084 model lithium-ion battery [46] was selected as storage system with the following specifications: The lowest expected lifespan of each rack is five years and each rack holds 11 modules. Each module has 22 series-connected cells. The capacity is 94 Ah and the rated capacity of each storage rack is 84 kWh. The area occupied by one rack is 0.31 m². The voltage rating is 774 1004 V and the lowest state of charge (SOC) is 20%. The C rate is 1 C and lifetime throughout is 337,123 kWh with 4000 cycles.

3.5. Power Conversion System

A dual-direction DC-AC converter is used for power conversion. In HES, when excess electricity is produced by diesel, PV and WG, the excess AC power will be changed into DC and stored in battery to supply it back to AC load when required. Conversion performance of 95% and 10 years of lifespan is assumed in the simulation.

3.6. Diesel Generators

The power supply system of the island discussed in this study uses diesel generators (DGs) to supply power 24 h a day. The main power supply, with a total capacity of 2850 kW, contains five DGs and the supply schedule is accordingly to meet the load demand.

To simplify the process of simulation analysis, set operation time of DG1 and DG2 to 0–11 a.m., operation time of DG3 and DG4 to 12–23 p.m., DG5 to backup, minimum load to 25%, and lifetime to 131,400 h. The simulation parameters of diesel generators are as shown in Table 2.

Table 2. Simulation parameters of the diesel generators.

Number	Prime Power	Type	Fuel Consumption Slope (Liter/kWh)	Operation Schedule
DG1	600 kW	Cummins QSK23-G3	0.2376	AM 00–11
DG2	600 kW	Cummins QSK23-G3	0.2376	AM 00–11
DG3	600 kW	Cummins KTA38-G1	0.2273	PM 12–23
DG4	600 kW	Denyo DCA series	0.2351	PM 12–23
DG5	450 kW	MAN B & W-8L20/27	0.2440	Backup

3.7. System Dispatch Strategy

System dispatch approach in HOMER is primarily for managing the operational rules of the power generation equipment and energy storage. The simulation software has two options: cycle charging strategy and load following strategy and [34]. HES operation in this case study is based on a DG power supply. PV and WG are used as auxiliary power supplies. Increasing the penetration of the PVs and WGs reduces fuel consumption. The DGs only generate sufficient power to fulfill load demand. DG has the limited base load output, so when PVs and WGs have excess power, the storage will charge, but if the energy storage is full and there is still excess energy that cannot be stored or used, the software will consider that to be excess electricity, as shown in (7). As in the actual operation of the HES, in the above scenario, the output of the PVs or WGs power converter must be controlled to reduce their output to maintain balance and/or damping load is required. Based on the operation mode, the dispatch mode of the power supply system must select the load following strategy and so as to calculate various costs such as fuel, operation and maintenance, and replacement.

4. Component Cost and Financial Assumption

4.1. System Component Cost

The summary of component costs used in this study is from the Taiwan System Integration Company, and is presented in Table 3. The cost consists of capital cost, replacement cost, operation and maintenance (O&M). In order to simplify the analysis, the capital and replacement cost is the same.

Capital cost of PVs is 4000 $/kW, including PV module, PV inverter, transportation, installation engineering. Capital cost of WGs is 5800 $/kW, including wind turbine, wind inverter, 30 m tower and cement foundation, transportation, installation engineering and so on. PVs, WGs and storage operate in parallel with the existing five DGs. To simplify the analysis, the capital cost of DG is not included in the calculation. The estimates of the cost of remaining components include transportation, and installation engineering.

Table 3. Summary of component costs.

Description	Data Description
PV system	
Capital cost ($/kW)	4000
Replacement cost ($/kW)	4000
Operation and maintenance cost ($/kW/yr)	20
WG system	
Capital cost ($/kW)	5800
Replacement cost ($/kW)	5800
Operation and maintenance cost ($/kW/yr)	40
Storage system	
Capital cost ($/kWh)	690
Replacement cost ($/kWh)	690
Operation and maintenance cost ($/kWh/yr)	5
Power converter	
Capital cost ($/kW)	700
Replacement cost ($/kW)	700
Operation and maintenance cost ($/kW/yr)	1
Diesel generator	
Capital cost ($/kW)	0
Replacement cost ($/kW)	400
Operation and maintenance ($/h)	0.03

4.2. Interest Rate and Inflation Rate

Figures 11 and 12 depict the interest rate and inflation rates of Bangko Sentral NG Pilipina (BSP) in recent years. Interest rates have been adjusted to 4.75% in 7 February2019 [47]. Refering to the BSP announcement in February 2019, the inflation rate is 4.4% [48]. Since January 2013 to February 2019, the highest interest rate was 4.75%, the lowest was 3%, so the average was 3.579%. The highest inflation rate was 6.7%, the lowest was −2%, and the average was 2.716% [48,49]. In this paper, the real discount rate is the average value in the statistical period, and the value calculated by (1) is 0.84%. The lifetime is set to 20 years.

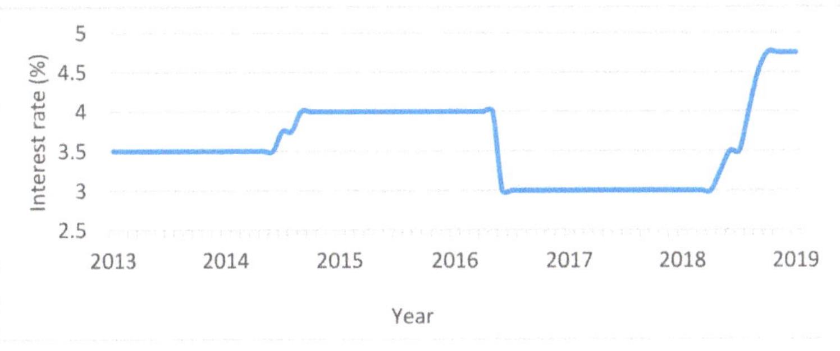

Figure 11. Interest rate information for Taiwan since January 2013 to February 2019.

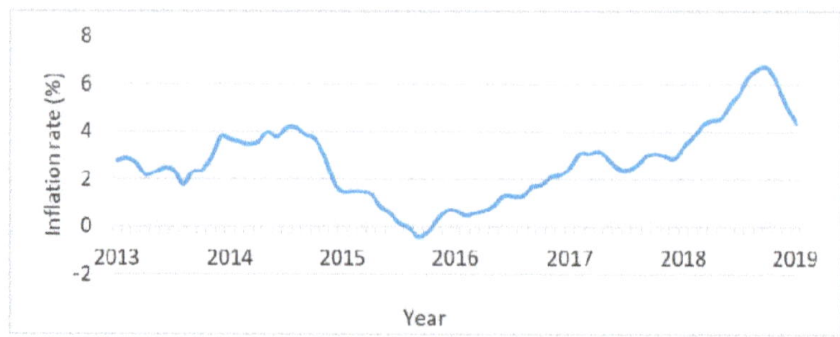

Figure 12. Inflation rate information for Taiwan since January 2013 to February 2019.

5. Simulation Results and Discussion

The flowchart of systems operation for cases 1 to 7 are presented in Figure A1 of Appendix A.

5.1. Case 1: Stand-Alone Diesel System

Simulation results are as shown in Table 4. Total annual fuel intake of the five DGs is 1,623,100 L/y (4447 L/d), the total annual power supply is 6,195,877 kWh/y, average fuel intake per kWh is 0.26 L/kWh, and annual total carbon dioxide (CO_2) emission is 4,274,156 kg/y.

Table 4. Electrical characteristics of stand-alone diesel system.

Component	Operation Schedule	Hours of Operation (h/yr)	Electrical Production (kWh/yr)	Fuel Consumption (L/yr)	Specific Fuel Consumption (L/kWh)
DG1	AM 0–11	4380	2,002,495	511,230	0.26
DG2	AM 0–11	4380	783,942	221,714	0.28
DG3	PM 12–23	4380	2,373,373	606,795	0.26
DG4	PM 12–23	4380	1,010,918	275,641	0.27
DG5	Backup	219	25,159	7720	0.31
System			6,195,887	1,623,100	0.26

No capacity shortage problem throughout the year and it meets the annual load requirements. DG5 is an auxiliary generator, which starts when the load demand reaches 90% of the rated capacity of two DGs to avoid any power supply shortage. The load ratio of DG5 is lower than that of other diesel generator sets, so the fuel consumption required per kWh (liter/kWh) is also relatively higher than that of other diesel generator sets. As indicated in Table 5, total NPC of the five DGs is $29,650,884 and the total COE is 0.2609 $/kWh. Among costs, the cost of oil is the largest and it takes 80% of the total NPC.

Table 5. Economic characteristics of stand-alone diesel system.

Component	Capital ($)	Replacement ($)	O&M ($)	Fuel ($)	Salvage ($)	NPC ($)	COE ($/kWh)
DG1	0	0	1,445,869	7,500,470	0	8,946,339	0.2436
DG2	0	0	1,445,869	3,252,855	0	4,698,724	0.3268
DG3	0	0	1,445,869	8,902,549	0	10,348,418	0.2377
DG4	0	0	1,445,869	4,044,047	0	5,489,916	0.2961
DG5	0	0	54,220	113,267	0	167,487	0.3630
System	0	0	5,837,696	23,813,188	0	29,650,884	0.2609

5.2. Hybrid Energy System without Storage

5.2.1. Case 2: PV-Diesel HES

The PV capacity configuration scheme for minimum COE of a PV-diesel HES is analyzed and the change of PV capacity is set from 50 to 2000 kWp and the step interval is 50 kWp in the simulation. The simulation results are shown in Figure 13. Because the analyzed system does not include an energy storage system, as the PV capacity rises, the percentage of renewable fraction (RF) and fuel saving will improve. The increment will reduce in accordance with the increase of PV capacity and the excess electricity fraction will increase relatively.

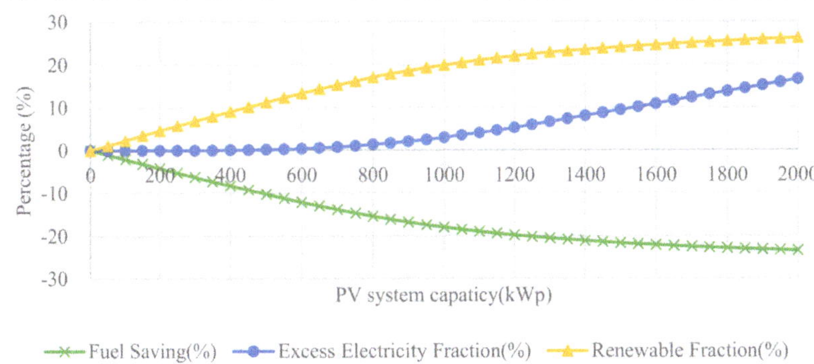

Figure 13. Effect of varying PV_S capacity on fuel saving, RF and excess electricity fraction.

The varying PV capacity on COE and RF are as shown in Figure 14. When the capacity of PVs is 550 kWp and PV operates with the DG in parallel, the lowest COE is 0.2583 \$/kWh, which is slightly lower than the COE (0.2609 \$/kWh) in the DG-only case. As shown in Table 6, the RF is 12.4%, the excess electricity fraction is 0.2% and the fuel saving is −11.17%. When the PV capacity continuously increases, the investment cost will also increase. Because the power generated by PVs is not entirely used by loads to minimize the DG fuel consumption, the COE will increase relatively. Table 7 shows economic characteristics of the lowest COE PV-diesel HES. Cost of oil is still the largest proportion, which is 72% of the total NPC. The value of NPC of PVs with 550 kWp total capacity is only 8% of the total NPC. Figure 15 shows the annual PV_S power production by the lowest COE PV-diesel HES. The output power of quarter 2 (April to June) and quarter 3 (July to September) can be observed, which is higher than quarter 1 (January to March) and quarter 4 (October to December).

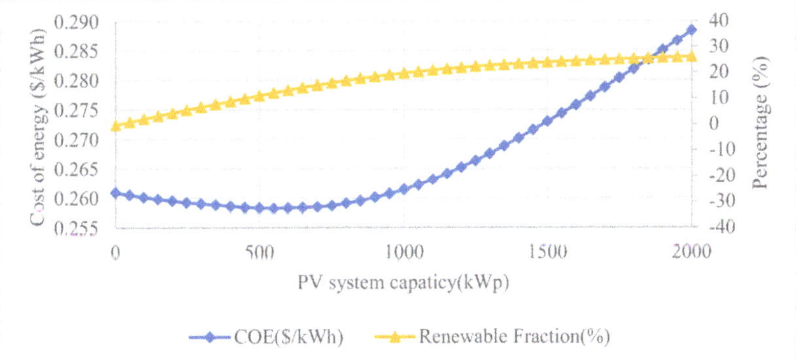

Figure 14. Effect of varying PV_S capacity on COE and RF.

Table 6. Electrical characteristics of lowest COE PV-diesel HES.

Genset	PV$_S$	Electrical Production	Excess Electricity	PV$_S$ Capacity Factor	Renewable Fraction	Fuel Consumption	Fuel Saving
(kW)	(kWp)	(kWh/yr)	(kWh/yr)	(%)	(%)	(liter/yr)	(%)
2850 (5 sets)	550	6,214,612	19,231.4 (0.3%)	16.32	12.4	1,441,803	−11.17

Table 7. Economic characteristics of the lowest COE PV-diesel HES.

Component	Capital	Replacement	O&M	Fuel	Salvage	NPC	COE
	($)	($)	($)	($)	($)	($)	($/kWh)
DG1	0	0	1,445,869	6,504,553	0	7,950,422	0.2525
DG2	0	0	1,445,869	2,892,353	0	4,338,222	0.3476
DG3	0	0	1,445,869	8,170,166	0	9,616,035	0.2434
DG4	0	0	1,445,869	3,552,188	0	4,998,057	0.3133
DG5	0	0	16,093	34,043	0	50,135	0.3603
PV$_S$	2,200,000	0	201,732	0	0	2,401,732	0.1666
System	2,200,000	0	6,001,301	21,153,303	0	29,354,604	0.2583

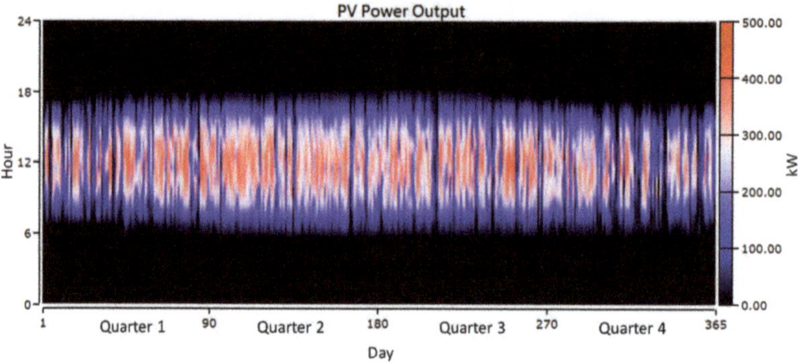

Figure 15. Annual PV$_S$ power production by lowest COE PV-diesel HES.

5.2.2. Case 3: Wind-Diesel HES

Using the analysis method mentioned in Section 5.2.1, we set the change of WG capacity between 200–800 kW and the interval at 10 kW to find the configuration with the lowest COE. As presented in Figure 16, trend of the curve is like the PV-diesel HES one. When the capacity of WGs is 380 kWp and with DG in parallel operation, the minimum COE calculated is 0.2541 $/kWh and this result is slightly less than COE in Case 1 and Case 2, as shown in Figure 17.

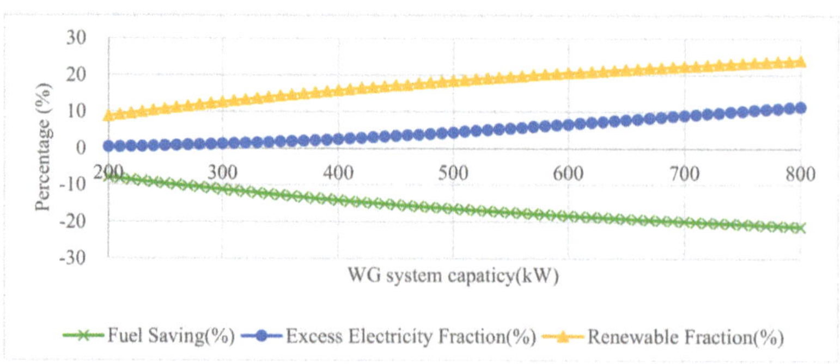

Figure 16. Effect of varying WG$_S$ capacity on fuel saving, RF and excess electricity fraction.

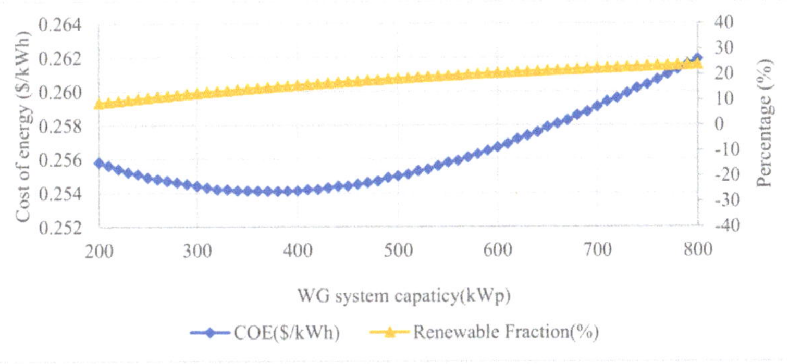

Figure 17. Effect of varying WG$_S$ capacity on COE and RF.

As shown in Table 8, the capacity factor of WGs is 32.28%, which is higher than 16.32% of PVs. Therefore, the installation capacity of WG is lower than PV but the system still has lower COE and higher RF: RF is 13.59%, the excess electricity fraction is 2.2%, and the fuel saving is −13.59%.

Table 8. Electrical characteristics of the lowest COE wind-diesel HES.

Genset	WG$_S$	Electrical Production	Excess Electricity	Capacity Factor	Renewable Fraction	Fuel Consumption	Fuel Saving
(kW)	(kW)	(kWh/yr)	(kWh/yr)	(%)	(%)	(liter/yr)	(%)
2850 (5 sets)	380	6,333,993	138,612.4 (2.2%)	32.48	15.2	1,402,500	−13.59

Table 9 shows the cost simulation of the lowest COE wind-diesel HES in 380 kW total of capacity. The COE is lower than the COE of DG 1–5 and that of PVs. Figure 18 shows the distribution of annual WGs output power and displays that output power of quarter 1 and 4 is higher than that of quarter 2 and 3. The trend is opposite to the result of the curve of quarter output power of PVs.

Table 9. Economic characteristics of the lowest COE wind-diesel HES.

Component	Capital	Replacement	O&M	Fuel	Salvage	NPC	COE
	($)	($)	($)	($)	($)	($)	($/kWh)
DG1	0	0	1,445,869	6,455,536	0	7,901,405	0.2530
DG2	0	0	1,445,869	3,077,750	0	4,523,619	0.3362
DG3	0	0	1,445,869	7,497,764	0	8,943,633	0.2498
DG4	0	0	1,445,869	3,491,624	0	4,937,493	0.3158
DG5	0	0	25,748	54,001	0	79,750	0.3621
WG$_S$	2,204,000	0	278,757	0	0	2,482,757	0.1252
System	2,204,000	0	6,087,982	20,576,675	0	28,868,656	0.2541

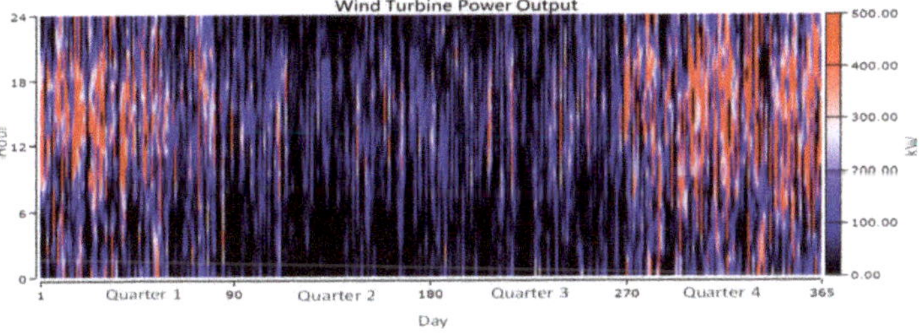

Figure 18. Annual WG$_S$ power production by the lowest COE wind-diesel HES.

5.2.3. Case 4: PV-Wind-Diesel HES

As shown in Figures 15 and 18, PVs and WGs have different seasonal power generation characteristics, so the hybrid power supply can be complementary to improve RF. After simulation analysis, the hybrid power system has PVs of 200 kWp and WGs of 340 kW, which is the lowest capacity configuration. As depicted in Table 10 the analysis of electrical characteristics, RF is 18.1%, fuel saving is −16.18%, which is better than Case 1 and 2. In Table 11, the economic characteristics analysis indicates that COE is 0.2539 $/kWh and result is slightly lower than Case 3. In the comparison, the result of Case 4 has the lowest COE in all of capacity configuration without the energy storage HES.

Table 10. Electrical characteristics of the lowest COE PV-wind-diesel HES.

Genset	PV$_S$	WG$_S$	Electrical Production	Excess Electricity	Renewable Fraction	Fuel Consumption	Fuel Saving
(kW)	(kWp)	(kW)	(kWh/yr)	(kWh/yr)	(%)	(liter/yr)	(%)
2850 (5 sets)	200	340	6,330,194	134,813.6 (2.1%)	18.1	1,360,550	−16.18

Table 11. Economic characteristics of the lowest COE PV-wind-diesel HES.

Component	Capital	Replacement	O&M	Fuel	Salvage	NPC	COE
	($)	($)	($)	($)	($)	($)	($/kWh)
DG1	0	0	1,445,869	6,225,212	0	7,671,081	0.2556
DG2	0	0	1,445,869	2,951,936	0	4,397,805	0.3437
DG3	0	0	1,445,869	7,395,589	0	8,841,458	0.2509
DG4	0	0	1,445,869	3,365,351	0	4,811,220	0.3214
DG5	0	0	10,894	23,121	0	34,014	0.3597
PV$_S$	800,000	0	73,357	0	0	873,357	0.1666
WG$_S$	1,972,000	0	249,414	0	0	2,221,414	0.1252
System	2,772,000	0	6,117,141	19,961,209	0	28,850,350	0.2539

5.3. Hybrid Energy System with Storage

5.3.1. Case 5: PV/Diesel/Storage HES

As the capacity of PV or WG increases, the RF will increase, but the excess electricity fraction will also increase. Although it will waste the energy, the problem could be enhanced by installing an energy storage. Capacity configuration for PV-diesel-storage HES of the lowest COE in each RF level addressed in this section. The setting parameters are as follows: RF is 25%–50%, the interval is 5% and the excess electricity fraction must be less than 5% considering the utilization rate of the energy.

Table 12 lists the electrical characteristics of PV-diesel-storage HES. The result displays that higher RF can let the effect of fuel saving better.

Table 12. Electrical characteristics of PV-diesel-storage HES.

Renewable Fraction	Genset	PV$_S$	Storage	Electrical Production	Excess Electricity	Fuel Consumption	Fuel Saving
(%)	(kW)	(kWp)	(kWh)	(kWh)	(kWh/yr)	(kWh/yr)	(liter/yr)
25		1200	1260	6,359,813	122,652.0 (1.9)	1,259,864	−22.38
30		1500	2100	6,478,906	200,845.9 (3.1)	1,188,383	−26.78
35	2850 (five	1750	3444	6,528,063	196,732.8 (3.0)	1,117,409	−31.16
40	sets)	2050	4872	6,630,449	239,512.0 (3.6)	1,041,868	−35.81
45		2300	6804	6,691,938	243,681.9 (3.6)	972,643	−40.07
50		2600	12,600	6,813,660	308,601.5 (4.5)	900,837	−44.50

Table 13 presents the economic characteristics of PV-diesel-storage HES. In RF 25% hybrid system configuration: PVs capacity is 1200 kWp, energy storage capacity is 1260 kWh and operates in parallel with DG. Above all, the minimum COE can be obtained is 0.2864 $/kWh. The cost and storage capacity have the significant difference in RF 45% and 50% because the simulation constraint should be satisfied: The capital cost difference is 1.35 times and the COE difference is 1.13 times.

Table 13. Economic characteristics of PV-diesel-storage HES.

Renewable Fraction	Capital	Replacement	O&M	Fuel	Salvage	NPC	COE
(%)	($)	($)	($)	($)	($)	($)	($/kWh)
25	6,716,295	965,724	6,379,540	18,484,004	0	32,545,563	0.2864
30	8,493,825	965,724	6,562,887	17,435,266	0	33,457,702	0.2945
35	10,417,873	965,724	6,775,842	16,393,978	0	34,553,418	0.3041
40	12,599,674	965,724	7,013,354	15,285,686	0	35,864,438	0.3157
45	14,927,993	965,724	7,280,970	14,270,060	0	37,444,747	0.3296
50	20,112,950	965,724	7,921,983	13,216,565	0	42,217,222	0.3716

5.3.2. Case 6: Wind-Diesel-Storage HES

Simulation analysis shows that WG power gen ration was concentrated in quarters 1 and 4, and the load power consumption during this period is lower than that of quarters 3 and 4. When the installed capacity of WGs and storage is more than RF 35%, the excess electric fraction has exceeded 5% and there is no system capacity configuration meets the set simulation conditions.

Tables 14 and 15 are the simulation results of the electrical and economic characteristics, respectively. In RF 25% hybrid system configuration: The capacity of WGs is 680 kW, storage capacity is 2520 kWh and operates in parallel with DG. Above all, the lowest COE can be obtained is 0.2874 $/kWh. The cost and storage capacity have the significant difference in RF 30% and 35% because the simulation constraint should be satisfied. The cost of the system configuration has increased significantly in RF 35%.

Table 14. Electrical characteristics of wind-diesel-storage HES.

Renewable Fraction	Genset	WG$_S$	Storage	Electrical Production	Excess Electricity	Fuel Consumption	Fuel Saving
(%)	(kW)	(kW)	(kWh)	(kWh)	(kWh/yr)	(kWh/yr)	(liter/yr)
25		670	2520	6,541,615	299,274.0 (4.6)	1,258,831	−22.44
30	2850 (five sets)	800	20,244	6,609,215	326,952.9 (4.9)	1,188,672	−26.77
35		1060	67,200	7,034,297	727,785.1 (10.3)	1,115,342	−31.28

Table 15. Economic characteristics of wind-diesel-storage HES.

Renewable Fraction	Capital	Replacement	O&M	Fuel	Salvage	NPC	COE
(%)	($)	($)	($)	($)	($)	($)	($/kWh)
25	6,668,590	965,724	6,552,369	18,468,839	0	32,655,522	0.2874
30	19,608,473	965,724	8,271,475	17,439,515	0	46,285,187	0.4074
35	53,400,400	965,724	12,765,424	16,363,651	0	83,495,199	0.7349

5.3.3. Case 7: PV-Wind-Diesel-Storage HES

The simulation results of the electrical and economic characteristics of the PV-wind-diesel-storage HES are shown in Tables 15 and 16, respectively. The capital cost, NPC and COE in case 7 are lower than in cases 5 and 6 when RF is 25–50%.

In RF 25% hybrid system configuration: The minimum capital cost is $5,317,506, the NPC is $31,100,367 and the COE is 0.2737 $/kWh. Because solar energy and wind energy have characteristics of seasonal power complementary, the required capacity of PVs, WGs and storage are lower than that of Cases 5 and 6, are 400 kWp, 440 kW and 168 kWh, respectively.

Table 16. Electrical characteristics of PV-wind-diesel-storage HES.

Renewable Fraction	Genset	PV$_S$	WG$_S$	Storage	Electrical Production	Excess Electricity	Fuel Consumption	Fuel Saving
(%)	(kW)	(kWp)	(kW)	(kWh)	(kWh/yr)	(kWh/yr)	(liter/yr)	(%)
25		400	440	168	6,468,587	264,437.6 (4.1)	1,259,850	−22.38
30		700	420	504	6,528,221	311,155.0 (4.8)	1,187,107	−26.86
35	2850 (five	950	410	1680	6,548,999	297,053.1 (4.5)	1,115,897	−31.25
40	sets)	1300	370	3108	6,623,216	322,638.7 (4.9)	1,043,925	−35.68
45		1750	280	5292	6,702,660	331,642.6 (4.9)	972,740	−40.07
50		2400	100	11,508	6,810,636	336101.2 (4.9)	900,511	−44.52

5.3.4. Optimal Capacity Configuration Analysis of the PV-Wind-Diesel-Storage HES

As mentioned in Section 5.3.3, if the cost is the main indicator, Case 7 has the lowest COE HES. This section uses the entropy weight and TOPSIS method and regards the fuel saving, capital cost, and COE as the indicators. Because the available space on the island is limited, the area occupied by the equipment is also considered. Above all, we evaluate the optimal hybrid system configuration for Case 7 at RF 25%–50%. Substituting the values from Table 17 into Equations (9)–(15), the information entropy and weighting of each index can be obtained. The weight of fuel saving is the highest, as shown in Table 18. Through the Equations (16)–(19), the relative closeness value and sorting result of each alternative can be obtained, as shown in Table 19. Therefore, the best system capacity configuration for this stud: The maximum relative closeness is the system capacity configuration in RF 35%, the PVs capacity is 950 kWp, the WGs capacity is 410 kW, and the energy storage capacity is 1680 kWh.

Table 17. Economic characteristics of the PV-wind-diesel-storage HES.

Renewable Fraction	Capital	Replacement	O&M	Fuel	Salvage	NPC	COE
(%)	($)	($)	($)	($)	($)	($)	($/kWh)
25	5,317,506	1,066,934	6,303,798	18,483,799	−71,670	31,100,367	0.2737
30	6,632,518	1,261,386	6,429,230	17,416,549	−277,216	31,462,467	0.2769
35	8,383,060	965,724	6,620,930	16,371,799	0	32,341,513	0.2846
40	10,532,861	965,724	6,849,419	15,315,875	0	33,663,880	0.2963
45	13,312,439	965,724	7,147,231	14,271,476	0	35,696,871	0.3142
50	19,142,161	965,724	7,821,603	13,211,782	0	41,141,271	0.3621

Table 18. Information entropy and weighting factors of each index.

Indices	Capital	COE	Fuel Saving	Occupied Area
Information entropy	0.87577	0.88614	0.83210	0.85829
Weighting factors (%)	22.68%	20.79%	30.66%	25.87%

Table 19. The relative closeness and ranking results.

	RF-25%	RF-30%	RF-35%	RF-40%	RF-45%	RF-50%
D+	0.30656	0.24911	0.20730	0.20175	0.25112	0.40201
D−	0.40201	0.36488	0.33507	0.30944	0.29627	0.30656
C	0.56736	0.59428	0.61779	0.60533	0.54124	0.43264
Ranking	4	3	1	2	5	6

5.4. Sensitivity Analysis

According to the analysis result in Section 5.3.4, the GHI, wind speed, diesel fuel price, and load consumption for RF 35% PV-wind-diesel-storage HES are used to discuss the economic and electrical of the system.

5.4.1. Global Horizontal Irradiation and Wind Speed

We set the value of GHI and wind speed 5.14 kW/m^2/d and 7.22 m/s in ±20% changes, respectively, i.e., GHI from 4.09 to 6.17 kW/m^2/d, and wind speed from 5.78 m/s to 8.66 m/s. According to Figures 19 and 20, when GHI or wind speed increases, the RF will continue to increase and this will let the DG fuel consumption and COE reduce. When GHI is 6.17 kW/m^2/d and wind speed is 8.66 m/s, RF can rise to 40.7% and COE can be reduced to 0.2747 $/kWh.

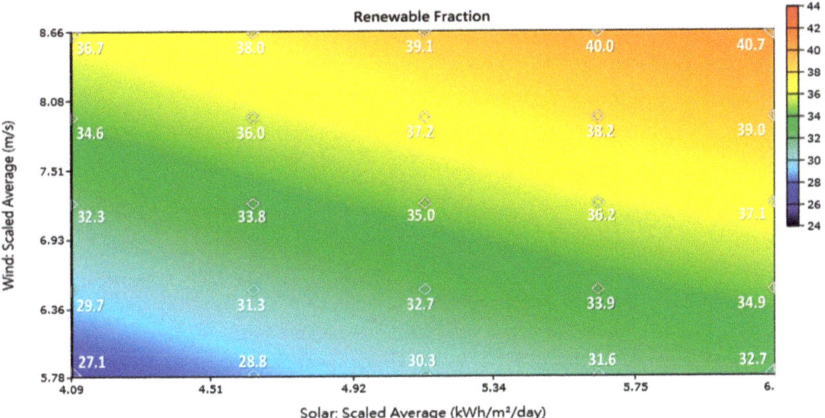

Figure 19. Effect of varying GHI and wind speed on RF generated by optimal PV-wind-diesel-storage HES.

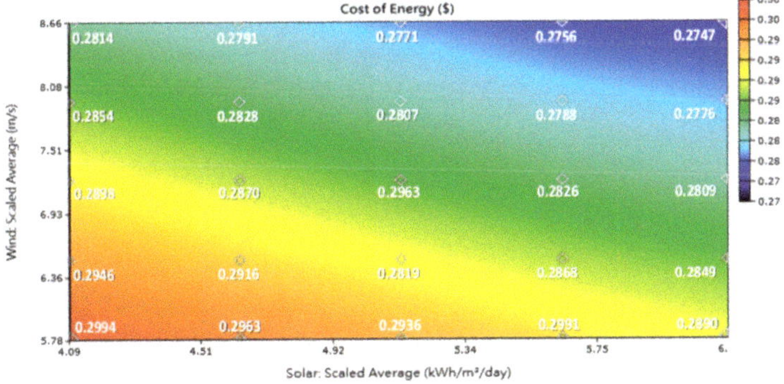

Figure 20. Effect of varying GHI and wind speed on COE generated by PV-wind-diesel-storage HES.

5.4.2. Diesel Fuel Price

We set the diesel fuel price change from 0.49–1.18 $/liter, which is 1.25 times the maximum and minimum values during the statistics period of oil price, as the change range of the oil price. Figure 21 show the impact on NPC and COE under different fuel price scenarios. When oil price is 1 $/L, NPC and COE are as high as $36,434,464 and 0.3207 $/kWh, respectively.

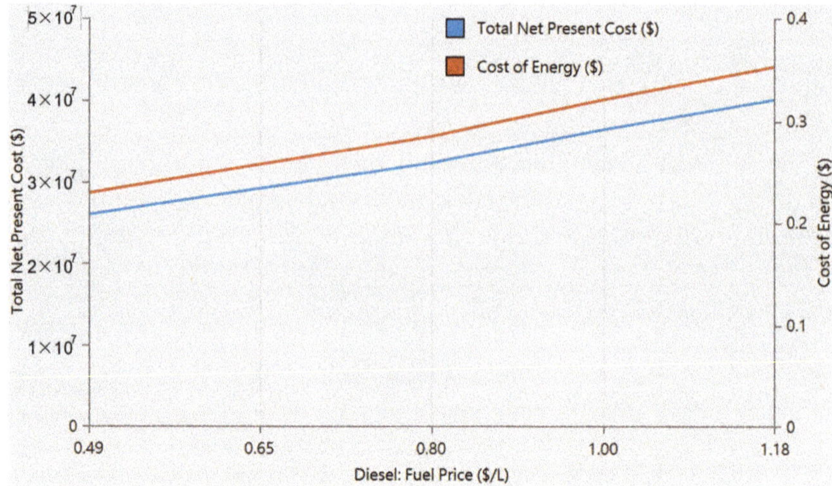

Figure 21. Effect of varying diesel fuel price on NPC and COE.

5.4.3. Load Consumption

We set the load consumption to vary from 16,974 to 20,000 kWh/d. As depicted in Figure 22, when the load consumption rises and the capacity configuration of PVs, WGs and storage do not change, the DGs will increase their power generation to meet the load demand, so the RF will decrease and the fuel usage will increase. As shown in Figure 23, when the load consumption increases, the fuel consumption, the fuel cost and the NPC rise because the power generation of DG increases. It can also be observed from (5) that the load consumption increases so the COE decreases.

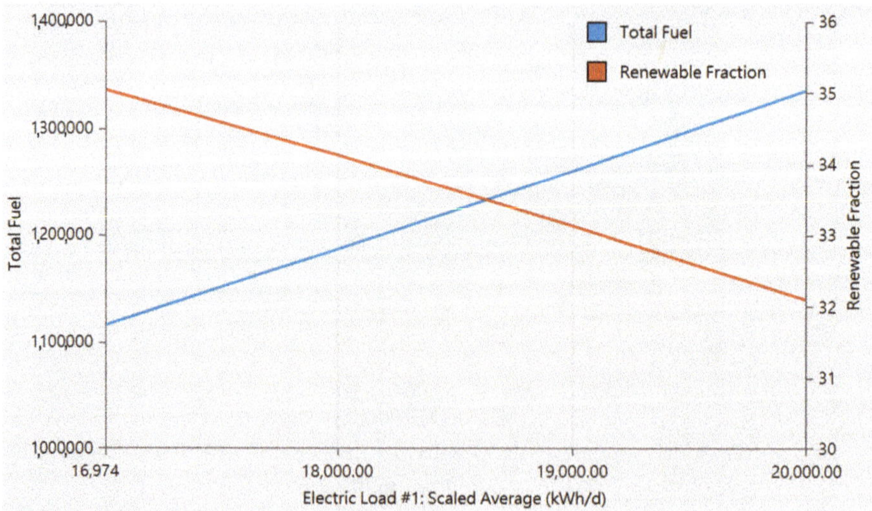

Figure 22. Effect of varying load consumption on fuel consumption and RF.

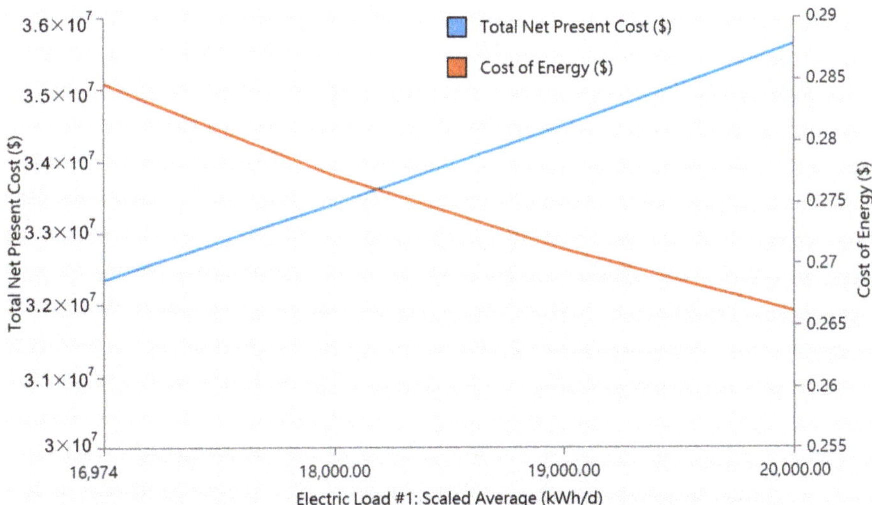

Figure 23. Effect of varying load consumption on NPC and COE.

6. Conclusions

Promoting the infrastructure of renewable energy is a global trend, especially in areas with insufficient power supply and low electrification. Renewable energy is regarded as a key to improving economic and development. This paper uses the Philippine islands, and a renewable energy and hybrid power supply to supplement existing diesel generators, to be the study case.

We use HOMER to simulate and investigate the techno-economic of different power supply systems' containing stand-alone diesel, PV-diesel HES, wind-diesel HES, PV-wind-diesel HES, PV-diesel-storage HES, wind-diesel-storage HES, PV-wind-diesel-storage HES. The study also uses entropy weight and TOPSIS method to evaluate optimal capacity configuration. The conclusions of the analysis are as follows:

- Based on the climate and load demand on the island, we simulate HES without an installing energy storage system such as PV-diesel HES, wind-diesel HES and PV-wind-diesel HES. Because solar energy and wind energy have characteristics of complementary seasonal power, the lowest COE obtained is 0.2539 \$/kWh in PV-wind-diesel HES. The result is lower than the COE in the DG-only case. The capacity of PVs and WGs are 200 kWp and 340 kW, respectively. In addition, RF is 18.1% and fuel saving is −16.18%.

- As the capacity of PVs or WGs increases, the RF will increase, but the excess electricity fraction will also increase. Although this will waste energy, the problem is mitigated by installing an energy storage system. In the three kinds of HES with the energy storage, discussing the capacity configuration of the lowest COE in each RF stage for PV-diesel-storage HES and the constraints are as follows: RF is 25%–50%, the interval is 5% and the excess electricity fraction must be less than 5% considering the utilization rate of the energy. In RF 25% hybrid system configuration: the minimum COE obtained is 0.2737 \$/kWh. Required capacity of PVs, WGs and storage are 400 kWp, 440 kW and 168 kWh, respectively. The fuel savings is −22.38%.

- For the system with energy storage, the result of RF 35% is the best capacity configuration in this research: PVs capacity is 950 kWp, WGs capacity is 410 kW, and energy storage capacity is 1680 kWh.

- RF 35% PV-wind-diesel-storage HES is used to analyze the sensitivity. When GHI and wind speed increase, RF will continuously increase, so the fuel consumption of the DGs will also decrease. Since the cost of fuel intake is the maximum percentage of all costs, oil price increase has a significant influence on COE. As electricity demand rises, DG will increase the power generation to fulfill the demand. This causes the RF to decline, the fuel consumption and NPC the rise, and COE conversely.

Author Contributions: Conceptualization, C.-C.K.; Data curation, W.-B.W.; Formal analysis, C.-T.T. and T.M.B.; Investigation, W.-B.W. and C.-C.K.; Methodology, C.-T.T. and T.M.B.; Resources, C.-T.T.; Supervision, C.-C.K.; Writing—original draft, W.-B.W.; Writing—review & editing, T.M.B. All authors have read and agreed to the published version of the manuscript.

Funding: Supported of this research by the Asian Productivity Organization Center of Excellence on Green Productivity (APO COE GP) Project and Ministry of Science and Technology of the Republic of China (Grant No. MOST108-3116-F-042A-004-CC2) are gratefully acknowledged.

Conflicts of Interest: The authors declare no conflicts of interest.

Appendix A

Table A1. Electrical parameters of the PV modules.

Electrical Parameters	
Rated power @ standard test conditions	300 Wp
Open circuit voltage, Voc	39.24 V
Maximum power point voltage, Vmp	31.51 V
Short circuit current, Isc	9.93 A
Maximum power point current, Imp	9.52 A
Temperature coefficient of maximum power	$-0.4003\%/^{\circ}\text{C}$
Temperature coefficient of open circuit voltage	$-0.2906\%/^{\circ}\text{C}$
Temperature coefficient of short circuit current	$0.0530\%/^{\circ}\text{C}$
Module efficiency	18.44%

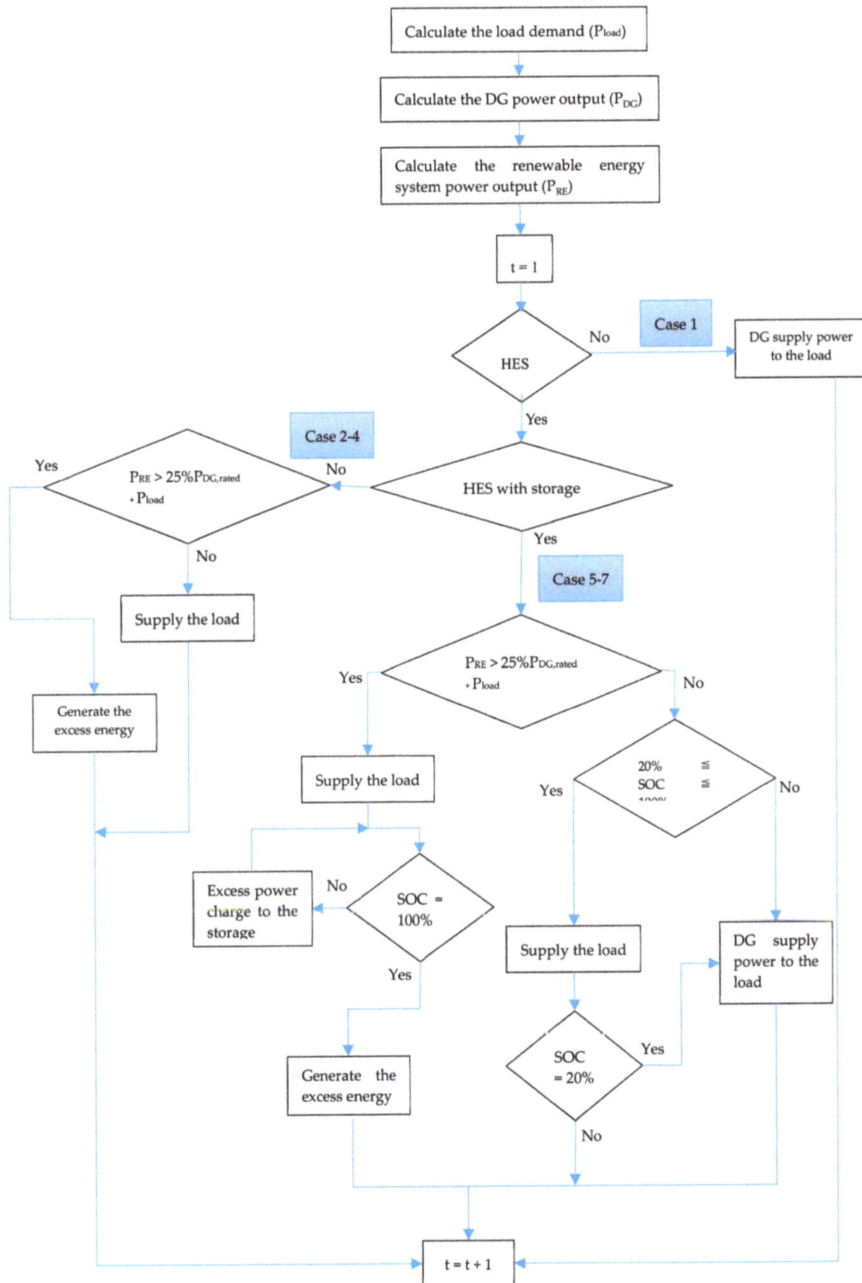

Figure A1. Flowchart of system operation in various cases 1–7.

References

1. Ma, T.; Javed, M.S. Integrated sizing of hybrid PV-wind-battery system for remote island considering the saturation of each renewable energy resource. *Energy Convers. Manag.* **2019**, *182*, 178–190. [CrossRef]

2. Krishan, O.; Suhag, S. Techno-economic analysis of a hybrid renewable energy system for an energy poor rural community. *J. Energy Storage* **2019**, *23*, 305–319. [CrossRef]

3. Lu, J.; Wang, W.; Zhang, Y.; Cheng, S. Multi-Objective Optimal Design of Stand-Alone Hybrid Energy System Using Entropy Weight Method Based on HOMER. *Energies* **2017**, *10*, 1664. [CrossRef]

4. Kartite, J.; Cherkaoui, M. Study of the different structures of hybrid systems in renewable energies: Cooling different structures of on hybrid systems renewable energies: A review Cherkaoui Assessing the Jihane feasibility using the heat demand-outdoor Cherkaoui temperature function for a long-term district heat demand forecast. *Energy Procedia* **2019**, *157*, 323–330.

5. Padrón, I.; Avila, D.; Marichal, G.N.; Rodríguez, J.A. Assessment of Hybrid Renewable Energy Systems to supplied energy to Autonomous Desalination Systems in two islands of the Canary Archipelago. *Renew. Sustain. Energy Rev.* **2019**, *101*, 221–230. [CrossRef]

6. *Solar Vision 2025*; Canadian Solar Industries Association: Ottawa, ON, Canada, 2010; Available online: http://www.pvinnovation.ca/files/CANSIA_Solar_Vision_2025_FINAL[1].pdf (accessed on 25 February 2019).

7. Brenna, M.; Longo, M.; Yaici, W.; Abegaz, T.D. Simulation and Optimization of Integration of Hybrid Renewable Energy Sources and Storages for Remote Communities Electrification. In Proceedings of the 2017 IEEE PES Innovative Smart Grid Technologies Conference Europe (ISGT-Europe), Torino, Italy, 26–29 September 2017; pp. 1–6.

8. Delucchi, M.A.; Jacobson, M.Z. Providing all global energy with wind, water, and solar power, Part II: Reliability, system and transmission costs, and policies. *Energy Policy* **2011**, *39*, 1170–1190. [CrossRef]

9. Valenciaga, F.; Puleston, P.F. Supervisor Control for a Stand-Alone Hybrid Generation System Using Wind and Photovoltaic Energy. *IEEE Trans. Energy Convers.* **2005**, *20*, 398–405. [CrossRef]

10. Wang, L.; Singh, C. Compromise between Cost and Reliability In Optimum Design of An Autonomous Hybrid Power System Using Mixed-Integer PSO Algorithm. In Proceedings of the IEEE 2007 International Conference on Clean Electrical Power, Capri, Italy, 21–23 May 2007.

11. Kadda, F.Z.; Zouggar, S.; el Hafyani, M.; Rabhi, A. Contribution to the Optimization of the Electrical Energy Production from a Hybrid Renewable Energy System. In Proceedings of the IEEE 2014 5th International Renewable Energy Congress (IREC), Hammamet, Tunisia, 25–27 March 2014; pp. 1–6.

12. Das, B.K.; Zaman, F. Performance analysis of a PV/Diesel hybrid system for a remote area in Bangladesh: Effects of dispatch strategies, batteries, and generator selection. *Energy* **2019**, *169*, 263–276. [CrossRef]

13. Halabi, L.M.; Mekhilef, S. Flexible hybrid renewable energy system design for a typical remote village located in Tropical climate. *J. Clean. Prod.* **2018**, *177*, 908–924. [CrossRef]

14. Yazici, M.S.; Yavasoglu, H.A.; Eroglu, M. A mobile off-grid platform powered with photovoltaic / wind / battery / fuel cell hybrid power systems. *Int. J. Hydrogen Energy* **2013**, *38*, 11639–11645. [CrossRef]

15. Zhou, X.; Feng, C. The impact of environmental regulation on fossil energy consumption in China: Direct and indirect effects. *J. Clean. Prod.* **2017**, *142*, 3174–3183. [CrossRef]

16. Ma, W.; Xue, X.; Liu, G. Techno-economic evaluation for hybrid renewable energy system: Application and merits. *Energy* **2018**, *159*, 385–409. [CrossRef]

17. Rehman, S.; Al-Hadhrami, L.M. Study of a solar PV-diesel-battery hybrid power system for a remotely located population near Rafha, Saudi Arabia. *Energy* **2010**, *35*, 4986–4995. [CrossRef]

18. Nayak, A.; Kasturi, K.; Nayak, M.R. Cycle-charging dispatch strategy based performance analysis for standalone PV system with DG & BESS. In Proceedings of the IEEE 2018 Technologies for Smart-City Energy Security and Power (ICSESP), Bhubaneswar, India, 28–30 March 2018; pp. 1–5.

19. Das, B.K.; Hoque, N.; Mandal, S.; Pal, T.K.; Raihan, M.A. A techno-economic feasibility of a stand-alone hybrid power generation for remote area application in Bangladesh. *Energy* **2017**, *134*, 775–788. [CrossRef]

20. Ghasemi, A.; Enayatzare, M. Optimal energy management of a renewable-based isolated microgrid with pumped-storage unit and demand response. *Renew. Energy* **2018**, *123*, 460–474. [CrossRef]

21. Ma, T.; Yang, H.; Lu, L. Electrical Power and Energy Systems Study on stand-alone power supply options for an isolated community. *Int. J. Electr. Power Energy Syst.* **2015**, *65*, 1–11. [CrossRef]

22. Ismail, M.S.; Moghavvemi, M.; Mahlia, T.M.I. Techno-economic analysis of an optimized photovoltaic and diesel generator hybrid power system for remote houses in a tropical climate. *Energy Convers. Manag.* **2013**, *69*, 163–173. [CrossRef]

23. Chauhan, A.; Saini, R.P. A review on Integrated Renewable Energy System based power generation for stand-alone applications: Con fi gurations, storage options, sizing methodologies and control. *Renew. Sustain. Energy Rev.* **2014**, *38*, 99–120. [CrossRef]

24. Guangqian, D.; Bekhrad, K.; Azarikhah, P.; Maleki, A. A hybrid algorithm based optimization on modeling of grid independent biodiesel-based hybrid solar/wind systems. *Renew. Energy* **2018**, *122*, 551–560. [CrossRef]

25. Akiki, H.; Eng, C.; Avenue, T. A decision support technique for the design of hybrid solar-wind power system. *IEEE Trans. Energy Convers.* **1998**, *13*, 76–83.

26. Venkataramanan, G.; Marnay, C.; Siddiqui, A.S.; Stadler, M.; Chandran, B.; Firestone, R. Optimal Technology Selection and Operation of Commercial-Building Microgrids. *IEEE Trans. Power Syst.* **2008**, *23*, 975–982.

27. Giraud, F.; Member, I.S.; Salameh, Z.M.; Member, S. Steady-State Performance of a Grid-Connected Rooftop Hybrid Power System with Battery Storage of Massachusetts, Lowell. *IEEE Power Eng. Rev.* **2001**, *21*, 54. [CrossRef]

28. Baneshi, M.; Hadianfard, F. Techno-economic feasibility of hybrid diesel/PV/wind/battery electricity generation systems for non-residential large electricity consumers under southern Iran climate conditions. *Energy Convers. Manag.* **2016**, *127*, 233–244. [CrossRef]

29. Murugaperumal, K.; Vimal, P.A.D. Feasibility design and techno-economic analysis of hybrid renewable energy system for rural electrification. *Sol. Energy* **2019**, *188*, 1068–1083. [CrossRef]

30. Batanes—Wikipedia. Available online: https://en.wikipedia.org/wiki/Batanes (accessed on 16 February 2019).

31. Welcome to HOMER. Available online: https://www.homerenergy.com/ (accessed on 25 February 2019).

32. Oil Monitor as of 16 October 2018 DOE Department of Energy Portal. Available online: https://www.doe.gov.ph/oil-monitor-16-october-2018 (accessed on 19 March 2019).

33. Halabi, L.M.; Mekhilef, S.; Olatomiwa, L.; Hazelton, J. Performance analysis of hybrid PV/diesel/battery system using HOMER: A case study Sabah, Malaysia. *Energy Convers. Manag.* **2017**, *144*, 322–339. [CrossRef]

34. Homer Energy, Welcome to HOMER. 2017. Available online: https://www.homerenergy.com/products/pro/docs/3.11/index.html (accessed on 16 February 2019).

35. Shannon, C.; Weaver, W. *The Mathematical Theory of Communication*; The University of Illinois Press: Champaign, IL, USA, 1964.

36. Shi, G.; Buffen, A.M.; Ma, H.; Hu, Z.; Sun, B.; Li, C.; Li, Y. Science Direct Distinguishing summertime atmospheric production of nitrate across the East Antarctic Ice Sheet. *Geochim. Cosmochim. Acta* **2018**, *231*, 1–14. [CrossRef]

37. Huang, J. Combining Entropy Weight and TOPSIS Method for Information System Selection. In Proceedings of the 2008 IEEE Conference on Cybernetics and Intelligent Systems, Chengdu, China, 21–24 September 2008; pp. 1281–1284.

38. Abusara, M.; Mallick, T.; al Badwawi, R.; Abusara, M.; Mallick, T. A Review of Hybrid Solar PV and Wind Energy System. *Smart Sci.* **2016**, *3*, 127–138.

39. Kolhe, M.L.; Ranaweera, K.M.I.U.; Gunawardana, A.G.B.S. Techno-economic sizing of off-grid hybrid renewable energy system for rural electrification in Sri Lanka. *Sustain. Energy Technol. Assess.* **2020**, *11*, 53–64. [CrossRef]

40. Li, C.; Ge, X.; Zheng, Y.; Xu, C.; Ren, Y.; Song, C. Techno-economic feasibility study of autonomous hybrid wind/PV/battery power system for a household in Urumqi, China. *Energy* **2013**, *55*, 263–272. [CrossRef]

41. Rohani, G.; Nour, M. Techno-economical analysis of stand-alone hybrid renewable power system for Ras Musherib in United Arab Emirates. *Energy* **2014**, *64*, 828–841. [CrossRef]

42. Sansone, G.; Hills, O.; Bohl, R. *Power Your Dream with the Wind*. 2001. Available online: www.bergey.com (accessed on 19 March 2019).

43. Nasser, Y.; Oumarou, H.; Lucien, M.; Benoit, N.; Jean, N. Analyzing of a Photovoltaic/Wind/Biogas/Pumped-Hydro Off-Grid Hybrid System for Rural Electrification in Sub-Saharan Africa—Case Study of Djoundé in Northern Cameroon. *Energies* **2018**, *11*, 2644.

44. Adaramola, M.S.; Agelin-chaab, M.; Paul, S.S. Analysis of hybrid energy systems for application in southern Ghana. *Energy Convers. Manag.* **2015**, *88*, 284–295. [CrossRef]

45. Adaramola, M.S.; Agelin-chaab, M.; Paul, S.S. Assessment of wind power generation along the coast of Ghana. *Energy Convers. Manag.* **2015**, *77*, 61–69. [CrossRef]

46. ESS Batteries by Samsung SDI Top Safety & Reliability Solutions. Available online: http://www.samsungsdi.com/upload/ess_brochure/201803_SamsungSDI%20ESS_EN.pdf (accessed on 16 February 2019).

47. Bangko Sentral ng Pilipinas—Publications & Research. Available online: http://www.bsp.gov.ph/publications/media.asp?id=4935 (accessed on 25 February 2019).
48. Philippines inflation Rates. Available online: http://www.bsp.gov.ph/statistics/spei_new/tab34_inf.htm (accessed on 25 February 2019).
49. Philippines Interest Rate 2019 Data Chart Calendar Forecast News. Available online: https://tradingeconomics.com/philippines/interest-rate (accessed on 25 February 2019).

Article

Integrated Analysis of Influence of Multiple Factors on Transmission Efficiency of Loader Drive Axle

Jianying Li [1,*], Tunglung Wu [1], Weimin Chi [1], Qingchun Hu [2] and Teenhang Meen [3,*]

[1] School of Mechanical and Automotive Engineering, ZhaoQing University, Zhaoqing 516260, China; tunglung216@gmail.com (T.W.); jeeweimin@gmail.com (W.C.)
[2] School of Mechanical and Automotive Engineering, South China University of Technology, Guangzhou 510640, China; huqc@scut.edu.cn
[3] Departments and Graduate Institute of Electronic Engineering, National Formosa University, Yulin 63202, Taiwan
* Correspondence: lijianying519@163.com (J.L.); thmeen@nfu.edu.tw (T.M.)

Received: 2 October 2019; Accepted: 22 November 2019; Published: 28 November 2019

Abstract: In this study, a loader drive axle digital model was built using 3D commercial software. On the basis of this model, the transmission efficiency of the main reducing gear, the differential planetary mechanism, and the wheel planetary reducing gear of the loader drive axle were studied. The functional relationship of the transmission efficiency of the loader drive axle was obtained, including multiple factors: the mesh friction coefficient, the mesh power loss coefficient, the normal pressure angle, the helix angle, the offset amount, the speed ratio, the gear ratio, and the characteristic parameters. This revealed the influence law of the loader drive axle by the mesh friction coefficient, mesh power loss coefficient, and speed ratio. The research results showed that the transmission efficiency of the loader drive axle increased with the speed ratio, decreased when the mesh friction coefficient and the mesh power loss coefficient increased, and that there was a greater influence difference on the transmission efficiency of the loader drive axle.

Keywords: drive axle; differential planetary mechanism; power loss coefficient; friction coefficient

---

1. Introduction

The loader is widely used as a construction machine, and the drive axle is an assembly of the loader drive system. The loader drive axle is mainly composed of a main reducing gear, a differential planetary mechanism, and a wheel planetary reducing gear. The transmission efficiency of the loader drive axle is a topic of interest among scholars. Xu and Kahraman [1] proposed a model of hypoid gears for predicting friction power loss, and their research showed that the friction power loss of a hypoid gear is affected by the gear smoothness, lubrication characteristics, and other factors. Fanghella et al. [2] discussed the functional characteristics of a nutating gearbox with bevel gears, for which the expressions of the transmission ratio and efficiency were obtained, including the main design. Kakavas et al. [3] proposed a thermal coupled transient model of a hypoid gear for power loss calculation and analyzed and compared numerical simulations and experimental results. Paouris [4] presented a tribodynamic model of differential hypoid gears, for which the lubricated contact of the meshing gear pair was integrated with lubricants of varying rheological properties, and the influence of lubricant formulation and gear geometry on system efficiency was examined. The results showed that there is a trend of increasing power loss when using lubricants with higher pressure–viscosity coefficients. Simon [5] found that the efficiency of a hypoid gear pair was improved by reducing the pressure and temperature in the oil film; the results were obtained based on the conditions of the mixed elastohydrodynamic lubrication (EHL) characteristics affected by the manufacturing parameters. To improve the transmission efficiency of a loader, Lyu et al. [6] designed an output power-split transmission system, in which a planetary gear set was applied as the dynamic coupling element to allow the output power of the loader to be split-transmitted. Xiong et al. [7] introduced a sizing approach for both hydrostatic–mechanical power-split transmission (PST) and hydrostatic transmission (HST) systems and presented a multidomain modeling approach for the powertrain of a wheel loader. The sizing method and simulation models should facilitate the development of powertrains for wheel loaders and other wheeled heavy vehicles. Barday et al. [8] analyzed the power losses of a truck drive axle, including the sliding and tooth friction of hypoid gear sets, rolling element bearings and oil churning, and then reviewed thermal exchanges and compared the bulk temperatures of the gear set.

The concept of virtual power was first proposed by Chen et al. [9] and was used to study the transmission efficiency of a simple planetary gear train. Then, on the basis of the virtual power theory, Chen et al. [10] analyzed the transmission efficiency of a dual-input, single-output planetary gear train. In Ref. [11–13], the virtual power theory was further developed to study the transmission efficiency of a compound planetary gear train. Li et al. [14] studied the power flow of a compound planet gear train with bevel gears via virtual power theory, and they analyzed the parameters' influence on transmission efficiency. In Ref. [15], Li et al. established a functional expression of transmission efficiency, including the characteristic parameters, the mesh friction coefficient, and the mesh angle of multistage microplanetary gearing transmission, and adopted the virtual power theory to study the influence of multiple factors on the transmission efficiency of a multistage microplanetary gearing transmission.

Although many studies have been carried out on the transmission efficiency of the main reducing gear, the differential planetary mechanism, and the wheel planetary reducing gear of the loader drive axle, the integrated analysis of the influence of multiple factors on the transmission efficiency of the loader drive axle assembly has not been reported. By establishing a three-dimensional digital model of a loader drive axle, the transmission efficiency of the main reducing gear, the differential planetary mechanism, and the wheel planetary reducing gear are studied, respectively. In doing so, the functional relationship between the transmission efficiency of the loader drive axle and the transmission efficiency of the main reducing gear, the differential planetary mechanism, and the wheel planetary reducing gear is obtained. This revealed the influence law between the transmission efficiency of the loader drive axle and the meshing friction coefficient, power loss coefficient, and speed ratio, which can provide a theoretical foundation for the design of heavy-duty automotive transmission systems.

2. Experimental

2.1. Digital Model of the Loader Drive Axle

According to the geometric characteristics of the main reducing gear, the differential planetary mechanism, and the wheel planetary reducing gear of the loader drive axle, the 3D commercial software parametric design method was used to establish the three-dimensional (3D) model of the main reducing gear, the differential planetary mechanism, and the wheel planetary reducing gear of the loader drive axle, as seen in Figure 1. The loader drive axle assembly is shown in Figure 2, in which the numbers 0–8 represent the loader drive axle frame, the main reducing driving gear, the main reducing driven gear, the differential planetary mechanism planet gear, the half shaft gear, the wheel planetary reducing planet gear, and the carrier, respectively. The basic parameters of the loader drive axle are listed in Table 1.

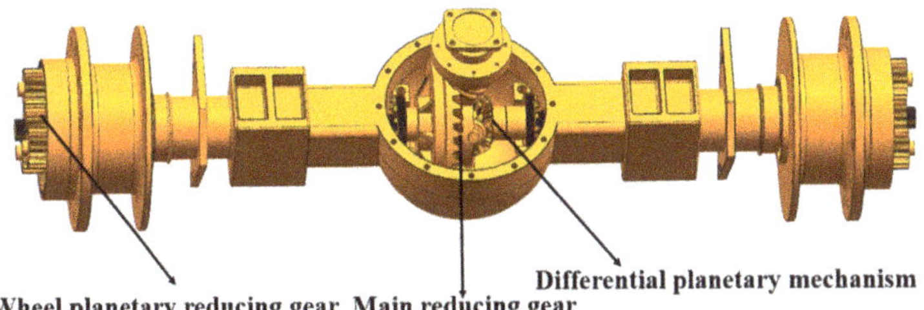

Figure 1. Three-dimensional (3D) digital model of the loader drive axle.

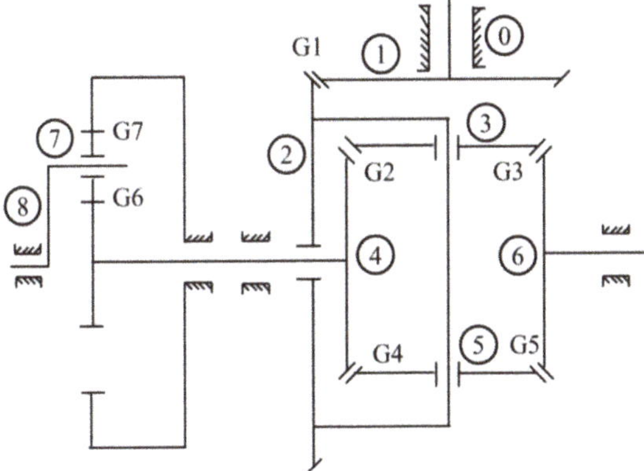

Figure 2. Schematic diagram of the loader drive axle.

Table 1. Basic parameters of the loader drive axle.

| Name | Main Reducing Gear | |
	Driving Gear	Driven Gear
Number of teeth	7	37
Transverse module (mm)	10.7	
Normal pressure angle	20° 30′	
Name	Differential Planetary Mechanism	
	Planet Gear	Half Shaft Gear
Number of teeth	10	18
Large transverse module (mm)	7	
Pressure angle	25°	
Name	Wheel Planetary Reducing Gear	
	Sun Gear	Planet Gear
Number of teeth	17	15
Modulus (mm)	5.5	
Pressure angle	20°	

2.2. Research on Transmission Efficiency of Loader Drive Axle

2.2.1. Research on Transmission Efficiency of the Main Reducing Gear

The theoretical transmission efficiency of the main reducing gear spiral bevel gear of the loader axle is shown in Equation (1) [16]:

$$\eta = \cfrac{1}{1 + \sqrt{\frac{T_{max}}{T}}\left[\frac{\mu}{\cos\phi}\sqrt{(\tan\psi_p - \tan\psi_g)^2 + \left(0.15\left(1 - \frac{E}{D}\right)^{16}\right)^2}\right]} \tag{1}$$

where η is the transmission efficiency; T is driven gear load torque; T_{max} is the driven gear peak torque at 2.75 times the fatigue limit bending stress; μ is the mesh friction coefficient; ϕ is the normal pressure angle of the driving gear surface; ψ_p is the spiral bevel gear driving gear nominal helix angle; ψ_g is the spiral bevel gear driven gear nominal helix angle; E is the offset of the hypoid gear driving gear; and D is the pitch diameter of the hypoid gear.

As reported by Saiki [17], the modified transmission efficiency of the spiral bevel gear does not depend on the load torque, as shown in Equation (2):

$$\eta_1 = \cfrac{1}{1 + \left[\frac{\mu}{\cos\phi}\sqrt{(\tan\psi_p - \tan\psi_g)^2 + \left(0.15\left(1 - \frac{E}{D}\right)^{16}\right)^2}\right]} \tag{2}$$

The nominal helix angle of the driving and driven spiral bevel gears is shown in Equations (3) and (4), respectively. F is the spiral bevel gear driven gear tooth width, and the rest of the symbol is in accordance with the meaning of Equation (1):

$$\psi_p = 25 + 5\sqrt{\frac{z_2}{z_1}} + 90(E/D) \tag{3}$$

$$\psi_g = \psi_p - \arcsin[2E/(D - F)] \tag{4}$$

2.2.2. Research on Transmission Efficiency of the Differential Planetary Mechanism

The diagram of the differential planetary mechanism of the loader drive axle shown in Figure 2 can be converted into the diagram shown in Figure 3 without considering the gear mesh power loss. The rules for the graphical conversion process are detailed in [14]. The virtual split power ratio β_3 and the split power ratio β'_3 of component 3 of the differential planetary mechanism of the loader drive axle are expressed by Equations (5) and (6), respectively:

$$\beta_3 = \frac{V - W}{W} = \frac{\omega_6 - \omega_2}{\omega_4 - \omega_2} = -1 \tag{5}$$

$$\beta'_3 = \frac{P}{M - P} = \frac{\omega_6}{\omega_4} = k \tag{6}$$

where the virtual split power ratio β_3 of component 3 of the differential planetary mechanism of the loader drive axle is defined as the ratio between the virtual power passing to component 6 and the virtual power passing to component 4 (Figure 3b). $V - W$ and W are the virtual power value flowing to member 6 and 4 and P and $M - P$ are the power value flowing to member 6 and 4. ω_i is the rotational angular velocity of component i. Specifically pointed out, the virtual power was measured by supposing that the observer stood on the carrier of a planetary gear train rotating at an angular velocity ω_i. The speed ratio k was defined as the ratio of the rotational angular velocity of component 6 (ω_6) to component 4 (ω_4). It can be found from Equations (5) and (6) that $V = 0$, $P = \frac{k}{1+k}M$.

The virtual power ratio α_4 of component 4 of the differential planetary mechanism of the loader drive axle is defined as the ratio between the virtual power and the power passing through component 4 [9] (Figure 3a,b), and established the relationship between relative angular velocity and angular velocity, as shown in Equation (7).

$$\alpha_4 = \frac{W}{M-P} = \frac{\omega_4 - \omega_2}{\omega_4} = \mathrm{sgn}\omega_4 \frac{\omega_3 z_3}{\omega_4 z_4} = \mathrm{sgn}\omega_4 hi \tag{7}$$

where $\mathrm{sgn}\omega_4 = \begin{cases} 1 & \text{Turn right} \\ 0 & \text{Straight} \\ -1 & \text{Turn left} \end{cases}$, $h = \frac{\omega_3}{\omega_4}, i = \frac{z_3}{z_4}$

It can be obtained from Equation (7) that

$$W = \mathrm{sgn}\omega_4 hi(M-P) = \mathrm{sgn}\omega_4 \frac{hi}{1+k}M \tag{8}$$

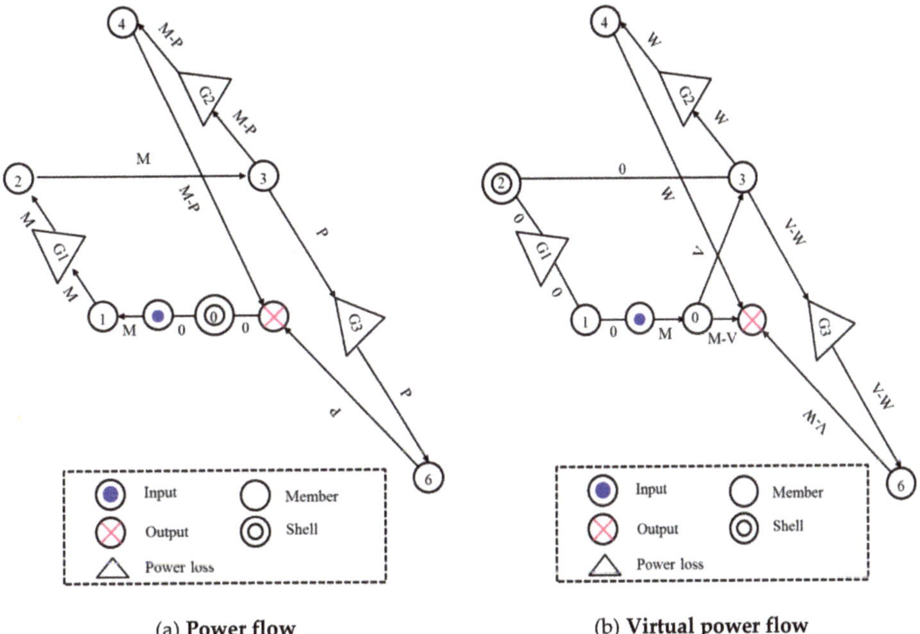

(a) **Power flow**   (b) **Virtual power flow**

Figure 3. Power flows of differential planetary mechanism without considering the gear mesh power loss.

The diagram of the differential planetary mechanism of the loader drive axle shown in Figure 2 can be converted into the diagram shown in Figure 4 when considering the gear mesh power loss. The gear mesh power loss is proportional to the input power, and the proportional ratio is called the mesh power loss coefficient, so L_1, L_2, and L_3 indicate the mesh power loss passing through the gear pairs G_1, G_2, and G_3, respectively, which are expressed by Equations (9) (11), respectively:

$$L_1 = \lambda_1 M \tag{9}$$

$$L_2 = \lambda_2 W \tag{10}$$

$$L_3 = \lambda_3 P \tag{11}$$

where λ_i represents the mesh power loss coefficient.

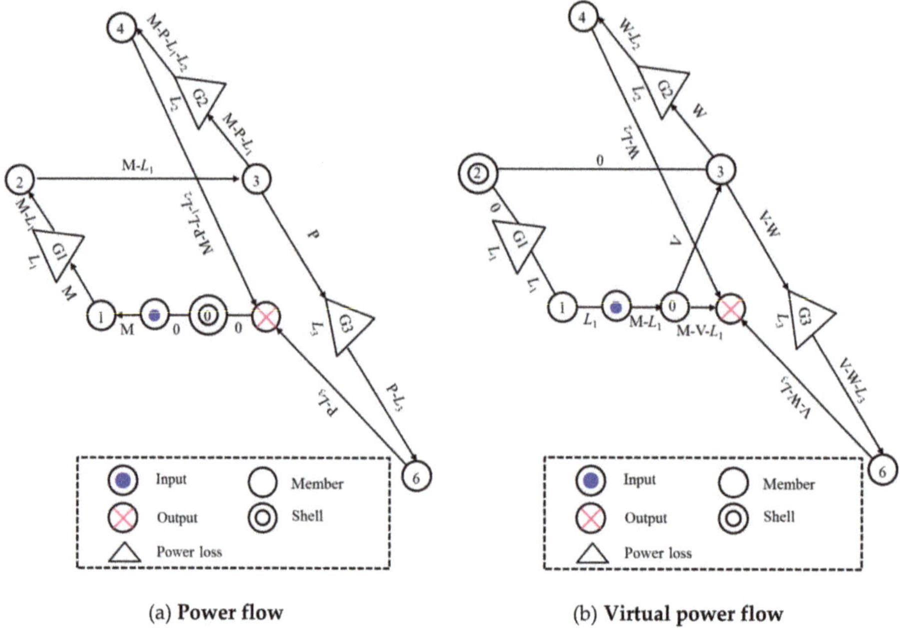

(a) **Power flow**　　　　　(b) **Virtual power flow**

Figure 4. Power flows of differential planetary mechanism considering the gear mesh power loss.

Considering the gear mesh power loss, the split power ratio β'_3 of component 3 and the virtual power ratio α_4 of component 4 of the differential planetary mechanism of the loader drive axle are expressed by Equations (12) and (13), respectively (Figure 4a,b):

$$\beta'_3 = \frac{P}{M - P - L_1} = k \tag{12}$$

$$\alpha_4 = \frac{W - L_2}{M - P - L_1 - L_2} = \mathrm{sgn}\omega_4 hi \tag{13}$$

It can be obtained from Equations (12) and (13) that

$$P = \frac{k}{1+k}(1 - \lambda_1)M \tag{14}$$

$$W = \left\{ \mathrm{sgn}\omega_4 \frac{hi(1 - \lambda_1)M}{1+k} \Big/ [1 + (\mathrm{sgn}\omega_4 hi - 1)\lambda_2] \right\} \tag{15}$$

From Equations (9)–(11), (14), and (15),

$$\frac{\sum L}{M} = \frac{L_1 + L_2 + L_3}{M} = \lambda_1 + \lambda_2 \left\{ \mathrm{sgn}\omega_4 \frac{hi(1 - \lambda_1)}{1+k} \Big/ [1 + (\mathrm{sgn}\omega_4 hi - 1)\lambda_2] \right\} + \lambda_3 \frac{k(1 - \lambda_1)}{1+k} \tag{16}$$

Accordingly, the transmission efficiency η_2 of the differential planetary mechanism of the loader drive axle is expressed as Equation (17):

$$\eta_2 = 1 - \frac{\sum L}{M} = 1 - \left\{ \lambda_1 + \lambda_2 \left\{ \mathrm{sgn}\omega_4 \frac{hi(1 - \lambda_1)}{1+k} \Big/ [1 + (\mathrm{sgn}\omega_4 hi - 1)\lambda_2] \right\} + \lambda_3 \frac{k(1 - \lambda_1)}{1+k} \right\} \tag{17}$$

2.2.3. Research on Transmission Efficiency of the Wheel Planetary Reducing Gear

The diagram of the wheel planetary reducing gear of the loader drive axle shown in Figure 2 can be converted into the diagram shown in Figure 5 without considering the gear mesh power loss. The virtual power ratio α_4 of component 4 of the wheel planetary reducing gear of the loader drive axle is defined as the ratio between the virtual power and the power passing through component 4 [9], and expressed by Equation (18) (Figure 5a,b):

$$\alpha_4 = \frac{M-V}{M} = \frac{\omega_4 - \omega_8}{\omega_4} = \frac{p}{1+p} \tag{18}$$

It can be obtained from Equation (18) that

$$V = \frac{1}{1+p}M \tag{19}$$

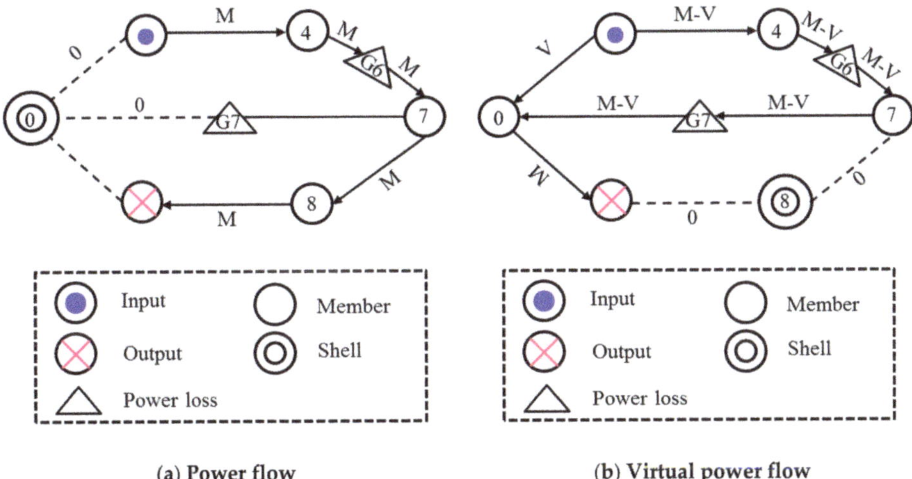

(a) Power flow   **(b) Virtual power flow**

Figure 5. Power flows of wheel planetary reducing gear without considering the gear mesh power loss.

The diagram of wheel planetary reducing gear of the loader drive axle shown in Figure 2 can be converted into the diagram shown in Figure 6 when considering the gear mesh power loss. L_6 and L_7 indicate the mesh power loss passing through the gear pairs G_6 and G_7, respectively, which are expressed by Equations (20) and (21), respectively:

$$L_6 = \lambda_6 M \tag{20}$$

$$L_7 = \lambda_7 (M - V - L_6) \tag{21}$$

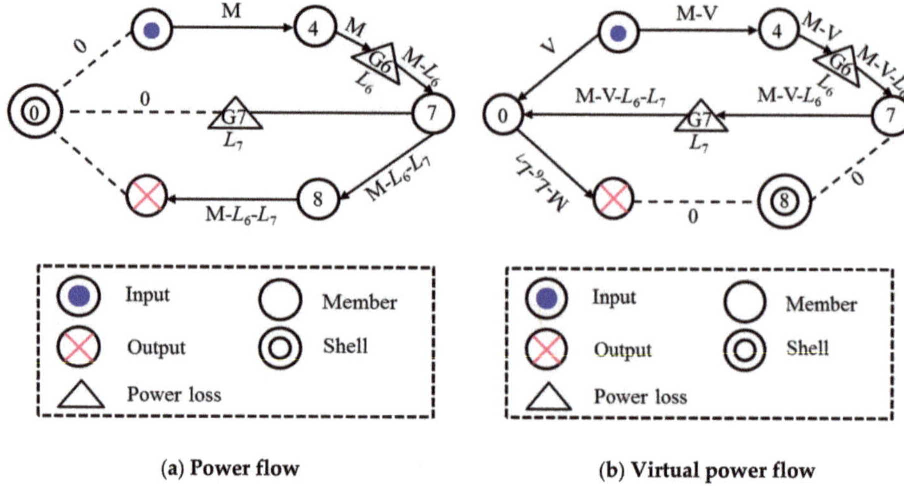

(a) **Power flow**　　　　　　　　(b) **Virtual power flow**

Figure 6. Power flows of wheel planetary reducing gear considering the gear mesh power loss.

Considering the gear mesh power loss, the virtual power ratio α_4 of component 4 of the wheel planetary reducing gear of the loader drive axle is expressed by Equation (22) (Figure 6a,b):

$$\alpha_4 = \frac{M - V}{M} = \frac{\omega_4 - \omega_8}{\omega_4} = \frac{p}{1 + p} \tag{22}$$

From Equation (22),

$$V = \frac{1}{1 + p} M \tag{23}$$

Accordingly, the transmission efficiency η_3 of the wheel planetary reducing gear of the loader drive axle is expressed by Equation (24):

$$\eta_3 = 1 - \frac{\sum L}{M} = 1 - \left(\lambda_6 + \lambda_7 \frac{p}{1 + p} - \lambda_6 \lambda_7 \right) \tag{24}$$

3. Results and Discussions

According to Equations (2), (17), and (24), the transmission efficiency of the main reducing gear, the differential planetary mechanism, and the wheel planetary reducing gear of the loader drive axle can be expressed by Equation (25):

$$\eta = \eta_1 \eta_2 \eta_3 = f(\mu, \phi, \psi, E, D, h, i, \lambda, k, p) \tag{25}$$

To analyze the influence of the mesh friction coefficient μ of the main reducing gear on the transmission efficiency η of the loader drive axle, the curve of the transmission efficiency η of the loader drive axle and the mesh friction coefficient μ of the gear teeth were drawn under the condition that the other parameters shown in Equation (25) were unchanged (Figure 7). According to Figure 7, the transmission efficiency η of the loader drive axle decreased when the mesh friction coefficient μ of the main reducing gear increased under the condition that the other parameters were unchanged. Simultaneously, the mesh friction coefficient μ of the main reducing gear was 0.05, and the transmission efficiency η of the loader drive axle was 0.707.

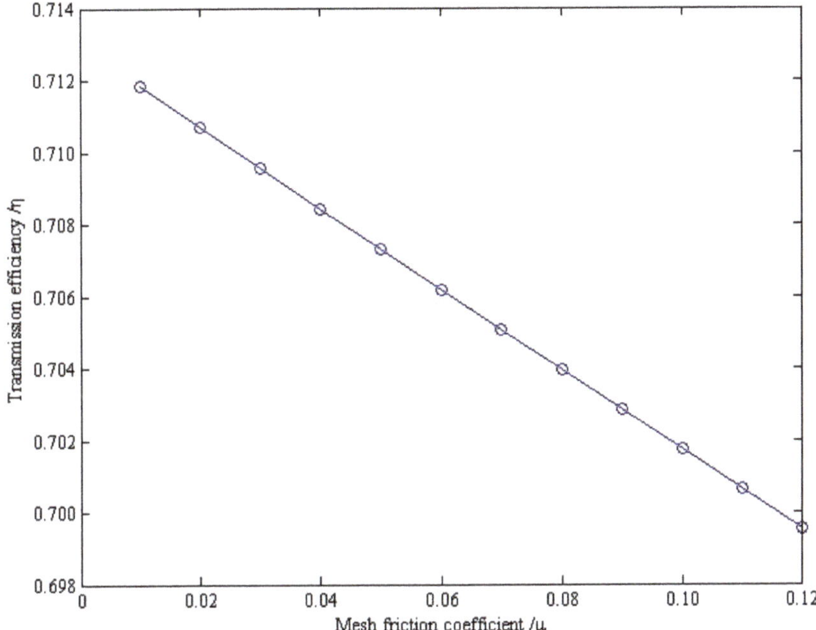

Figure 7. Relationship of the transmission efficiency η of the loader drive axle and the mesh friction coefficient μ.

To determine if the transmission efficiency η of the loader drive axle is affected by the mesh power loss coefficient λ_i, the curve of the transmission efficiency η of the loader drive axle and the mesh power loss coefficient λ_i was drawn under the condition that the other parameters shown in Equation (25) were unchanged (Figure 8). According to Figure 8, the transmission efficiency η of the loader drive axle decreased as the mesh power loss coefficient λ_i increased. That is, the transmission efficiency η of the loader drive axle decreased as the mesh power loss coefficients λ_1 and λ_6 increased, and it also decreased as the mesh power loss coefficients λ_3 and λ_7 increased. As the mesh power loss coefficients λ_3 and λ_7 increased, the transmission efficiency η of the loader drive axle decreased, but the degree of decrease was slower than λ_1 and λ_6; as the mesh power loss coefficient λ_2 increased, it initially decreased slowly and then sharply.

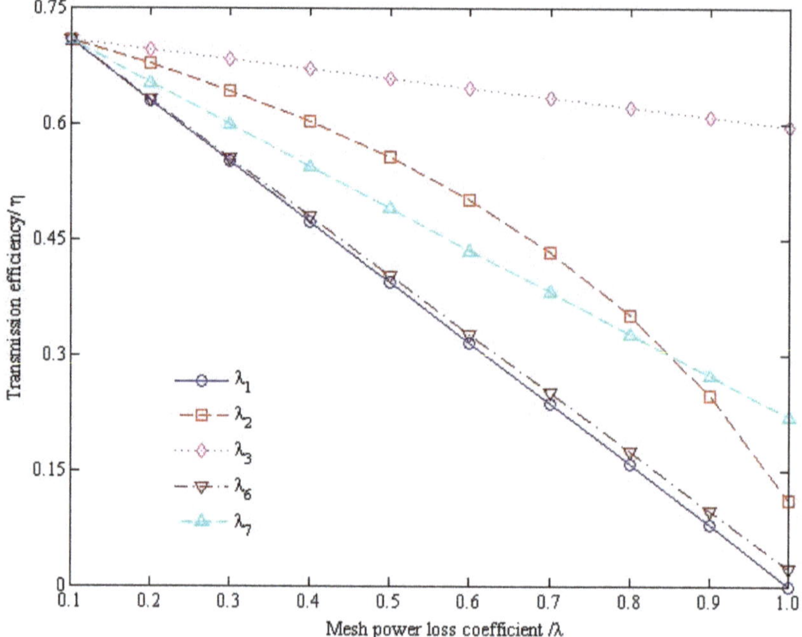

Figure 8. Relationship of the transmission efficiency η of the loader drive axle and the mesh power loss coefficient λ_i.

Here, the speed ratio γ was defined as the ratio of the rotational angular velocity of component 2 (ω_2) to component 4 (ω_4), as shown in Equation (26):

$$\gamma = \frac{\omega_2}{\omega_4} \tag{26}$$

To determine if the transmission efficiency η of the loader drive axle is affected by the speed ratio γ, the curve of the transmission efficiency η of the loader drive axle and the speed ratio γ were drawn under the condition that the other parameters shown in Equation (25) were unchanged (Figure 9). According to Figure 9, the transmission efficiency η of the loader drive axle increased with the increase of the speed ratio γ when the other parameters were unchanged, but the transmission efficiency η ranged from 0.7055 to 0.7090; that is, the modification of the speed ratio γ had no obvious effect on the transmission efficiency η of the loader drive axle.

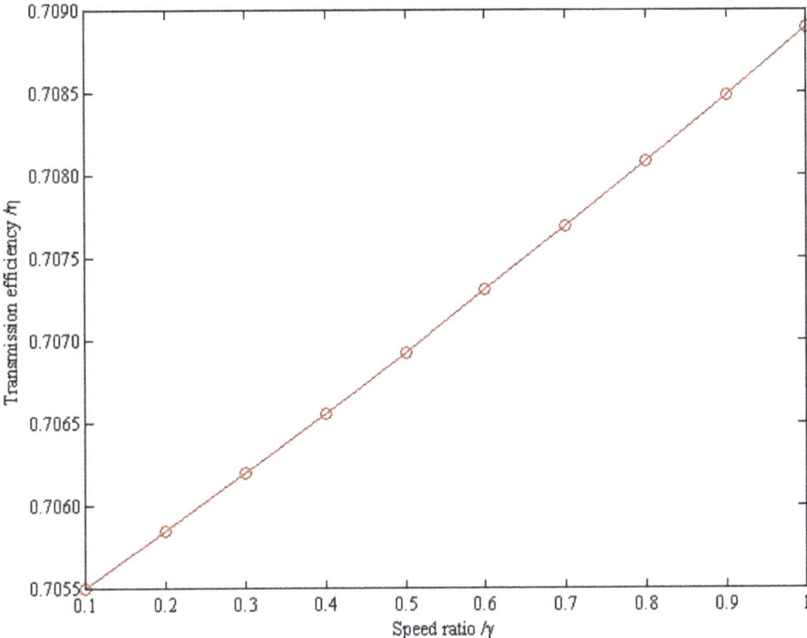

Figure 9. Relationship of the transmission efficiency η of the loader drive axle and the speed ratio γ.

4. Conclusions

(1) By establishing a three-dimensional digital model of a loader drive axle, the transmission efficiency of the main reducing gear, the differential planetary mechanism, and the wheel planetary reducing gear of the loader drive axle are studied. The transmission efficiency functional relationship between the loader drive axle and the main reducing gear, differential planetary mechanism, and wheel planetary reducing gear were obtained; that is, a functional relationship between the transmission efficiency of the loader drive axle and the gear mesh friction coefficient, the mesh power loss coefficient, the normal pressure angle, the helix angle, the offset amount, the speed ratio, and the characteristic parameters.

(2) The transmission efficiency of the loader drive axle increased with speed ratio and decreased when the gear teeth mesh friction coefficient and the mesh power loss coefficient increased. However, compared with the mesh power loss coefficient, the degree of influence of the transmission efficiency of the loader drive axle was different than the gear teeth friction coefficient, and the mesh power loss coefficient of each gear pair had a different influence on the transmission efficiency of the loader drive axle.

Based on the above research of this paper, the influence of rolling element bearings and temperatures of the gears on the transmission efficiency of the loader drive axle was carried out, so that the influence of multiple factors on the transmission efficiency of the loader drive axle can be studied more comprehensively.

Author Contributions: J.L. performed the formula derivation, data analyses, and wrote the manuscript; T.M. and Q.H. proposed the concept of the study; T.W. and W.C. contributed to writing, reviewing, and editing.

Funding: This research was funded by the Research Project of Guangdong Higher Education Institute (19GYB014), the Zhaoqing University Quality Engineering and Teaching Reform Project (zlgc201751 and zlgc201829), and the Guangdong Youth Innovation Talents Project (2018KQNCX290).

Acknowledgments: The authors wish to express their gratitude to the School of Mechanical and Automotive Engineering, Zhaoqing University, and express their gratitude for the funding provided by the Guangdong Youth Innovation Talents Project.

Conflicts of Interest: The authors declare no conflicts of interest.

References

1. Xu, H.; Kahraman, A. Prediction of friction-related power losses of hypoid gear pairs. *Proc. Inst. Mech. Eng. Part K J. Multi-Body Dyn.* **2007**, *221*, 387–400. [CrossRef]
2. Fanghella, P.; Bruzzone, L.; Ellero, S.; Landò, R. Kinematics, Efficiency and Dynamic Balancing of a Planetary Gear Train Based on Nutating Bevel Gears. *Mech. Based Des. Struct. Mach.* **2015**, *44*, 72–85. [CrossRef]
3. Kakavasa, I.; Olvera, A.V.; Dinia, D. Hypoid Gear Vehicle Axle Efficiency. *Tribol. Int.* **2016**, *101*, 314–323. [CrossRef]
4. Paouris, L.; Theodossiades, S.; Rahmani, R.; Rahnejat, H.; Hunt, G.; Barton, W. Effect of Lubricant Rheology on Hypoid Gear Pair Efficiency. In Proceedings of the 3rd Biennial International Conference on Powertrain Modelling and Control, Loughborough, UK, 7–9 September 2016; pp. 1–25.
5. Simon, V. Improvements in the Mixed Elastohydrodynamic Lubrication and in the Efficiency of Hypoid Gears. *Proc. Inst. Mech. Eng. Part J J. Eng. Tribol.* **2019**. [CrossRef]
6. Lyu, C.; Zhao, Y.; Lyu, M. Loader Power-Split Transmission System Based on a Planetary Gear Set. *Adv. Mech. Eng.* **2018**, *10*, 1–8. [CrossRef]
7. Xiong, S.; Wilfong, G.; Lumkes Jr., J. Components Sizing and Performance Analysis of Hydro-Mechanical Power Split Transmission Applied to a Wheel Loader. *Energies* **2019**, *12*, 1613. [CrossRef]
8. Bardav, D.; Fossier, C.; Changenet, C.; Ville, F.; Berier, V. Investigations on Drive Axle Thermal Behaviour: Power Loss and Heat-Transfer Estimations. *SAE Int. J. Eng.* **2018**, *11*, 55–66.
9. Chen, C.; Angeles, J. Virtual power flow and mechanical gear-mesh power losses of epicyclic gear trains. *J. Mech. Des.* **2007**, *129*, 107–113. [CrossRef]
10. Chen, C.; Liang, T.T. Theoretic study of efficiency of two DOFs of epicyclic gear transmission via virtual power. *J. Mech. Des.* **2011**, *133*, 031007–031013. [CrossRef]
11. Chen, C. Power flow analysis of compound epicyclic gear transmission: Simpson gear train. *J. Mech. Des.* **2011**, *133*, 1–5. [CrossRef]
12. Chen, C. Power Analysis of Epicyclic Transmissions Based on Constraints. *J. Mech. Robot.* **2012**, *4*, 041004. [CrossRef]
13. Chen, C. Power flow and efficiency analysis of epicyclic gear transmission with split power. *Mech. Mach. Theory* **2013**, *59*, 96–106. [CrossRef]
14. Hu, Q.C.; Li, J.Y.; Duan, F.H. Power Flow and Efficiency Analysis of Compound Planetary Gears Transmission with Bevel Gears. *J. Mech. Eng.* **2015**, *51*, 42–48. [CrossRef]
15. Li, J.Y.; Wu, T.L.; Zhu, T.J.; Hu, Q.C.; Cai, C.M.; Zong, C.F. Influence of Multi-Factor on the Transmission Efficiency of Multistage Micro-Planetary Gearing Transmission. *Microsyst. Technol.* **2019**. [CrossRef]
16. Beijing Gear Factory. *Gleason Technical Information Translation*, 2nd ed.; Mechanical Industry Press: Beijing, China, 1983; pp. 137–139.
17. Saiki, K.; Fukamachi, S. Research on High Efficiency Hypoid Gear for Automotive Application. In Proceedings of the 14th IFToMM World Congress, Taipei, Taiwan, 25–30 October 2015; pp. 293–297.

Article

Management and Distribution Strategies for Dynamic Power in a Ship's Micro-Grid System Based on Photovoltaic Cell, Diesel Generator, and Lithium Battery

Wanneng Yu [1,2,*], Suwen Li [1], Yonghuai Zhu [1] and Cheng-Fu Yang [3,*]

[1] School of Marine Engineering, Jimei University, Xiamen 361021, China; suwenli@jmu.edu.cn (S.L.); 201711824012@jmu.edu.cn (Y.Z.)

[2] Provincial Key Laboratory of Naval Architecture & Ocean Engineering, Xiamen 361021, China

[3] Department of Chemical and Materials Engineering, National University of Kaohsiung, Kaohsiung 811, Taiwan

* Correspondence: wnyu2007@jmu.edu.cn (W.Y.); cfyang@nuk.edu.tw (C.-F.Y.)

Received: 14 October 2019; Accepted: 24 November 2019; Published: 27 November 2019

Abstract: Combining new energy technology with electric propulsion technology is an effective way to decrease the pollution of water resources caused by cruise ships. This study examines the stable parallel operation of a ship's micro-grid system through a dynamic power management strategy involving a step change in load. With cruise ships in mind, we construct a micro-grid system consisting of photovoltaics (PV), a diesel generator (DG), and a lithium battery and establish a corresponding simulation model. We then analyze the system's operating characteristics under different working conditions and present the mechanisms that influence the power quality of the ship's micro-grid system. Based on an analysis of the power distribution requirements under different working conditions, we design a power allocation strategy for the micro-grid system. We then propose an optimization allocation strategy for dynamic power based on fuzzy control and a load current feed-forward method, and finally, we simulate the whole system. Through this study we prove that the proposed power management strategy not only verifies the feasibility and correctness of the ship's micro-grid structure and control strategy, but also greatly improves the reliability and stability of the ship's operation.

Keywords: photovoltaics; lithium battery; ship micro-grid system; fuzzy logic; power management

1. Introduction

To comply with the requirements of the energy efficiency design index (EEDI), the shipping industry must find ways to reduce its emissions of greenhouse gases. Low-carbon-emission shipping strategies and technologies that recently have been applied on ships in many developed countries include ship form optimization, electrical propulsion, nuclear propulsion, fuel cells, biomass fuels, and renewable energy sources [1,2].

Solar energy is an important natural resource, utilized mainly through photo-thermal and photovoltaic (PV) techniques, which are widely applied in aerospace, architecture, electrical power generation, and elsewhere [3–12]. Solar PV technology has been introduced into ships' power systems to reduce their greenhouse gas emissions, improve energy efficiency, and reinforce power system stability [13]. PV technology is mainly used in four ways on ships: as a propulsion power supply for small ships, a power supply for a ship's auxiliary machinery, a supply for general power consumption on large and medium passenger ships, and a heating source for large oil tankers [3]. Figure 1 shows the solar cruise ship "Shangdeguosheng", which was built for the Shanghai World Expo in 2010.

Passengers can enjoy a life of leisure on the ship, such as dining and entertainment, and the ship can be used to hold important business meetings and for other functions. In 2011, the German ship "Planet Solar" was constructed (Figure 2), with its owner claiming it to be the world's largest solar-powered ship. In 2013, the sightseeing ship "Yun Dang No. 1" was navigated in Xiamen Yun Dang Lake, powered by solar energy (Figure 3). The "Cosco Tengfei" was remodeled to become China's first large automobile ro-ro ship with solar PV, shown in Figure 4.

Figure 1. "ShangDeGuoSheng" solar cruise ship.

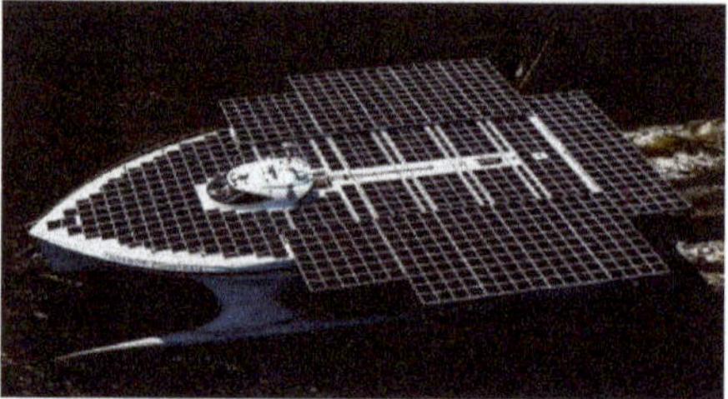

Figure 2. "Planet Solar" solar ship.

Figure 3. "Yun Dang No.1" solar cruise ship.

Figure 4. "Cosco Tengfei" ro-ro ship.

Although PV technology is widely utilized in power systems on land, to the best of our knowledge, hybrid PV/diesel/battery power systems on ships have not been extensively investigated and developed, especially the use of marine micro-grid technology to manage and control hybrid power systems on ships. Glykaset et al. discussed PV systems applied to merchant marine vessels to reduce fuel costs [14]. A stability assessment and economic analysis of a hybrid PV/DG ship system are presented by Tsekouras et al. [15]. Adamo et al. offered a preliminary analysis of reducing the emissions from an electrical ship in a berth [16]. Thus, Wen et al. performed a cost analysis using a particle swarm optimization algorithm to find the optimal size and capacity for various types of energy storage systems [17]. And, Lan et al. proposed a method for determining the ideal size for a PV power generation system; a diesel generator and an energy storage system were designed in a ship's stand-alone power system with the aim of minimizing investment cost, fuel cost, and CO_2 emissions [13].

Tang [18] proposed a novel structure for a large-scale PV array and its controls, designed for the illumination unit of a ship's power grid. Lee et al. [19] reported on the experimental results from the operation of a prototype green ship on Geoje Island, South Korea. In [20], Sun and colleagues built a simulation model of a PV-driven power system for a ship, based on a PSCAD/EMTDC platform. The use of new energy technologies in ships has so far been limited to the development of "green" touring vessels.

Yu et al. had developed multi-energy ship micro-grid energy control system based on solar lithium battery and diesel generator set for cruise ships and proposed the energy distribution control strategy according to the requirements of safe and stable operation, and they had verified the feasibilities on the actual ships [21,22]. Banaei et al. conducted a study on all-electrical propulsion ships, which consisted of fuel cells, batteries, PV, and two diesel generators [23]. The battery state of charge (SOC) was divided into three different states, in which $SOC > SOC_{high}$, $SOC_{low} < SOC < SOC_{high}$, and $SOC < SOC_{low}$. They also proposed the different energy management strategies in different states, for that the diesel generators, energy storage systems, and fuel cells were rationally utilized. For the ship's micro-grid composing of diesel generator and other energy systems, Kuo et al. had proposed a storage for the ship's power supply mode. The proposed storage could be determined according to the SOC state of energy storage system and it achieved the effects of reducing the fuel consumptions and exhausted emissions of the diesel engine [24].

However, all of above statements were investigated for the managements of energy sources in static situations, and no dynamic situations were studied. Until now, only few studies were focused on the dynamic control of ship's micro-grid, but there were many related researches on terrestrial micro-grid. Kong et al. proposed a three-stage proportional plus derivative control (PD) preemptive dynamic scheduling strategy, which could overcome the problem of dynamic control and the PD control was easy to cause oscillation, but the settings of three time constants increased the difficulty of system design [25]. Because the different micro-grid systems have different time constants, so it is not an almighty method. Ding et al. proposed a method based on fuzzy control strategy for battery energy storage system to improve the transient stability of power grid at when the input of fuzzy controller had the problem of frequency difference [26]. Above two methods were investigated to suppress the frequency difference, but both would cause a delay in control. Li analyzed the mathematical model of diesel generator, and he obtained the reason of frequency fluctuation during load abrupt change [27]. Also, Li had proposed the active compensation strategy of current feed forward compensation and PD control algorithm, but this method was easy to cause the high frequency oscillation.

These applications often have shortcomings, such as frequency fluctuation, high cost, and complicated control systems. We therefore propose management and distribution strategies for dynamic power by combining PV, diesel generator, and lithium battery. Via the fuzzy control and compensation method of the ship's micro-grid system, we were able to obtain optimal compensation power. After taking into consideration the balance between sailing mileage and economic efficiency, as well as how to integrate PV, DG, and lithium battery characteristics, we investigated a ship's micro-grid system to combine those three power sources. Therefore, this investigated article was based on the dynamic research of terrestrial micro-grid, which combined with the characteristics of the static power allocation strategy and dynamic power allocation strategy of ship's micro-grid. The strategies for the frequency fluctuation during the sudden change of load in the PV-diesel generator-battery operation mode were analyzed and investigated. The characteristics of the energy storage system could be utilized to respond quickly, and the output or developed power could be used to lower down the frequency fluctuation, and a dynamic energy scheduling strategy was put forward based on current the feed-forward and fuzzy control.

2. Modelling of the Ship's Micro-Grids System

Unlike in previous studies, we considered step changes in strong coupling, the great complexity of the random changes that occur in a ship's power system, and step changes in loads. We then present a novel dynamic power distribution strategy for a ship's micro-grid system. The topological structure is shown in Figure 5 and can be described as follows [28,29]:

(1)   The ship's micro-grid system is composed of a diesel generator, a lithium battery, and PV solar cells, and the PV and lithium-ion-based battery are connected by inverters to the AC grid bus, which can supply power in parallel with the diesel generator.

(2)   To use the renewable energy more effectively, the micro-grid power system is mainly operated by the inverter, while the diesel generators are used as an auxiliary or emergency power source.

(3)   The micro-grid system has many operation modes, depending on the ship's navigation state. For example, when the ship is in berth with sufficient sunlight, solar power will charge the lithium battery and supply power for the daily living equipment, and the diesel generator will be idle. When the ship is under normal sailing conditions and the voyage is short, propeller power and daily living power will be jointly supplied by the solar and lithium batteries, and the diesel generator will remain idle. When the ship is under normal sailing conditions and the voyage is long, propeller power and living power will be jointly supplied by the solar and lithium batteries, and the diesel generator and energy management control strategy for the ship's micro-grid system will work together, which is the primary working mode investigated in this paper.

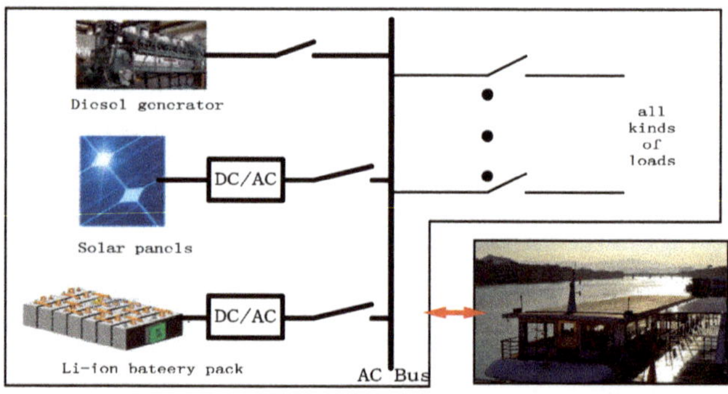

Figure 5. Proposed structure of ship's micro-grid system.

The ship's micro-grid system is very different from a traditional land micro-grid system, as the ship's has no unified dispatch center and the operating conditions are more complicated. In the topology described above, one of the frequent operation modes is very complex because the multiple power sources are arranged in parallel. To better analyze the micro-grid system's performance in this mode, using mathematical analysis of the power supply mechanisms for the various energy sources, we established a simulation model of the micro-grid system. This provides the foundation for further research on the strategy for controlling the micro-grid system under changing dynamic loads.

2.1. The Dynamic Characteristics of the Diesel Generator

The electromagnetic torque formula of for the synchronous generator can be expressed as Equation (1) as when the salient effect is ignored:

$$T_e = E_p' i_p \tag{1}$$

In this formula, E_p' is the instantaneous electromotive force (EMF) of the generator and i_p is the stator current of the p axis. The diesel motor drives the synchronous generator's rotation, and the stator winding of the synchronous generator takes the induction EMF by cutting the magnetic line motion in the magnetic field, providing electrical energy in the case of an external load. When the load alters suddenly, so does its active current. According to Equation (1), the electromagnetic torque also changes.

According to [30], the functioning of an output diesel-engine speed-control system can be expressed as:

$$\omega(s) = \frac{K(1 + T_2 s)(1 + T_3 s)}{M} \omega_{ref}(s) - \frac{s(1 + T_1 s)(1 + T_4 s)(1 + T_5 s)(1 + T_d s)}{M} T_e(s) \tag{2}$$

where $M = T_j s^2 (1 + T_1 s)(1 + T_4 s)(1 + T_5 s)(1 + T_d s) + K(1 + T_2 s)(1 + T_3 s)$, and T_1, T_4, T_5, and T_d are the small time constants, T_j is a time constant in the diesel engine model, K refers to the parameter of transfer function, subscript symbol s refers to the Laplace transform of model, and Te refers to the electromagnetic torque. If we set $T = T_1 + T_4 + T_5 + T_d$, then $M = T_j s^2 (1 + Ts) + K(1 + T_2 s)(1 + T_3 s)$. Thus, the output of the system can be expressed as:

$$\omega(s) = \frac{K(1 + T_2 s)(1 + T_3 s)}{M} \omega_{ref}(s) - \frac{s(1 + Ts)}{M} T_e(s) \tag{3}$$

From Equation (3) we see that the output is determined by the system input volume $\omega_{ref}(s)$ and $T_e(s)$. The transfer function $\omega_{ref}(s) = 0$ can be obtained under load disturbance. The transfer function $G(s)$ can be then expressed as:

$$G(s) = \frac{\omega(s)}{T_e(s)} = \frac{-s(1 + Ts)}{M} \tag{4}$$

2.2. The Model of the PV Cells

PV cells are semiconductor devices based on the PV effect of semiconducting materials with a PN junction and have the function of directly converting sunlight into electrical energy. According to [31], when sunlight shines on PV cells, the electrons in the P region move to the N region, resulting in a large amount of electrons accumulating on the receivers.

The PV battery equivalent circuit model is shown in Figure 6, where I_{ph} is a photo-generated current whose magnitude depends on the light intensity S and temperature T, and which acts as a controlled current source; I_d is a voltage generated by a reactive PN junction of voltage V; I_{sh} is the short-circuit current caused by leakage from the PV cells; R_s and R_{sh} are the series resistance and parallel resistance of the PV cells; V is the PV output voltage; and I is the PV output current. The output current of the PV cells can be obtained from the equivalent circuit diagram and Equation (5):

$$I = I_{ph} - I_{os}\left\{exp\left[\frac{q(V + IR_s)}{AKT}\right] - 1\right\} - \frac{V + R_s I}{R_{sh}} \tag{5}$$

I_{ph} is determined thus:

$$I_{ph} = I_{sc} = I_{scr}\left[\frac{S}{S_r} + k_i(T - T_r)\right] \tag{6}$$

I_{os} is determined thus:

$$I_{os} = I_{or}\left(\frac{T}{T_r}\right)^3 \exp\left[\frac{qE_g}{nKT_r} - \frac{qE_g}{nKT}\right] \tag{7}$$

In Equation (5), I_{ph} is the photo current; I_{os} is the diode reverse saturation current; q is the electron constant, $1.602 \times e^{-19}$(C); V and I are the PV output voltage and current; R_s is the series resistance; R_{sh} is the parallel resistance; A is the diode characteristic fitting coefficient; K is the Boltzmann constant, with a value of $1.831 \times e^{-23}$(J/K); T is the PV cell working temperature in Kelvin; I_{scr} is the short-circuit current under the conditions of solar radiation intensity $S_r = 1000$W/m^2 and temperature $T_r = 25$ °C; k_i is the short-circuit current temperature coefficient; I_{or} is the dark saturation current under T_r; E_g is the forbidden band width of the semiconductor material; and N is the junction constant, which is 1.5. The electrical characteristics of the PV cells are shown in Figure 7.

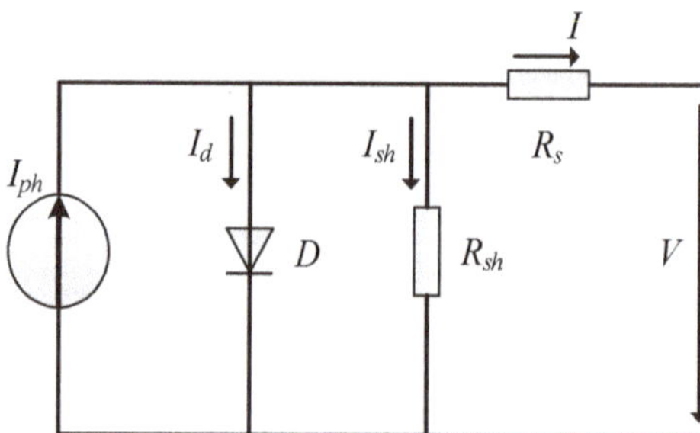

Figure 6. Equivalent circuit of photovoltaic cells.

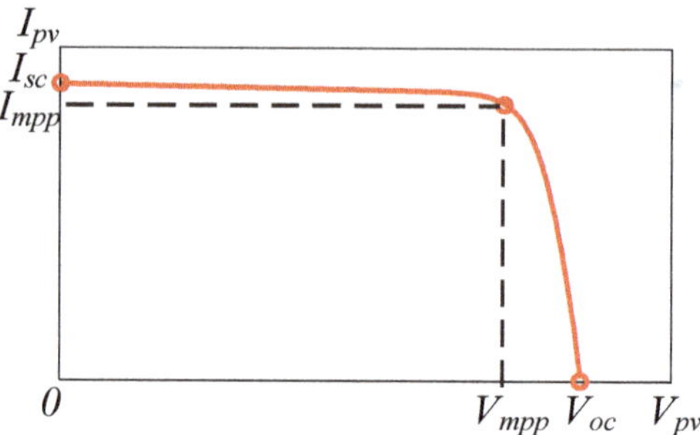

Figure 7. Characteristic of photovoltaic cells.

According to the actual needs and electric parameters, the PV cells should be connected in series or in parallel. The output current of each PV cell can then be expressed as:

$$I = N_p I_{sc} - N_p I_{os}\left\{\exp\left[\frac{q(V + IR_{se})}{N_s nKT}\right] - 1\right\} \tag{8}$$

In Equation (8), where N_p and N_s are the number of the PV cells' strings and parallel connections, and R_{se} is the equivalent series resistance of the PV cells after series and parallel connection.

2.3. The Model of the Lithium Iron Phosphate Battery

Recently, battery technologies based on lithium compounds have reduced cost and improved performance in terms of increased energy density. We used a LiFePO$_4$ battery as the power battery for the system, a type widely used in automobiles. Ships are generally less sensitive to weight restrictions than cars are, so more energy can be stored in a ship, which has boosted interest in using purely electric cruise ships based on lithium batteries. According to [32], models for the charge or discharge processes of ships can use the improved Shepherd equation, expressed in Equations (9) and (10), which disregards the influence of temperature on batteries. When batteries are in the discharge and charge processes, their voltage can be expressed thus:

$$V_{batt} = E_0 - K\frac{Q}{it - 0.1Q} \times i^* - K\frac{Q}{Q - it} \times it + A\exp(-B \times it) \tag{9}$$

$$V_{batt} = E_0 - K\frac{Q}{Q - it} \times i^* - K\frac{Q}{Q - it} \times it + A\exp(-B \times it) \tag{10}$$

where V_{batt} is the battery terminal voltage, E_0 is the nominal voltage, K is the polarization constant, Q is the battery capacity, i^* is the filter current, A is the exponential voltage, and B is the exponential time constant.

2.4. Working Principle and Model of Inverter Power

Considering that the ship's distribution mode is mainly a three-phase, three-wire system, a relatively simple three-phase full bridge topology is applied for the main power circuit. This is shown in Figure 8, where the LC filter is selected for the filter and U_{dc} is used as the DC voltage; u_{ia}, u_{ib}, and u_{ic} act as the inverter output voltages; i_{ia}, i_{ib}, and i_{ic} act as the inverter output currents; u_{sa}, u_{sb}, and u_{sc} are the bus voltages; i_{sa}, i_{sb}, and i_{sc} are the bus currents; L is the filtering inductance; and C is the filtering capacitance.

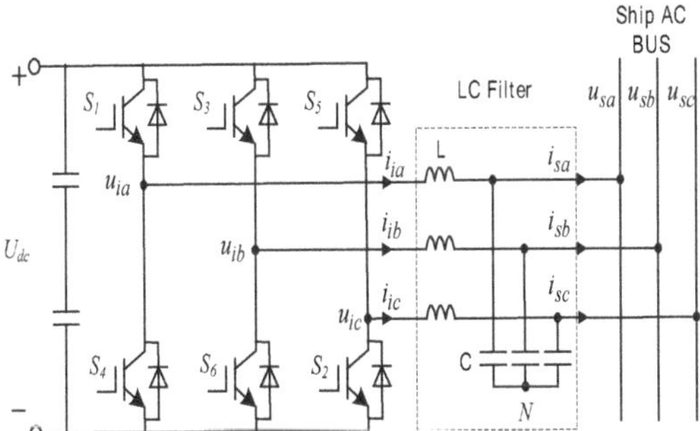

Figure 8. Circuit of the investigated inverter.

The inverter control strategy can be divided into three control modes: constant power control (PQ), constant voltage and frequency control (V/f), and dropping control. Depending on the requirements of different operation modes, different control algorithms can be selected to match each mode. The main purpose of this paper is to investigate the best operation mode for parallel inverters and then better utilize the solar energy of PV cells, so we adopted the PQ control method. The voltage/current values on the dq axis can be obtained by PARK transformation, and according to instantaneous power theory, the active and reactive power outputs from the inverter can be expressed as:

$$\begin{cases} P = u_{sd}i_{sd} + u_{sq}i_{sq} \\ Q = u_{sq}i_{sd} - u_{sd}i_{sq} \end{cases} \tag{11}$$

Selecting the d axis and the voltage/current vector in the same direction, the following formula can be obtained.

$$\begin{cases} P = u_{sd}i_{sd} \\ Q = -u_{sd}i_{sq} \end{cases} \Rightarrow \begin{cases} i_{sdref} = \dfrac{P_{ref}}{u_{sd}} \\ i_{sqref} = \dfrac{-Q_{ref}}{u_{sd}} \end{cases} \tag{12}$$

According to Equation (12) and under the dp axis, the active power P is determined by u_{sd} and i_{sd}, while the reactive power Q is determined by u_{sd} and i_{sq}. The decoupling control of active power P_{ref} and reactive power Q_{ref} is realized by controlling i_{sdref} and i_{sqref}, respectively, so that the voltage of the micro-grid system is unchanged. To control the current, the circuit equation of the inverter can be obtained from Figure 9:

$$\begin{cases} L\frac{di_{sd}}{dt} = u_{ld} - u_{sd} + wLi_{sq} \\ L\frac{di_{sq}}{dt} = u_{lq} - u_{sq} + wLi_{sd} \end{cases} \tag{13}$$

Using Equations (12) and (13) we can obtain the structural diagram of the PQ control, as shown in Figure 9. From this figure the inner ring current references, i_{dref} and i_{qref}, are obtained from the power references, P_{ref} and Q_{ref}, to divide the voltage u_{sd} in the outer ring. PI control is carried out using the difference between the currents references and the actual output currents i_{sd} and i_{sq}, and the modulated signals u_{dref} and u_{qref} are obtained through feed-forward compensation and cross-coupling compensation.

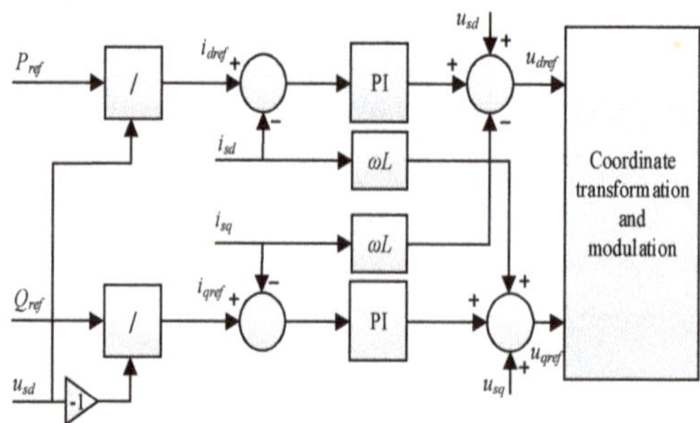

Figure 9. Power control (PQ) control strategy of inverter.

2.5. The Simulation Model of the Ship's Micro-Grid System

Combined with the proposed topological structure, the simulation model of the ship's micro-grid system is established using the MATLAB Simscape component libraries and modules, as presented in Figure 10.

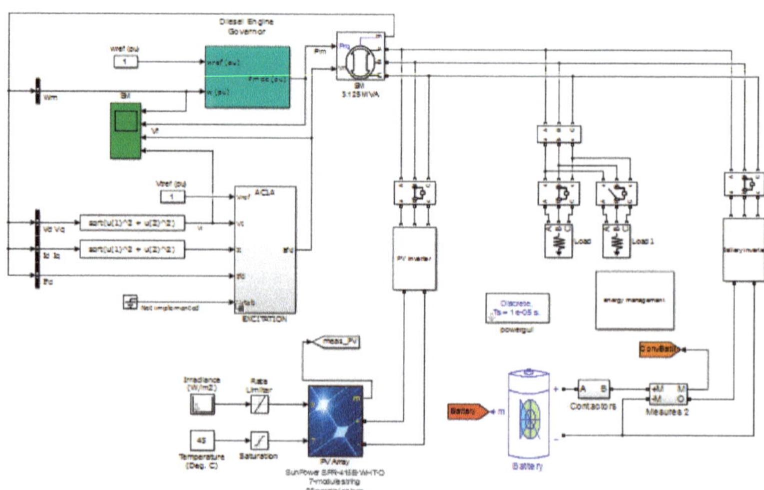

Figure 10. Simulation model of the proposed ship's micro-grid system.

3. Analysis of the Operation Characteristics in Parallel of the Ship's Micro-Grid System

According to the design of ship's micro-grid system, the system's voltage and frequency were supported by the diesel generator, which provided the main power supply, and the sources from the inverter of solar PV cells and lithium batteries both were used according the following algorithm. Assuming that the solar power output was maximum and constant, the power distribution of the micro-grid system could be determined by controlling the power output of the lithium battery pack. Based on the characteristics of the marine electrical propulsion system, this article studies the micro-grid energy management strategy for the sudden change in load, which can improve the power quality (frequency) of the ship's micro-grid under dynamic load. This strategy is based on the assumption that the output of PV power generation system remains constant, and we do not explore the complex dynamics of PV power generation system. Since this study is based on the output of different power generation units equal to the load power (conservation of energy), the PV output variation can also be considered as sudden load change, so the research conclusion can also be used for PV output variation. From this point, the proposed system can still control the actual PV channels (practical PV panels).

The governor characteristic (power-frequency characteristic curve) of the diesel generator is shown in curve 1 of Figure 11 [30]; this is also called the first adjustment rule between output power and power frequency. To stabilize the frequency around the ratings value, the throttle can be adjusted according to the output power, which shifts the characteristic curve up and down; this is called the second adjustment rule. The output frequency of the lithium battery inverter follows the frequency of the main power because the current-following algorithm is adopted. The output frequency is independent of the output power, which is determined by a given PQ value. The power-frequency characteristic of the lithium battery inverter is shown in curve 2 of Figure 11 [31]. The output frequency of the solar inverter is also followed by the main power frequency, and it is independent of the output power. The output power is determined by the maximum power point tracking (MPPT) algorithm, and the power-frequency characteristic of the solar inverter is shown in curve 3 of Figure 11. The effect of load change on frequency fluctuation in the ship's micro-grid system can be analyzed as follows.

Assuming that the ship's micro-grid system is in stable operation at A point (rated frequency f_n) under the certain working conditions in Figure 11, the solar output power is P_1, the output power of the lithium battery pack is P_2, and the output power of the diesel generator is P_3. Load increase $\Delta P = P_4 - P_3$ always happens suddenly. When we use the PQ algorithm for inverter operation, ΔP will be borne by the diesel generators only because the power outputs of the solar PV and batteries are constant values. From Figure 11, we can describe the regulation process for static system frequency. At first, the power frequency will adjust according to curve 1 and the system operation point from A to B. Then, the diesel throttle will be controlled to change the system's operation point from B to C (curve 4). For better operation, we postulate the initial conditions in which the system load is 80 kW, the power output of the PV array controlled by the MPPT algorithm is 15 kW, the power output of the energy storage battery is 20 kW, the remaining 45 kW of power is borne by the diesel generators, and the system reaches stable operation status2 s after the simulation start. When t = 4 s, the load increases from 0 to 30 kW with a step increase of 5 kW. The simulation results of the system's corresponding dynamic frequency response are shown in curve 7 to curve 1 of Figure 12.

The above analysis of system static output shows that when the load power increases, the system's frequency drops from f_n to f_1. If the system is not controlled, the frequency fluctuation will inevitably affect the system's power quality. Especially during dynamic system processes, when the mutation load increases, the harmonic increase in inverter power output is largely due to the frequency fluctuation overshoot. That condition not only will reduce the power grid's power quality but also will make the grid collapse in a serious situation (such as the propeller touching bottom).

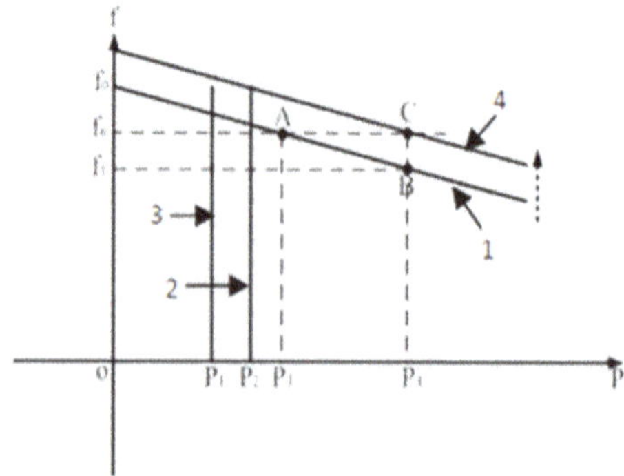

Figure 11. Frequency static regulation process.

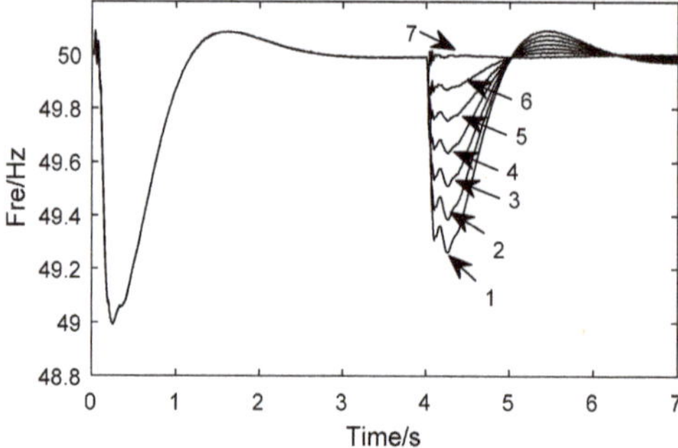

Figure 12. The system's corresponding frequency dynamic response.

4. The Dynamic Power Distribution Strategy of the Ship's Micro-Grid System

4.1. Analysis of the Dynamic Power Distribution Strategy

Because of the transient nature of the step load, the operation of impulsive power demands very high amplitude. Diesel generators are not capable of supporting transient load requirements because they need to operate constantly for a relatively long time to control the inertia of the fuel valves, combustors, and other mechanical components [33]. For supplying transient power, the lithium battery system is capable of supporting sudden changes in load and can help maintain the frequency stability of the micro-grid bus with its high power density. The main control progress of the micro-grid system is shown in Figure 13 and can be described as follows.

First, we assume that the system operates stably at point A and the load increases suddenly by $\Delta P = P_4 - P_3$ at a certain moment. If the reference value for the output power of the lithium battery inverter regulated using the PQ algorithm is changed from P_2 (curve 2) to P_{21} (curve 5) at that time, and the increase in output power of the lithium battery is controlled by $P_{21} - P_2 = P_4 - P_3 = \Delta P$, the working state of the micro-grid system will remain unchanged for the power output of the solar and diesel generators, and the load of the mutation will be borne entirely by the lithium battery pack. The frequency of the micro-grid system will thereby be kept stable throughout the load mutation. However, the output power P_2 from the package of lithium batteries is the optimized conditions of the micro-grid system. If the micro-grid system continues in this situation, a large current will be discharged from the lithium battery, negatively affecting its performance and lifetime. Therefore, the load of the lithium battery pack must be transferred to the diesel generator sets gradually.

According to the working characteristics of lithium battery, the optimal range of output capacity is 30–70%, and the maximum short-term discharge current cannot exceed 3C. In this study, the capacity of lithium battery pack is 53 kWh and the maximum output is set as 53 kW in simulation. If the load of system is suddenly increased to 80 kW, then 27 kW can be directly generated by the diesel generator. Curve 5 represents the characteristic curve of output power-frequency, which is generated by the reverse power supply of the lithium battery pack. The output power-frequency characteristic of the lithium battery changes from curve 5 to curve 2, and the frequency characteristic of the output power of the diesel generator set changes to curve 4 from curve 1. The output power of the lithium battery pack decreases from P_{21} back to P_2, and the output power assumed by the diesel generator increases from P_3 to P_4. Through this method of control, the total sudden increase in load is borne by the diesel generators, while the output power of the lithium battery pack returns to the original working-state value, and the micro-grid system reaches stable functioning at point C.

We were able to conclude that the frequency of the micro-grid system can be kept stable throughout load changes if an effective control strategy is developed. According to the above analysis, the key to the system's dynamic control process is the precise measurement of the output power reference value $P_{21} = P_2 + \Delta P$ and the algorithm for power transfer control in the lithium battery inverter source. The first part of the output power reference to value P_2 is the output power of the lithium battery inverter when the micro-grid system is operated at point A, before the power changes. The reference value is given by optimizing the precision predictions for the PV output power, battery pack capacity, load demand, and real-time output power of the diesel generators. The second part of the output power reference value ΔP needed to be tested and calculated. To do this, we proposed a dynamic control strategy based on current compensation and fuzzy control, which assumed that load power was dependent on and proportional to the current and that the response speed of the current was faster than the response speed of the diesel engine. Compared the results in Figures 11 and 13, Figure 11 shows that if the proposed algorithm is not used as a control strategy, the frequency will have large fluctuation. Nevertheless, Figure 13 shows that as the proposed control strategy is used to control the energy system, the frequency fluctuation will be greatly reduced and the power quality of the grid will be greatly enhanced. The frequency of the micro-grid system will thereby be kept stable even the load has apparent frequency fluctuation.

This strategy can improve the dynamic and static response performances for frequency feedback and give the lithium battery inverter system better dynamic response performance and precision. The characteristics of the rapid output power of the lithium battery energy storage system suggest that the frequency fluctuation of the system under load mutation is significantly reduced, and the power quality of the ship's micro-grid system is improved. The results are also applicable if there is a sudden decrease in the output of the solar power generation system, meaning the strategy can effectively reduce the influence of intermittent solar characteristics on the system and greatly improve the stability of system operation.

The control strategy for the compensation current is dependent on the actual change in current value to control the output power reference ΔP_1 of the whole system. The deviation problem is solved by frequency feedback, which can be achieved using a fuzzy control algorithm. This keeps the input at Δf and $d\Delta f/dt$, the output power is reference ΔP_2 of the whole system, and better dynamic performance is achieved. The input parameters of the fuzzy control algorithm are the difference Δf between real frequency f and a given reference frequency f_{ref} and the change rate in the difference $d\Delta f/dt$. The output power of the system will be referred to ΔP_2 to realize feedback control over the frequency. This will further improve the dynamic and static performance of the system. In this situation, the reference output power of the system is ΔP_2; feedback control over the frequency is implemented, and the system's dynamic and static performance are further improved. Given the limited output capacity of the battery pack, we adopt the amplitude limitation control links in the system control strategy. A schematic diagram of the dynamic control strategy for the ship's micro-grid system is shown in Figure 14.

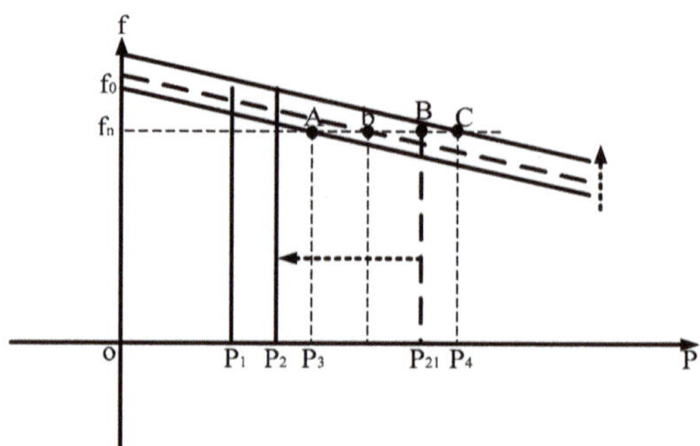

Figure 13. The proposed static frequency regulation process.

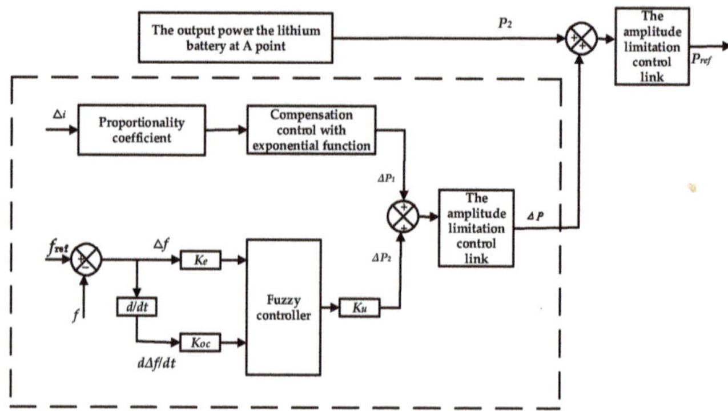

Figure 14. Proposed dynamic control strategy.

4.2. Controller Design

To meet the requirements for dynamic control and smooth load transfer, the current compensation power reference value is changed according to an exponential function. The time domain expression of this exponential function is shown in Equation (14):

$$\Delta p_1(t) = A \times e^{-\frac{t}{\tau}} \tag{14}$$

where A is the initial value and τ is the time constant ($\tau = 4$ for simulation purposes). The waveform of the function is shown in Figure 15. The function value $\Delta P_1(t)$ is reduced from the beginning A to zero, and the time constant τ determines the attenuation speed of the value $\Delta P_1(t)$. The fuzzy controller is equivalent to a PD controller, having the characteristics of a PD controller and significant robustness [34]. First, the fuzzy controller is used to blur the input value, then fuzzy reasoning is carried out according to the designed fuzzy control rules, and the fuzzy quantity is ambiguity resolved to obtain a clear value. In this study, the capacity of lithium battery pack is 53 kWh, which is set as the maximum output in simulation. When the increased load is greater than the difference value between 70% rated capacity of the lithium battery and the current capacity of the assembled lithium battery, the output of the assembled lithium battery will be controlled as the difference value, and the other part of the capacity will be loaded by the diesel generator set. Therefore, the value of A varies according to the situation, which is related to the battery discharge current and load variation. In general, the value of A is linearly related to the discharge current value of lithium battery.

The basic domain of input Δf and dΔf/dt is [−2.5 + 2.5], [−15 + 15], and the quantifying factors K_e and K_{ec} are 30 and 2, respectively. According to the control's requirements, the input value is divided into seven fuzzy sets: negative big {NB}, negative middle {NM}, negative small {NS}, zero

{ZO}, positive small {PS}, positive middle {PM}, and positive big {PB}. When the errors of {NB} and {PB} are large, the Gaussian subordinating degree function is used to improve stability. The rest value is used with the triangle membership function to improve resolution and control sensitivity. The basic domain of output is [−3000,3000], and the proportion factor is 3, which is divided into seven fuzzy subsets: negative big {NB}, negative middle {NM}, negative small {NS}, zero {ZO}, positive small {PS}, positive middle {PM}, and positive big {PB}, and the triangle membership function is applied. The input and output functions of the membership function are shown in Figures 16–18. Based on the control requirement, when the frequency change and frequency change rate are larger, the power reference value $\Delta P_2(t)$ should be increased to improve dynamic performance and reduce error. In other circumstances, the reference value $\Delta P_2(t)$ should be reduced to avoid overtones, and then the fuzzy control rule in Table 1 is available. The control quantity of power output $\Delta P_2(t)$ must be determined. According to a fuzzy subset of the output in Table 1, the gravity method is used to solve the ambiguity [35,36], and $\Delta P_2(t)$ is controlled by timing the proportional factor K_u.

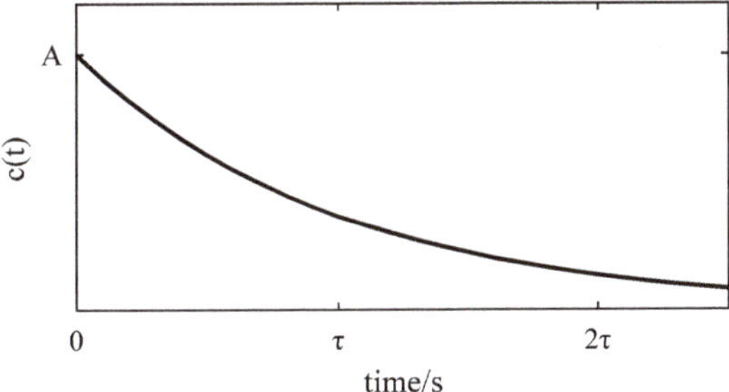

Figure 15. Waveform of the e exponential function.

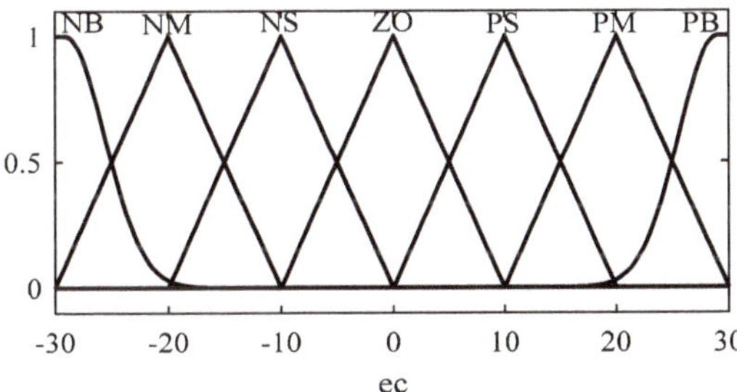

Figure 16. Δf of the membership function.

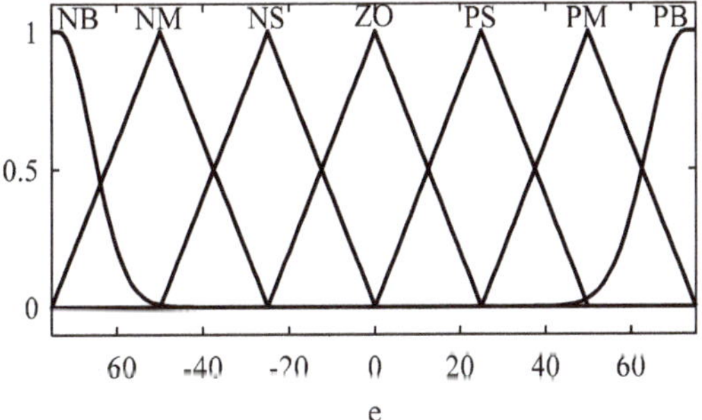

Figure 17. dΔf/dt of the membership function.

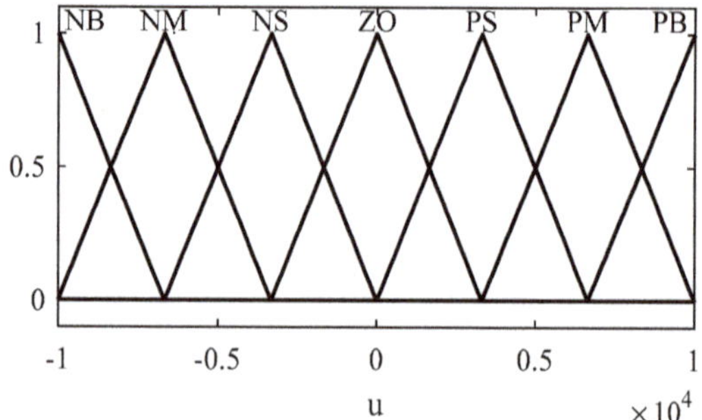

Figure 18. Output ΔP_2 of the membership function.

Table 1. Fuzzy logic rules.

Δf \\ $d\Delta f/dt$	NB	NM	NS	ZO	PS	PM	PB
NB	PB	PB	PM	PS	ZO	ZO	ZO
NM	PB	PB	PM	PS	ZO	ZO	ZO
NS	PM	PM	PM	ZO	ZO	NS	NS
ZO	PM	PM	PS	ZO	NS	NM	NM
PS	PS	PS	ZO	NS	NM	NM	NM
PM	ZO	ZO	ZO	NS	NM	NB	NB
PB	ZO	ZO	ZO	NB	NB	NB	NB

5. Analysis of Simulation and Results

Based on the overall model of the ship's micro-grid system developed by MATLAB/SIMULINK, we ran a simulation to evaluate the effectiveness of the power distribution and control allocation strategy and the system's operational stability. The input parameters of the simulation were as follows:

(1) The specified outputs of the diesel generator were 100 kW, 380 V, and 50 Hz.
(2) The maximum power output of the PV was 15 kW.
(3) The rated capacity of the battery was 200 Ah, 537 V.
(4) Resistors were used to simulate the system load.

The initial conditions were that the system was operated in parallel on stable status: initial total load 45 kW, PV output 15 kW, diesel generator output 10 kW, and battery output 20 kW. At five seconds, the load was suddenly increased to 80 kW. The frequency response curve of the micro-grid system is shown in Figure 19, and the operational profiles of the different energy modules are shown in Figure 20.

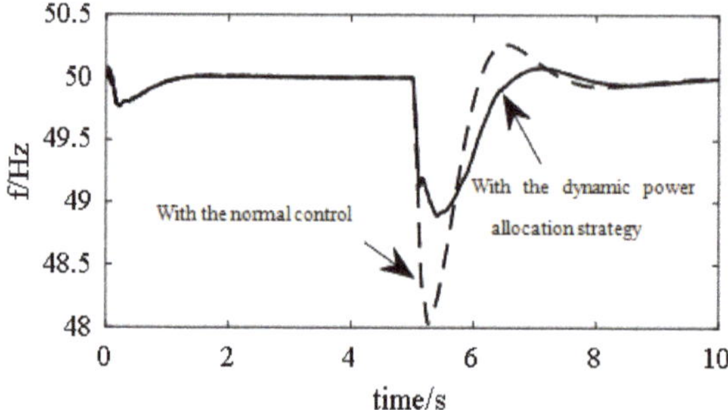

Figure 19. Frequency response curves of the micro-grid system.

The dotted line in Figure 20 shows that the maximum frequency deviation was 2 Hz when the load was suddenly increased to 80 kW. The real line indicates the frequency response curve of the micro-grid system with the dynamic power allocation strategy. From the real line, the maximum frequency deviation was 1 Hz when the load variation suddenly increased to 80 kW. Compared with the two curves, the system frequency fluctuation was decreased by [(2 − 1)/2] × 100% = 50%using the proposed control strategy, proving it could efficiently improve the system's dynamic performance.

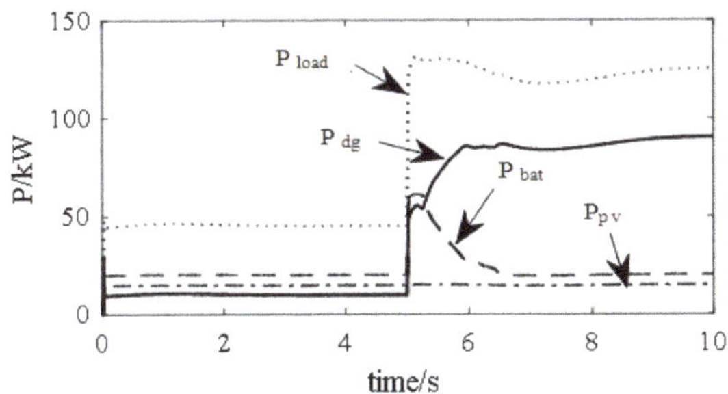

Figure 20. Operational profiles of different energy modules.

As can be seen from the load and power simulations for each power generation unit, presented in Figure 20, at the load mutation moment, the dynamic power reference value of the battery system quickly increased and the power diesel generator was minimized in the working state. At the same time, the dynamic power reference of the battery system decreased and the output power of the diesel generators increased. Finally, the dynamic power reference value of the battery system decreased to zero and all of the increased loads were borne by the diesel generators. Because the output power of the battery system decreased gradually according to the design of the *e* exponential function, the diesel generators had enough response time for speed regulation, so they did not produce larger frequency fluctuations in the micro-grid system. The whole process of the power distribution system enabled full use of the lithium battery's good instantaneous discharge performance, improved the power quality in the dynamic process of the micro-grid system, and efficiently ensured the system's stability.

The combination of new energy technology and electrical propulsion technology is an effective way to solve the problems of energy saving and greenhouse gas emission reduction of ships. But as a relatively independent unit and regardless of actual structures or grid characteristics, the power grids of ships have the typical characteristics different from the ones of terrestrial regions. Investigation of the coordinated control strategy for different power generation units under different situations is the essential scientific problem that must be developed to ensure the stable economic operation of the ship's micro-grid. This paper investigated a useful strategy to improve the utilization rate of renewable energy under static conditions, reduce the frequency fluctuation of the system under dynamic conditions, and improve the power quality and system stability of the ship's micro-grid.

In the past, Guo et al. had combined the battery with SC as the energy storage system and they investigated the control strategy of the power balance and the optimal energy management in MVDC power system [37]. The system could be used to maintain the MVDC bus voltage within desired margin, which was usually 10% around the nominal MVDC voltage. The objective of this research was to investigate the battery as the main power supply of cruise ships and investigate the control strategy of the power quality (constant frequency), which could have the optimal energy management in LVAC power system. For the widely used alternating current (AC) electrical propulsion ships, if the above micro-grid system architecture is adopted, a large number of power converters need to be connected to the direct current (DC) grid side to realize the conversion of the electrical energy form. Then the electrical energy utilization rate is low and the cost is high, although the voyage of large ships can be far away, the application prospect is not good. Therefore, the investigated management and distribution strategies are used, the ship's AC grid composed of the generator and the new energy is more advantageous, practical, and popularized.

6. Conclusions

The combination of new energy technology and electrical propulsion technology is an effective way to solve energy saving and greenhouse gas emission reduction of ships. But as a relatively independent unit, regardless of the actual structures or grid characteristics, ships have typical characteristics different from terrestrial power grids. The coordinated control strategy of different power generation units under different situations is the essential scientific problem that must be solved to ensure the stable economic operation of the ship's micro-grid. This paper investigated a strategy for the ship's micro-grid, which consisted of PV, DG, and a lithium battery. For that, the characteristics of the grid-connected operation of the new energy inverter power system, the ship's power station, and the corresponding dynamic power allocation strategy were proposed. In this study, we successfully constructed a simulation model to integrate the characteristics of a photovoltaic cell, a diesel generator, and a lithium battery (PV/DG/BAT), then investigated using a ship's micro-grid system to combine the three power sources. The constructed model allowed us to explore frequency fluctuations in the multi-energy micro-grid system under dynamic conditions. The dynamic power reference value was easily obtained by fuzzy control and a current compensation algorithm. Via the proposed distribution strategy for dynamic power, under the same conditions the diesel generator had enough response time to regulate its rotating speed, and the output powers had reasonable distribution. When the load was suddenly changed, the proposed strategies reduced the frequency fluctuation by 50%. The proposed strategies could also be used to reduce the frequency fluctuation caused by sudden changes because of the intermittent characteristics of PV power, greatly improving the stability of PV/DG/BAT-based dynamic power.

Author Contributions: Methodology, S.L. and W.Y.; software, S.L., Y.Z., and C.-F.Y.; formal analysis, W.Y., and C.-F.Y.; investigation, W.Y., S.L. and Y.Z.; data curation, Y.Z. and C.-F.Y.; original draft preparation, S.L., and W.Y.; review and editing, W.Y., Y.Z., and C.-F.Y.

Funding: The authors acknowledge the support provided by the National Natural Science Foundation of China (NO. 51679106) and the major science and technology project of Fujian province (NO. 2018H6014). This work was also supported by projects under No. MOST 108-2221-E-390-005 and MOST 108-2622-E-390-002-CC3.

Conflicts of Interest: The authors declare no conflict of interest.

References

1. O'Rourke, R. *Nuclear and Fossil Fuel Powered Surface Ships, Quick Look Analysis: U.S. Naval Nuclear Propulsion Program-CRS Report RL33946*; Congressional Research Service: Washington, WA, USA, 2007.

2. O'Rourke, R. *Navy Ship Propulsion Technologies: Options for Reducing Oil Use (Background for Congress)-CRS Report RL33360*; Congressional Research Service: Washington, WA, USA, 2006.

3. Zhang, H.; Chen, H.J.; Du, X.Z.; Wen, D.S. Photothermal conversion characteristics of gold nanoparticle dispersions. *Sol. Energy* **2014**, *100*, 141–147. [CrossRef]

4. Savchuk, O.A.; Carvajal, J.J.; Massons, J.; Aguilo, M.; Diza, F. Determination of photo-thermal conversion efficiency of grapheme and grapheme oxide through an integrating sphere method. *Carbon* **2016**, *103*, 134–141. [CrossRef]

5. Ayad, M.; Aissat, A. Photo-stability of the photons converter applied in the photovoltaic conversion systems. *Sol. Energy* **2016**, *133*, 221–225. [CrossRef]

6. Rosa-Clot, M.; Rosa-Clot, P.; Tina, G.M.; Ventura, C. Experimental photovoltaic thermal power plants based on TESPI panel. *Sol. Energy* **2016**, *133*, 305–314. [CrossRef]

7. Lee, D.Y.; Cutler, J.W.; Mancewicz, J.; Ridley, A.J. Maximizing photovoltaic power generation of a space-dart configured satellite. *Acta Astronaut.* **2015**, *111*, 283–299. [CrossRef]

8. Girish, T.E.; Aranya, S.; Nisha, N.G. Photovoltaic power generation using albedo and thermal radiations in the satellite orbits around planetary bodies. *Sol. Energy Mater. Sol. Cells* **2007**, *91*, 1503–1504. [CrossRef]

9. Wu, Z.; Xia, X.H. Optimal switching renewable energy system for demand side management. *Sol. Energy* **2015**, *114*, 278–288. [CrossRef]

10. Kibria, M.A.; Saidur, R.; AI-Sulaiman, F.A.; Aziz, M.M.A. Development of a thermal model for a hybrid photovoltaic module and phase change materials storage integrated in buildings. *Sol. Energy* **2016**, *124*, 114–123. [CrossRef]

11. Atmaca, E. Energy management of solar car in circuit race. *Turk. J. Electr. Eng.Comput. Sci.* **2015**, *23*, 1142–1158. [CrossRef]

12. Vinnichenko, N.A.; Uvarov, A.V.; Znamenskaya, I.A.; Ay, H.; Wang, T.H. Solar car aerodynamic design for optimal cooling and high efficiency. *Sol. Energy* **2014**, *103*, 183–190. [CrossRef]

13. Lan, H.; Wen, S.; Yu, D.C. Optimal sizing of hybrid PV/diesel/battery in ship power system. *Appl. Energy* **2015**, *158*, 26–34. [CrossRef]

14. Glykas, A.; Papaioannou, G.; Perissakis, S. Application and cost-benefit analysis of solar hybrid power installation on merchant marine vessels. *Ocean Eng.* **2010**, *37*, 592–602. [CrossRef]

15. Tsekouras, G.J.; Kanellos, F.D.; Prousalidis, J. Simplified method for the assessment of ship electric power systems operation cost reduction from energy storage and renewable energy sources integration. *IET Electr. Syst. Transp.* **2015**, *5*, 61–69. [CrossRef]

16. Adamo, F.; Andria, G.; Cavone, G.; De Capua, C.; Lanzolla, A.M.L.; Morello, R.; Morello, R.; Spadavecchia, M. Estimation of ship emissions in the port of Taranto. *Measurement* **2014**, *47*, 982–988. [CrossRef]

17. Wen, S.; Lan, H.; Yu, D.C.; Fu, Q. Optimal sizing of hybrid energy storage sub-systems in PV/Diesel ship power system using frequency analysis. *Energy* **2017**, *140*, 198–208. [CrossRef]

18. Ruoli, T. Large-scale photovoltaic system on green ship and its MPPT controlling. *Sol. Energy* **2017**, *157*, 614–628.

19. Lee, K.J.; Shin, D.; Yoo, D.W.; Choi, H.K.; Kim, H.J. Hybrid photovoltaic/diesel green ship operating in standalone and grid-connected mode-experimental investigation. *Energy* **2013**, *49*, 475–483. [CrossRef]

20. Sun, Y.; Qiu, Y.; Yuan, C.; Tang, X.; Wang, Y.; Jiang, Q. Research on the transient characteristic of photovoltaics-ship power system based on PSCAD/EMTDC. In Proceedings of the International Conference on Renewable Energy Research and Applications (ICRERA), Palermo, Italy, 22–25 November 2015; pp. 397–402.

21. Yu, W.N.; Li, D.; Zheng, W.M. Research and development of energy control system for solar tour vessels. *China Shipbuild.* **2013**, *54*, 177–183.

22. Yu, W.N.; Liao, W.Q.; Yang, R.F.; Li, S.W. Research and development of multi-energy ship micro-grid energy control system based on solar lithium battery and diesel generator set. *China Shipbuild.* **2017**, *58*, 170–176.

23. Banaei, M.R.; Alizadeh, R. Simulation-Based Modeling and Power Management of All-Electric Ships Based on Renewable Energy Generation Using Model Predictive Control Strategy. *IEEE Intell. Transp. Syst. Mag.* **2016**, *8*, 90–103. [CrossRef]

24. Gao, D.Z.; Shen, A.D.; Yan, J.X.; Huang, X.X. Energy Management and Control Strategy for Hybrid Electric Ships. *J. Shanghai Marit. Univ.* **2015**, *36*, 70–74.

25. Kong, L.Z.; Tang, X.S.; Qi, Z.P.; Lin, N. Research on energy storage dynamic energy scheduling strategy of Photovoltaic-Diesel Generator-Battery micro-grid system. *Power Syst. Prot. Control* **2012**, *40*, 6–12.

26. Ding, D.; Liu, Z.Q.; Yang, S.L.; Wu, X.G.; Li, T.T. Auxiliary AGC Frequency Modulation Method for Battery Energy Storage System Based on Fuzzy Control. *Power Syst. Protec. Control* **2015**, *43*, 81–87.

27. Li, W.K. *Mathematical Modeling of Diesel Generator Sets and Its Power Compensation Technology*; Harbin Institute of Technology: Harbin, China, 2013.

28. Zhai, D.; An, L.W.; Dong, L.X.; Zhang, Q.L. *Robust Adaptive Fuzzy Control of a Class of Uncertain Nonlinear Systems with Unstable Dynamics and Mismatched Disturbances*; IEEE Transactions on Cybernetics: Piscataway, NJ, USA, 2017.

29. Liao, W.Q.; Zhang, R.C.; Yu, W.N.; Wang, G.L. Prediction of output power of photovoltaic based on similar samples and principal component analysis. *J. Sol. Energy* **2016**, *37*, 2377–2385.

30. Benhamed, S.; Ibrahim, H.; Belmokhtar, K.; Hosni, H. Dynamic Modelling of Diesel Generator Based on Electrical and Mechanical Aspects. In Proceedings of the IEEE Electrical Power and Energy Conference (EPEC), Ottawa, ON, Canada, 12–14 October 2016.

31. Joga Rao, G.; Shrivastava, S.K.; Mangal, D.K. Coordinated V-f and P-Q Control of Solar Photovoltaic Generators with MPPT and Battery Storage in Microgrids. *IEEE Trans. Smart Grid* **2014**, *5*, 1270–1281.

32. Abada, S.; Marlair, G.; Lecocq, A.; Petit, M.; Sauvant-Moynot, V.; Huet, F. Safety focused modelling of lithium-ion batteries: A. review. *J. Power Sources* **2016**, *306*, 178–192. [CrossRef]
33. Khan, M.M.S.; Faruque, M.O.; Newaz, A. *Fuzzy Logic Based Energy Storage Management System for MVDC Power System of All Electric Ship*; IEEE Transactions on Energy Conversion: Piscataway, NJ, USA, 2017.
34. Sun, L.; Yue, Y.T. Research of fuzzy PD control algorithm based on DSP. *China Electro-Tech. Soc.* **2015**, *30*, 465–468.
35. Song, Q.; Wu, Y. Study on the Robustness Based on PID Fuzzy Controller. In Proceedings of the 2017 International Conference on Computing Intelligence and Information System (CIIS), Nanjing, China, 21–23 April 2017; pp. 142–145.
36. Qiang, G.; Junfeng, H.; Wei, P. PMSM Servo Control System Design Based on Fuzzy PID". In Proceedings of the 2017 2nd International Conference on Cybernetics, Robotics and Control (CRC), Chengdu, China, 21–23 July 2017; pp. 85–88.
37. Guo, Y.; Khan, M.M.S.; Faruque, M.O.; Sun, K. Fuzzy logic based energy storage supervision and control strategy for mvdc power system of all electric ship. In Proceedings of the 2016 IEEE Power and Energy Society General Meeting (PESGM), Boston, MA, USA, 17–21 July 2016.

Article

Analysis of Energy Flux Vector on Natural Convection Heat Transfer in Porous Wavy-Wall Square Cavity with Partially-Heated Surface

Yan-Ting Lin [1] and Ching-Chang Cho [2],*

[1] Institute of Nuclear Energy Research, Atomic Energy Council, Taoyuan 32546, Taiwan; yantinglin@iner.gov.tw

[2] Department of Vehicle Engineering, National Formosa University, No.64, Wunhua Rd., Huwei Township, Yunlin County 632, Taiwan

* Correspondence: cccho@nfu.edu.tw; Tel.: +886-5-6315699

Received: 16 October 2019; Accepted: 20 November 2019; Published: 22 November 2019

Abstract: The study utilizes the energy-flux-vector method to analyze the heat transfer characteristics of natural convection in a wavy-wall porous square cavity with a partially-heated bottom surface. The effects of the modified Darcy number, modified Rayleigh number, modified Prandtl number, and length of the partially-heated bottom surface on the energy-flux-vector distribution and mean Nusselt number are examined. The results show that when a low modified Darcy number with any value of modified Rayleigh number is given, the recirculation regions are not formed in the energy-flux-vector distribution within the porous cavity. Therefore, a low mean Nusselt number is presented. The recirculation regions do still not form, and thus the mean Nusselt number has a low value when a low modified Darcy number with a high modified Rayleigh number is given. However, when the values of the modified Darcy number and modified Rayleigh number are high, the energy flux vectors generate recirculation regions, and thus a high mean Nusselt number is obtained. In addition, in a convection-dominated region, the mean Nusselt number increases with an increasing modified Prandtl number. Furthermore, as the length of the partially-heated bottom surface lengthens, a higher mean Nusselt number is presented.

Keywords: energy flux vector; porous cavity; natural convection; wavy-wall; heat transfer enhancement; visualization technique

---

1. Introduction

The plot of the heat flow paths is important since it can provide physical insights into the energy transport process in detail. To achieve this purpose, Kimura and Bejan [1] have suggested a heatline visualization technique. Following the study of Kimura and Bejan [1], numerous researchers have explored the process of heat transport within thermal-fluid systems by utilizing the technique [2–4].

Hooman [5,6] has presented an energy-flux-vector method, which is basically similar to the heatline technique, for visualizing the heat flow paths. Comparing the two visualization methods, Hooman [5,6] has pointed out that the energy-flux-vector method is simpler than the heatline visualization technique since the algebraic equations do not require solving. In the literature, numerous researchers [7–9] have demonstrated that the process of heat transport can be completely explained by using the energy-flux-vector method.

Natural convection in porous cavities has numerous practical applications in engineering fields, including biomedical engineering, chemical and material processing, fluidized beds, geothermal engineering, thermal insulation, solar collection, and so on [10–14]. To achieve the purpose of heat transfer enhancement, wavy-surface geometries are often imposed [7,8]. Note that the practical applications for the natural convection in a wavy-wall cavity include heat exchangers, solar collectors, condensers in refrigerators, geothermal engineering, and so on [15–17]. In the literature, the investigation into the natural convection heat transfer behavior in a porous cavity with wavy surfaces has been widely discussed [18–20]. The results have demonstrated that the use of the wavy-surface geometries can improve the heat transfer effect. Recently, Biswal et al. [21] have utilized the energy-flux-vector method to explain the process of heat transport of natural convection within a porous cavity with curved side walls. Their results have shown that given suitable curved-side wall forms with appropriate flow conditions, the heat transfer performance can be enhanced.

As discussed above, the analysis of the energy-flux-vector method on natural convection heat transfer in a porous cavity with wavy surfaces has attracted little attention. Accordingly, in the current study, the energy-flux-vector method is utilized to analyze the heat transfer characteristics of natural convection within a porous square cavity bounded by a partially-heated flat bottom wall, low temperature left and right wavy-walls, and an insulated flat top wall. The simulations focus particularly on the effects of the modified Darcy number, modified Rayleigh number, modified Prandtl number, and length of the partially-heated bottom surface on the energy-flux-vector distribution and mean Nusselt number, respectively.

2. Mathematical Formulation

2.1. Governing Equations and Boundary Conditions

Figure 1 illustrates the studied porous square wavy-wall cavity with a partially-heated bottom surface and a characteristic length of L_c. As shown, the partially-heated wall surface has a length of L_H, and it is placed on the center of the bottom wall. Meanwhile, the left and right walls are assumed to have a constant low temperature and a complex-wavy surface.

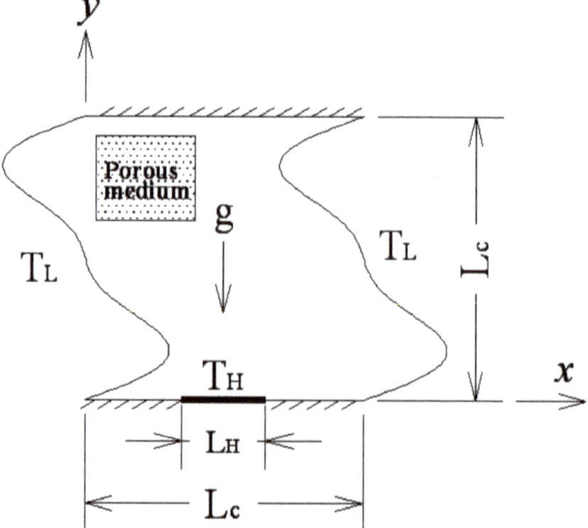

Figure 1. Geometry of partially-heated wavy-wall porous cavity.

It is assumed that the working fluid is Newtonian and incompressible, and the flow and temperature fields are two-dimensional and steady state. In addition, it is also assumed that the Rayleigh number is less than 10^9 and thus, the assumption of laminar flow is valid. Furthermore, the Boussinesq approximation [7,8,22] is imposed, and the local thermal equilibrium condition is assumed to be achieved.

Defining the non-dimensional quantities are as follows [14]:

$$x^* = \frac{x}{L_c}, \; y^* = \frac{y}{L_c}, \; \alpha_{eff} = \frac{k_{eff}}{\varepsilon \rho C_p}, \; u^* = \frac{u L_c}{\alpha_{eff}}, \; v^* = \frac{v L_c}{\alpha_{eff}}, \; p^* = \frac{p L_c^2}{\rho \alpha_{eff}^2},$$

$$\theta = \frac{T - T_L}{T_H - T_L}, \; K_m = \frac{K}{\varepsilon}, \; \Pr_m = \frac{\mu}{\rho \alpha_{eff}}, \; Da_m = \frac{K_m}{L_c^2}, \; Ra_m = \frac{\rho g \beta L_c^3 (T_H - T_L)}{\mu \alpha_{eff}}$$

(1)

where ρ, C_p, and p represent the density, specific heat, and pressure, respectively; u and v are the velocity components along x- and y-axes, respectively; T_H and T_L signify the high temperature and low temperature, respectively; subscript m indicates the modified value; and superscript $*$ denotes the non-dimensional quantity. After the viscous dissipation and thermal radiation effects are ignored, the continuity, momentum and energy conservation equations described the flow behavior and heat transfer characteristics of natural convection within the porous cavity is written in the following non-dimensionalized forms [14]:

Continuity equation:

$$\frac{\partial u^*}{\partial x^*} + \frac{\partial v^*}{\partial y^*} = 0$$

(2)

x-direction momentum equation:

$$u^* \frac{\partial u^*}{\partial x^*} + v^* \frac{\partial u^*}{\partial y^*} = -\frac{\partial p^*}{\partial x^*} - \frac{\Pr_m}{Da_m} u^* + \Pr_m \left(\frac{\partial^2 u^*}{\partial x^{*2}} + \frac{\partial^2 u^*}{\partial y^{*2}} \right) - \frac{1.75}{\sqrt{150}} \frac{\sqrt{u^{*2} + v^{*2}}}{\sqrt{Da_m}} u^*$$

(3)

y-direction momentum equation:

$$u^* \frac{\partial v^*}{\partial x^*} + v^* \frac{\partial v^*}{\partial y^*} = -\frac{\partial p^*}{\partial y^*} - \frac{\Pr_m}{Da_m} v^* + \Pr_m \left(\frac{\partial^2 v^*}{\partial x^{*2}} + \frac{\partial^2 v^*}{\partial y^{*2}} \right) - \frac{1.75}{\sqrt{150}} \frac{\sqrt{u^{*2} + v^{*2}}}{\sqrt{Da_m}} v^* + Ra_m \cdot \Pr_m \cdot \theta$$

(4)

Energy conservation equation:

$$u^* \frac{\partial \theta}{\partial x^*} + v^* \frac{\partial \theta}{\partial y^*} = \frac{\partial^2 \theta}{\partial x^{*2}} + \frac{\partial^2 \theta}{\partial y^{*2}}$$

(5)

The dimensionless boundary conditions are expressed as follows:

Bottom partially-heat wall: $\quad u^* = v^* = 0, \quad \theta = 1$

Bottom other walls: $\quad u^* = v^* = 0, \quad \partial \theta / \partial \vec{n}^* = 0$

Left and right wavy walls: $\quad u^* = v^* = 0, \quad \theta = 0$

Top wall: $\quad u^* = v^* = 0, \quad \partial \theta / \partial \vec{n}^* = 0$

where $\vec{n}^*$ is the normal vector.

2.2. Energy Flux Vectors and Nusselt Number

The energy flux vectors ($\vec{E}$) are defined as follows [5–9]:

$$\vec{E} = \frac{\partial H^*}{\partial y^*} \vec{i} - \frac{\partial H^*}{\partial x^*} \vec{j}$$

(6)

$$\frac{\partial H^*}{\partial y^*} = u^* \theta - \frac{\partial \theta}{\partial x^*},$$

(7)

$$-\frac{\partial H^*}{\partial x^*} = v^* \theta - \frac{\partial \theta}{\partial y^*}.$$

(8)

Note that H^* is the non-dimensional heat function, and $\vec{i}$ and $\vec{j}$ are the unit components in x- and y-directions, respectively.

The mean Nusselt number (Nu_m) along the partially-heated bottom surface is defined as follows:

$$Nu_m = \int Nud\xi \tag{9}$$

where $Nu(= \frac{hL_c}{k_{eff}} = -\left(\frac{\partial\theta}{\partial\overline{n}}\right))$ is the Nusselt number, and h is the convection heat transfer coefficient.

2.3. Geometric Description and Numerical Method

In this study, the left and right walls of the porous cavity are assumed to have a complex-wavy surface, which is formulated as follows [7,8]:

$$y^* = \alpha_w Sin(2\pi x^*) + \frac{\alpha_w}{2.5} Sin(4\pi x^*) \tag{10}$$

where α_w is the amplitude of the wavy surface.

In the current study, the generalized coordinate transform technique, finite volume method, and SIMPLE algorithm are used to solve the governing equations presented in Equations (2)–(5). The numerical methods are identical to those used in our previous studies [7,8]. A detailed description of the numerical methods applied in this work can be found in [7,8].

2.4. Numerical Validation and Grid Independence Evaluation

The currently used numerical models and numerical methods were valid by comparing the current results with those presented by Singh et al. [14]. Table 1 shows these results. Note that the Rayleigh number of $Ra = 10^5$, porosity of $\varepsilon = 0.6$, and Prandtl number of $Pr = 1$ are given in the case. It is shown that the current results are identical to those presented in Singh et al. [14].

Table 1. Comparison of current results for mean Nusselt number with Singh et al. [14] for two different Darcy numbers.

	$Da = 10^{-2}$	$Da = 10^{-4}$
Current results	3.441	1.067
Singh et al. [14]	3.461	1.067
Error (%)	0.6	0.0

The mesh sizes of 101×201, 101×501, 101×1001, 201×1001 have been examined for the variation of mean Nusselt number, respectively, and the results showed that the mesh size of 101×1001 has a grid-independent solution.

3. Results and Discussion

In the study, the ranges of the non-dimensional parameters were set as follows: modified Darcy number: $Da_m = 10^{-5}$ to $Da_m = 10^{-2}$; modified Rayleigh number: $Ra_m = 10^2$ to $Ra_m = 10^5$; modified Prandtl number: $Pr_m = 0.1$ to $Pr_m = 10$; amplitude of wavy surface: $\alpha_w = 0.25$; and length of the partially-heated bottom surface: $L_H^* = 0.3$ to $L_H^* = 0.9$.

Figures 2 and 3 illustrate the distributions of the energy flux vectors and isotherms within the porous cavity for various modified Darcy numbers and modified Rayleigh numbers. Figure 4 shows the effects of the modified Darcy number and modified Rayleigh number on the mean Nusselt number. According to the definition of the energy flux vectors given in Equation (6), the conduction mechanism dominates if the flow of energy flux vectors is directly from the bottom partially-heated high-temperature wall to the left and right low-temperature wavy-walls, while the convection mechanism dominates if the energy flux vectors generate closed recirculation regions [5–9]. Therefore, as shown in Figure 2a, when the values of the modified Darcy number and modified Rayleigh number are both low, the heat transfer effect within the porous partially-heated cavity is dominated by a pure conduction mechanism since the closed recirculation regions in the energy-flux-vector distribution are not created. In addition, a low modified Darcy number represents a low permeability due to the presence of a high flow resistance [14]. Therefore, although a larger buoyancy effect induced by giving a high modified Rayleigh number is presented, a high flow resistance is also generated within the porous partially-heated cavity. As a result, the energy flux vectors do still not form closed recirculation regions, and thus the conduction mechanism continually dominates heat transfer behavior (see Figure 2b). Under conduction-domination, the high-temperature fluid heated by the partially-heated bottom surface is slowly diffused to left and right low-temperature wavy-wall surfaces to be dissipated (see the isotherms in Figure 2). Consequently, under a low modified Darcy number, a low mean Nusselt number is presented, irrespective of the value assigned to the modified Rayleigh number (see Figure 4).

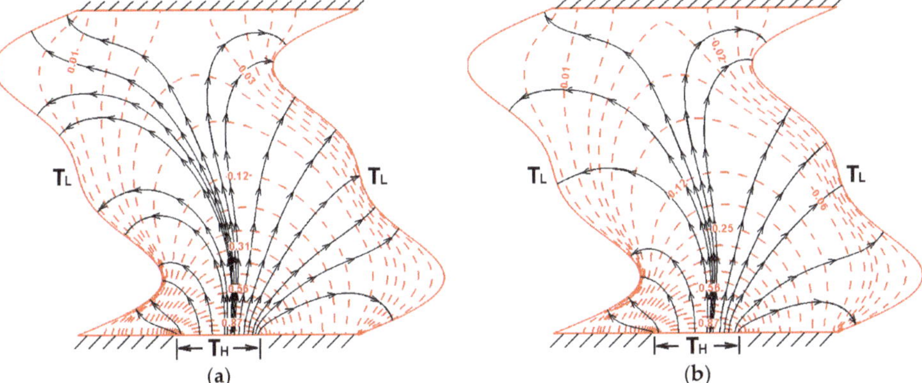

Figure 2. Distributions of energy flux vectors and isotherms within a partially-heated porous cavity given a modified Darcy number of $Da_m = 10^{-5}$ and modified Rayleigh numbers of: (a) $Ra_m = 10^2$ and (b) $Ra_m = 10^5$. Note that $Pr_m = 1$ and $L_H^* = 0.3$. Note also that the black solid lines indicate the flow direction of the energy-flux-vector, while the red dashed lines indicate the isothermal contours.

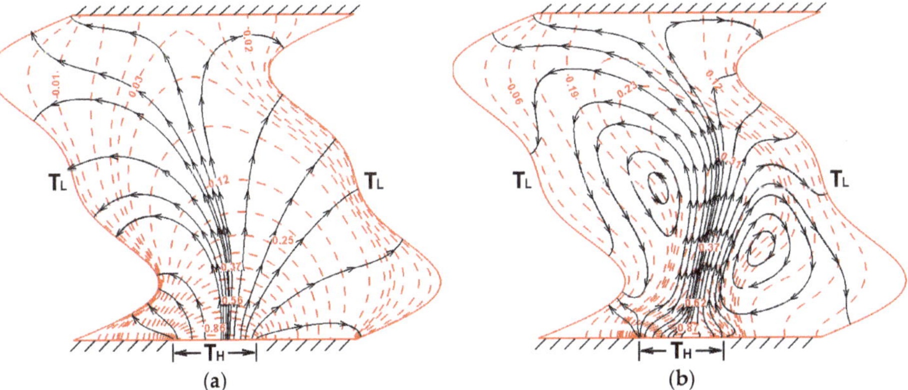

Figure 3. Distributions of energy flux vectors and isotherms within a partially-heated porous cavity given a modified Darcy number of $Da_m = 10^{-7}$ and modified Rayleigh numbers of: (a) $Ra_m = 10^2$ and (b) $Ra_m = 10^5$. Note that $Pr_m = 1$ and $L_H^* = 0.3$. Note also that the black solid lines indicate the flow direction of the energy-flux-vector, while the red dashed lines indicate the isothermal contours.

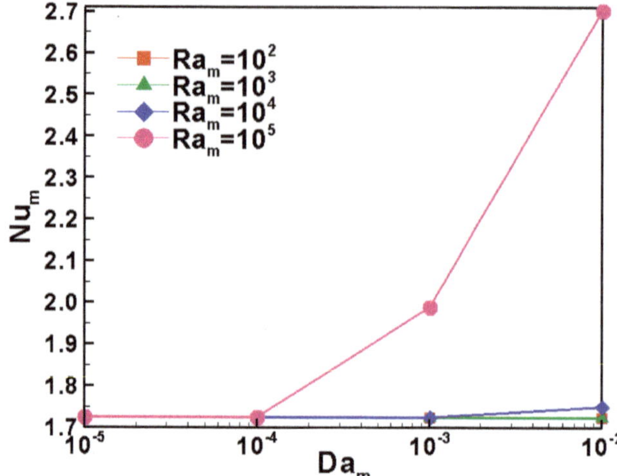

Figure 4. Variation of mean Nusselt number with modified Darcy number as a function of modified Rayleigh number. Note that $Pr_m = 1$ and $L_H^* = 0.3$.

When a high modified Darcy number and a low modified Rayleigh number is given, the closed recirculation regions of the energy flux vectors are still not formed within the porous partially-heated cavity (see Figure 3a). Although a high modified Darcy number has a high permeability and a low flow resistance, a low modified Rayleigh number induces a small buoyancy effect. Therefore, the convection effect is still weak, and the conduction mechanism continues to dominate the heat transfer effect. As a result, a low mean Nusselt number is presented (see Figure 4).

When the modified Darcy number and the modified Rayleigh number are both high, a low flow resistance and a high buoyancy effect occur within the porous partially-heated cavity. Therefore, the closed energy-flux-vector recirculations are formed, resulting in enhancement of the convection effect (see Figure 3b). The high-temperature fluid on the partially-heated bottom wall is promptly driven to the left and right low-temperature walls to be dissipated (see the isotherms in Figure 3b). As a result, a high mean Nusselt number is obtained (see Figure 4).

Figures 5 and 6 illustrate the distributions of the energy flux vectors and isotherms within the porous partially-heated cavity for various modified Prandtl numbers and modified Rayleigh numbers. Figure 7 shows the effects of the modified Prandtl number and modified Rayleigh number on the mean Nusselt number. Note that the modified Darcy number is set as $Da_m = 10^{-2}$ in the cases. It shows that given a low modified Rayleigh number, the distributions of the energy flux vectors and isotherms are similar for various modified Prandtl numbers (see Figure 5a,b). The recirculation regions of energy flux vectors are not created, and thus the conduction heat transfer dominates. In other words, the effect of the modified Prandtl number on the heat transfer is insignificant. Therefore, a low and approximately constant mean Nusselt number is presented (see Figure 7). When a high modified Rayleigh number is given, it shows that the size of closed energy-flux-vector recirculation regions enlarges as the modified Prandtl number is increased (see Figure 6a,b). In other words, a higher modified Prandtl number has a stronger convection effect under high modified Rayleigh numbers. Consequently, a higher mean Nusselt number is presented (see Figure 7).

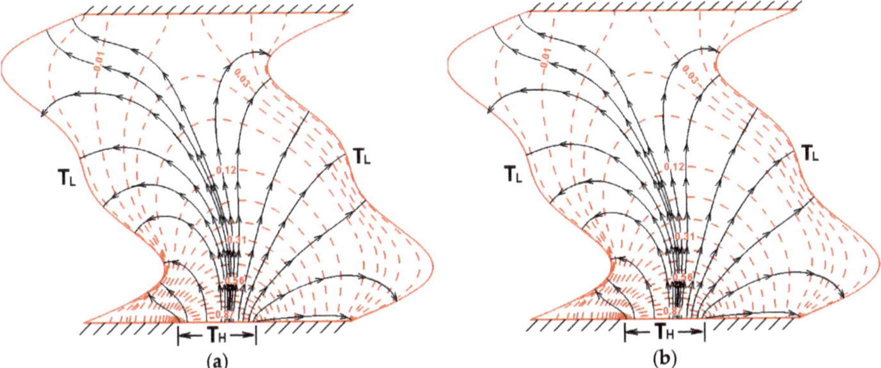

Figure 5. Distributions of energy flux vectors and isotherms within a partially-heated porous cavity given a modified Rayleigh number of $Ra_m = 10^2$ and modified Prandtl numbers of: (**a**) $Pr_m = 0.1$ and (**b**) $Pr_m = 10$. Note that $Da_m = 10^{-2}$ and $L_H^* = 0.3$. Note also that the black solid lines indicate the flow direction of energy flux vectors, while the red dashed lines indicate the isothermal contours.

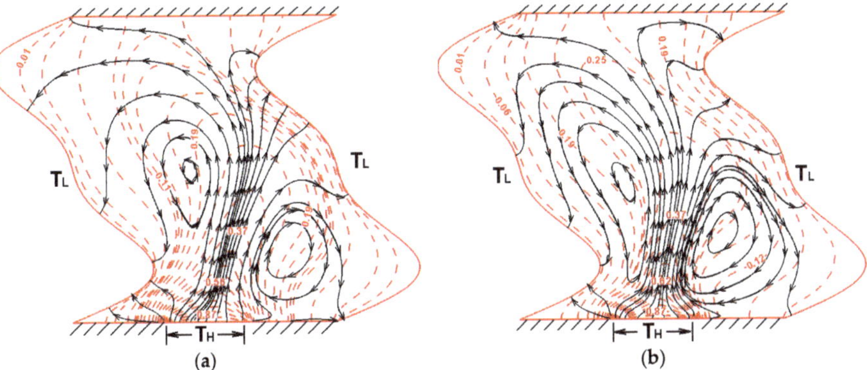

Figure 6. Distributions of energy flux vectors and isotherms within a partially-heated porous cavity given a modified Rayleigh number of $Ra_m = 10^5$ and modified Prandtl numbers of: (**a**) $Pr_m = 0.1$ and (**b**) $Pr_m = 10$. Note that $Da_m = 10^{-2}$ and $L_H^* = 0.3$. Note also that the black solid lines indicate the flow direction of energy flux vectors, while the red dashed lines indicate the isothermal contours.

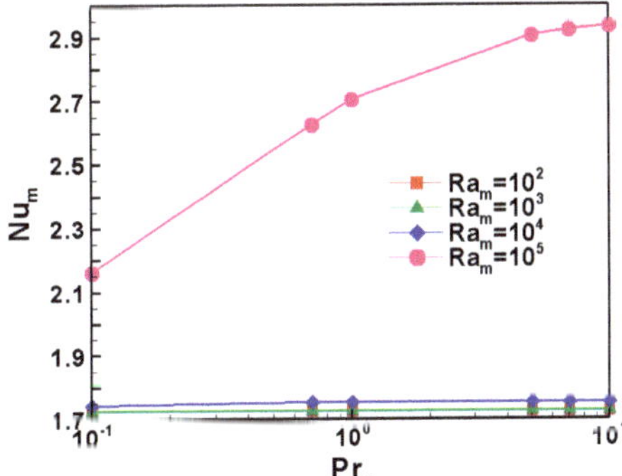

Figure 7. Variation of mean Nusselt number with modified Prandtl number as a function of a modified Rayleigh number. Note that $Da_m = 10^{-2}$ and $L_H^* = 0.3$.

Figure 8 illustrates the distributions of the energy flux vectors within the porous cavity for various partially-heated lengths and modified Rayleigh numbers. Figure 9 shows the effects of the partially-heated length and modified Rayleigh number on the mean Nusselt number. Note that the modified Darcy number is set as $Da_m = 10^{-2}$ in these cases. For a low modified Rayleigh number, the conduction heat transfer dominates since the recirculation regions of the energy flux vectors are not formed within the porous cavity for various partially-heated lengths (see Figure 8a). Therefore, a low mean Nusselt number is presented. As the length of the partially-heated bottom surface is lengthened, a larger bottom heated area is presented. Consequently, a higher mean Nusselt number is obtained (see Figure 9). Given a high modified Rayleigh number, the energy flux vectors form closed recirculation regions. The size of the recirculation regions enlarges as the partially-heated length is extended (see Figure 8b). As a result, the strength of the convection effect enhances. Consequently, the mean Nusselt number rises with the increasing length of the partially-heated bottom surface (see Figure 9).

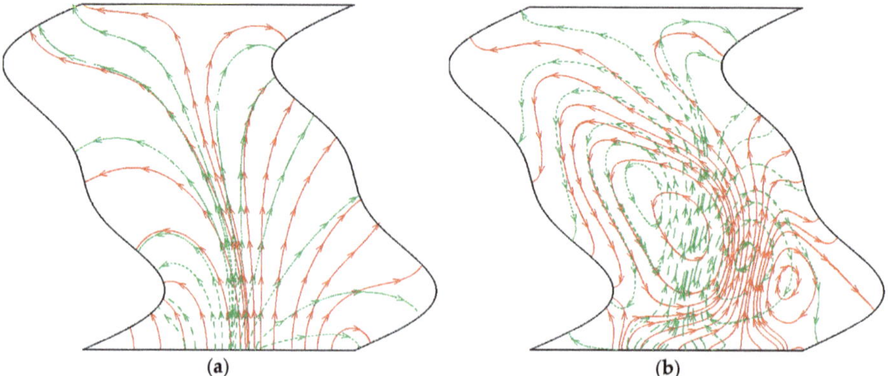

(a)   (b)

Figure 8. Distributions of energy flux vectors within a porous cavity given partially-heated lengths of $L_H^* = 0.3$ and $L_H^* = 0.9$ and modified Rayleigh numbers of: (**a**) $Ra_m = 10^2$ and (**b**) $Ra_m = 10^5$. Note that $Da_m = 10^{-2}$ and $\Pr_m = 1$. Note also that the red solid lines indicate the partially-heated length of $L_H^* = 0.9$, and the green dashed lines indicate the partially-heated length of $L_H^* = 0.3$.

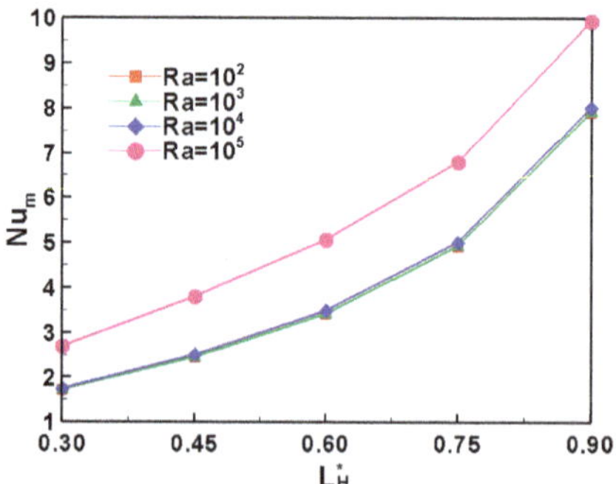

Figure 9. Variation of mean Nusselt number with partially-heated length as a function of a modified Rayleigh number. Note that $Da_m = 10^{-2}$ and $\Pr_m = 1$.

4. Conclusions

This paper has analyzed the heat transfer behavior of natural convection in a porous square wavy-wall cavity with a partially-heated bottom surface using the energy-flux-vector method. The effects of the flow parameters and length of a partially-heated bottom surface on the energy-flux-vector distribution and mean Nusselt number have been discussed. The studied results are summarized as follows:

1. Given a low modified Darcy number, the energy flux vectors did not generate recirculation regions within the porous cavity, irrespective of the value assigned to the modified Rayleigh number. The conduction heat transfer dominated, and thus the mean Nusselt number was low.
2. Given a high modified Darcy number and a low modified Rayleigh number, the heat transfer effect was dominated by the conduction mechanism since no recirculation region was formed in the energy flux vectors. Therefore, a low mean Nusselt number was obtained.
3. Given a high modified Darcy number with a high modified Rayleigh number, recirculation regions in the energy-flux-vector distribution were produced, resulting in a convection-domination. Consequently, a high mean Nusselt number was presented.
4. In conduction-dominated region, the effect of the modified Prandtl number on the energy-flux-vector distribution and mean Nusselt number was insignificant. However, in convection-dominated region, the size of the closed energy-flux-vector recirculation region enlarged, and the value of the mean Nusselt number raised as the modified Prandtl number was increased.
5. In conduction-dominated or convection-dominated regions, the mean Nusselt number was raised as the length of the partially-heated bottom surface lengthened.

Author Contributions: Conceptualization, C.-C.C.; Methodology, C.-C.C.; Software, C.-C.C. and Y.-T.L.; Validation, C.-C.C. and Y.-T.L.; Formal Analysis, C.-C.C.; Investigation, C.-C.C.; Resources, C.-C.C. and Y.-T.L.; Data Curation, C.-C.C. and Y.-T.L.; Writing-Original Draft Preparation, C.-C.C.; Writing-Review & Editing, C.-C.C.; Visualization, C.-C.C.; Supervision, C.-C.C.; Project Administration, C.-C.C.; Funding Acquisition, C.-C.C. and Y.-T.L.

Funding: The authors would like to thank the Ministry of Science and Technology, Taiwan, for the financial support of this study under Contract Nos. MOST 108-2221-E-150-011, MOST 108-3116-F-042A-006 and MOST 107-2221-E-150-035.

Conflicts of Interest: The authors declare no conflict of interest.

Nomenclature

Da_m	modified Darcy number, $Da_m = \frac{K_m}{L_c^2}$
$\vec{E}$	energy flux vector
k_{eff}	effective thermal conductivity, $Wm^{-1}K^{-1}$
L_c	characteristic length of square cavity, m
L_H	length of partially-heated bottom surface, m
Nu	Nusselt number, $Nu = \frac{hL_c}{k_{eff}} = -\left(\frac{\partial\theta}{\partial \vec{n}}\right)$
Nu_m	mean Nusselt number, $Nu_m = \int Nud\xi$
Pr_m	modified Prandtl number, $Pr_m = \frac{\mu}{\rho\alpha_{eff}}$
Ra_m	modified Rayleigh number, $Ra_m = \frac{\rho g\beta L_c^3(T_H-T_L)}{\mu\alpha_{eff}}$
T	temperature, K

Greek symbols

α_{eff}	effective thermal diffusivity, m^2s^{-1}
α_w	amplitude of wavy surface
β	thermal expansion coefficient, K^{-1}
ε	porosity
K	permeability, m^2
θ	dimensionless temperature, $\theta = \frac{T-T_L}{T_H-T_L}$

References

1. Kimura, S.; Bejan, A. The "Heatline" visualization of convective heat transfer. *J. Heat Transf.-Trans. ASME* **1983**, *105*, 916–919. [CrossRef]
2. Trevisan, O.V.; Bejan, A. Combined heat and mass transfer by natural convection in a vertical enclosure. *J. Heat Transf.-Trans. ASME* **1984**, *109*, 104–112. [CrossRef]

3. Ramakrishna, D.; Basak, T.; Roy, S.; Pop, I. A complete heatline analysis on mixed convection within a square cavity: Effects of thermal boundary conditions via thermal aspect ratio. *Int. J. Therm. Sci.* **2012**, *57*, 98–111. [CrossRef]

4. Biswal, P.; Nag, A.; Basak, T. Analysis of thermal management during natural convection within porous tilted square cavities via heatline and entropy generation. *Int. J. Mech. Sci.* **2016**, *115–116*, 596–615. [CrossRef]

5. Hooman, K.; Gurgenci, H.; Dincer, I. Heatline and energy-flux-vector visualization of natural convection in a porous cavity occupied by a fluid with temperature-dependent viscosity. *J. Porous Media* **2009**, *12*, 265–275. [CrossRef]

6. Hooman, H. Energy flux vectors as a new tool for convection visualization. *Int. J. Numer. Methods Heat Fluid Flow* **2010**, *20*, 240–249. [CrossRef]

7. Cho, C.C. Heat transfer and entropy generation of mixed convection flow in Cu-water nanofluid-filled lid-driven cavity with wavy surface. *Int. J. Heat Mass Transf.* **2018**, *119*, 163–174. [CrossRef]

8. Cho, C.C. Mixed convection heat transfer and entropy generation of Cu-water nanofluid in wavy-wall lid-driven cavity in presence of inclined magnetic field. *Int. J. Mech. Sci.* **2019**, *151*, 703–714. [CrossRef]

9. Nayak, R.K.; Bhattacharyya, S.; Pop, I. Numerical study on mixed convection and entropy generation of a nanofluid in a lid-driven square enclosure. *J. Heat Transf.-Trans. ASME* **2016**, *138*, 012503. [CrossRef]

10. Lu, D.A.; Flamant, G.; Snabre, P. Towards a generalized model for vertical walls to gas-solid fluidized beds heat transfer-I. Particle convection and gas convection. *Chem. Eng. Sci.* **1993**, *48*, 2479–2492. [CrossRef]

11. Nield, D.A.; Bejan, A. *Convection in Porous Media*; Springer: New York, NY, USA, 2006.

12. Prud'homme, M.; Jasmin, S. Inverse solution for a biochemical heat source in a porous medium in the presence of natural convection. *Chem. Eng. Sci.* **2006**, *61*, 1667–1675. [CrossRef]

13. Al-Amiri, A.; Khanafer, K.; Pop, I. Steady-state conjugate natural convection in a fluid-saturated porous cavity. *Int. J. Heat Mass Transf.* **2008**, *51*, 4260–4275. [CrossRef]

14. Singh, A.K.; Basak, T.; Nag, A.; Roy, S. Heatlines and thermal management analysis for natural convection within inclined porous square cavities. *Int. J. Heat Mass Transf.* **2015**, *87*, 583–597. [CrossRef]

15. Misirlioglu, A.; Baytas, A.C.; Pop, I. Free convection in a wavy cavity filled with a porous medium. *Int. J. Heat Mass Transf.* **2005**, *48*, 1840–1850. [CrossRef]

16. Sultana, Z.; Hyder, M.N. Non-darcy free convection inside a wavy enclosure. *Int. Commun. Heat Mass Transf.* **2007**, *34*, 136–146. [CrossRef]

17. Cho, C.C.; Chiu, C.H.; Lai, C.Y. Natural convection and entropy generation of Al_2O_3-water nanofluid in an inclined wavy-wall cavity. *Int. J. Heat Mass Transf.* **2016**, *97*, 511–520. [CrossRef]

18. Kumar, B.V.R.; Singh, P.; Murthy, P.V.S.N. Effect of surface undulations on natural convection in a porous square cavity. *J. Heat Transf.-Trans. ASME* **1997**, *119*, 848–851. [CrossRef]

19. Chen, X.B.; Yu, P.; Winoto, S.H.; Low, H.T. Free convection in a porous wavy cavity based on the Darcy-Brinkman-Forchheimer extended model. *Numer. Heat Tranf. A-Appl.* **2007**, *52*, 377–397. [CrossRef]

20. Khanafer, K.; Al-Azmi, B.; Marafie, A.; Pop, I. Non-Darcian effects on natural convection heat transfer in a wavy porous enclosure. *Int. J. Heat Mass Transf.* **2009**, *52*, 1887–1896. [CrossRef]

21. Biswal, P.; Basak, T. Heatlines visualization of convective heat flow during differential heating of porous enclosures with concave/convex side walls. *Int. J. Numer. Methods Heat Fluid Flow* **2018**, *28*, 1506–1538. [CrossRef]

22. Singh, A.K.; Basak, T.; Nag, A.; Roy, S. Role of entropy generation on thermal management during natural convection in tilted porous square cavities. *J. Taiwan Inst. Chem. Eng.* **2015**, *50*, 153–172. [CrossRef]

Article

Performance Enhancement of Hybrid Solid Desiccant Cooling Systems by Integrating Solar Water Collectors in Taiwan

Win-Jet Luo [1,*], **Dini Faridah** [1], **Fikri Rahmat Fasya** [2], **Yu-Sheng Chen** [2], **Fikri Hizbul Mulki** [2] and **Utami Nuri Adilah** [3]

[1] Graduate Institute of Precision Manufacturing Engineering, National Chin-Yi University of Technology, Zhongshan Rd., Taiping Dist., Taichung 41170, Taiwan
[2] Department of Refrigeration, Air Conditioning and Energy Engineering, National Chin-Yi University of Technology, Zhongshan Rd., Taiping Dist., Taichung 41170, Taiwan
[3] Department of Refrigeration and Air Conditioning Engineering, Politeknik Negeri Bandung, Gegerkalong Hilir Rd., Parongpong Dist., Bandung 40012, Indonesia
* Correspondence: wjluo@ncut.edu.tw; Tel.: +886-423-924-505 (ext. 5110)

Received: 30 July 2019; Accepted: 4 September 2019; Published: 9 September 2019

Abstract: A hybrid solid desiccant cooling system (SDCS), which combines a solid desiccant system and a vapor compression system, is considered to be an excellent alternative for commercial and residential air conditioning systems. In this study, a solar-assisted hybrid SDCS system was developed in which solar-heated water is used as an additional heat source for the regeneration process, in addition to recovering heat from the condenser of an integrated heat pump. A solar thermal collector sub-system is used to generate solar regeneration water. Experiments were conducted in the typically hot and humid weather of Taichung, Taiwan, from the spring to fall seasons. The experimental results show that the overall performance of the system in terms of power consumption can be enhanced by approximately 10% by integrating a solar-heated water heat exchanger in comparison to the hybrid SDCS system. The results show that the system performs better when the outdoor humidity ratio is large. In addition, regarding the effect of ambient temperature on the coefficient of performance (COP) of the systems, a critical value of outdoor temperature exists. The COP of the systems gradually rises with the increase in ambient temperature. However, when the ambient temperature is greater than the critical value, the COP gradually decreases with the increase in ambient temperature. The critical outdoor temperature of the hybrid SDCS is from 26 °C to 27 °C, and the critical temperature of the solar-assisted hybrid SDCS is from 27 °C to 30 °C.

Keywords: hybrid solid desiccant cooling system; regeneration process; solar thermal collector; coefficient of performance

---

1. Introduction

Heating, ventilating and air conditioning (HVAC) systems are designed to maintain specific indoor conditions, which vary depending on the application. The main factors that influence the thermal comfort of occupants are metabolic rate, clothing insulation, air temperature, mean radiant temperature, air velocity and relative humidity [1]. The main purpose of the HVAC system is to provide good indoor air quality that meets the criteria for hygienic air conditions and satisfies the thermal comfort of occupants or products in a building or space. Moreover, the working environment can influence the productivity of workers, which is an economic reason for installing HVAC systems in buildings [2,3].

Generally, HVAC systems require a large amount of energy, especially in large capacity applications. HVAC systems make a significant contribution to carbon-based energy consumption and greenhouse gases emissions [4]. In the USA, nearly 50% of the energy consumption of buildings is used for HVAC systems [5]. Furthermore, conventional HVAC systems cause pollution, not only due to consuming a large amount of energy, but also due to the use of hydro-chlorofluorocarbon (HCFC), hydrofluorocarbon (HFC) and other refrigerants that produce greenhouse gases [6].

Cooling-based dehumidification systems are the most popular systems of recent decades [7,8]. Such systems provide cooling and dehumidification by utilizing a single vapor-compression unit. Cooling coils are used to cool the air below its dew point so moisture can be removed from the air. Therefore, low humidity and low temperature air can be generated and a reheat coil is required to avoid overcooling, which consumes a large amount of energy and is difficult to control. Current studies are developing new approaches that are more energy efficient and more environmentally friendly.

Recently, novel modern air conditioning systems have been proposed, most notably utilizing split processes of cooling and dehumidification. Instead of the cooling coil unit used in conventional cooling dehumidification systems, a single unit that can handle latent heat is applied, so unnecessary energy use can be avoided. One developed method is a solid desiccant cooling system (SDCS) in which refrigerants of HCFCs are unnecessary and low-grade thermal energy can be used for regeneration [9,10]. In an SDCS system, solid desiccants, such as silica gel, activated carbon, molecular sieves, alumina gel and other materials with strong hygroscopic ability, are used to dehumidify the process air. The solid desiccant itself has many pores, and the inner surface of each pore is concave. When the process air passes through the solid desiccant, because of the lower partial pressure of water vapor on the concave surface with a small radius of curvature, the vapor may migrate from the air to the concave surface of the pores. Then, the vapor condenses on the concave surface and releases adsorption heat to the desiccant. The most widely used solid desiccant dehumidification equipment is the rotary dehumidification wheel, in which solid desiccant is coated on the surface of wheel. The rotary dehumidification wheel can realize continuous dehumidification and regeneration through periodical rotation. The high-humidity process air passes through a portion of the desiccant wheel, and the vapor in the process air is absorbed by the solid desiccant of the wheel due to the vapor pressure difference between the air and the desiccant. Then, the temperature of the process air rises due to adsorption heat, and humidity is reduced after passing through the wheel. When the desiccant in the process air stream absorbs enough vapor from the process air, the vapor adsorption portion of the desiccant wheel is rotated to the high temperature regeneration air stream. While the regeneration air passes through the portion of the desiccant wheel containing a greater amount of vapor in the desiccant, the heat from the high temperature air stream leads to higher vapor pressure in the desiccant in comparison to the vapor pressure in the air stream. Then, vapor in the desiccant is ejected into the regenerated air stream until the lower vapor pressure condition in the desiccant is attained. The desiccant wheel periodically and dynamically rotates between the process and regeneration air streams. Dehumidification and regeneration processes are conducted in the corresponding air streams, and a low-humidity process air stream can be continuously obtained using the rotary wheel. Narayanan et al. [11] analyzed the performance characteristics of a solid-desiccant evaporative cooling system with TRNSYS software. The results show that the system could provide thermal comfort, however, the capability of the system to provide suitable air temperature and humidity depends on the performance of the evaporative cooling, energy-recovery, and heat-generation systems. Therefore, the availability of a cheap or waste-heat source is essential in making this system economically viable. Narayanan et al. [12] numerically investigated the dehumidification potential of a solid desiccant-based evaporative cooling system with an enthalpy exchanger operating in subtropical and tropical climates with TRNSYS software. The results show that in hot and humid conditions, the system thermal comfort capability drops to around 54% to 63%.

Solar energy utilization is a new approach developed in recent years specifically for space heating applications. The solar energy source is limitless and safer for the environment. Recently, heat from solar energy was developed for desiccant cooling systems. The heat generated by a solar thermal collector can be used for the regeneration process of the dehumidification wheel. Guidara et al. [13]

proposed the use of evaporative coolers for pre-cooling and re-cooling of the process air before and after the dehumidification wheel in order to satisfy the load demand of an air-conditioned space. From their numerical analysis, they indicated that the pre-cooling design is suitable to be applied in drier ambient conditions. Enteria et al. [14] investigated the performance of a solar-desiccant cooling system with a silica gel and titanium dioxide desiccant wheel in East Asia. From numerical simulations, it was found that using solar desiccant cooling systems has great potential for East Asian countries. The study pointed out that, in the tropical region, a larger area and capacity of the solar thermal collector and greater regeneration air flow rate are required, and the operational performance of the system is in the range of 1.5 to 3. The major operational energy loss of the proposed system comes from solar collectors, water pipes, electric heaters and thermal storage tanks. Speerforck et al. [15] numerically investigated the performance of a solar-desiccant cooling system incorporating borehole heat exchangers for direct cooling and solar energy for desiccant regeneration. They indicated that the proposed system allows electricity saving of 50% and reduces CO_2 equivalent emissions by 91%. White et al. [16] proposed a solar-assisted SDCS incorporating direct and indirect evaporative coolers in a serial arrangement to cool process air after the dehumidification wheel. From numerical analysis, it was indicated that the sensible heat removal capability of the process air can be enhanced by a two-stage evaporative cooling design, which is suitable for cities with low-humidity climatic conditions.

Environmental conditions in different regions may affect the performance of a desiccant cooling system (DCS). In hot and humid regions, the use of a hybrid SDCS that is integrated with a heat pump is suggested, instead of a conventional SDCS that utilizes an evaporative cooler and can generate a more humid air supply. Jani et al. [17] proposed a hybrid cooling system integrated with a heat pump (hybrid SDCS), and indicated the proposed system has good performance in hot and humid climate conditions. The dehumidification performance of the system is also highly sensitive to the outdoor ambient conditions. The performance of the hybrid SDCS is better for cases where ambient humidity is high [18]. It was found that the use of a hybrid SDCS can decrease the total power consumption by 20–30% and increase the cooling capacity by 40–60% [19]. Jani et al. [20] used the numerical software TRNSYS to simulate the performance of a solid desiccant-assisted hybrid space cooling system. The results show that the system achieves a good performance in hot humid climates. The humidity ratio of a room's process air was substantially lowered, from 0.014 kg of water/kg of dry air to 0.006 kg of water/kg of dry air, by use of a solid desiccant-based rotary dehumidifier.

In addition to environmental conditions, regeneration temperature is one of the main parameters used to determine the performance of a SDCS. The performance of an SDCS is sensitive to changes in humidity and regeneration temperature [21,22]. However, if the regeneration temperature in the hybrid SDCS is too high, the system performance may be reduced due to high condensation pressure and an increased amount of work performed by the compressor. Several heat sources can be used for the regeneration process, such as an electric heater, gas, a low-grade thermal energy such as solar energy, and waste heat. A further cogeneration system can also be considered as the heat source for the regeneration process.

Beccali [23,24] developed a new solar-assisted hybrid SDCS in which the condensing heat from the condenser of the incorporated heat pump is recovered to preheat the regeneration air stream before solar heating. Long-term measurements of the developed system were conducted in southern Italy. Five control modes based on the temperature and humidity of the outdoor air conditions were designed. From their performance analysis, it was pointed out that preheating of the recovered heat for the regeneration process can reduce the required heat from the solar collector by approximately 30%. In other words, the required area of the solar collector can also be moderately reduced. Fong et al. [25,26] developed a solar assisted SDCS incorporating an adsorption chiller and analyzed the performance of the system by numerical analysis. The outdoor air was dehumidified by a rotary dehumidification wheel and passed through a radiant cooling coil in an air-conditioned space. The chilled water from the adsorption chiller was provided to the radiant cooling coil for handling the sensible heat of the air-conditioned space. Both required regeneration heating for the rotary dehumidification wheel and the adsorption chiller supplied by the solar collector. Compared with the traditional centralized air conditioning system, the energy saving potential of the integrated system could reach 36.5%. As alternatives to silica gel, Bareschino et al. [27] proposed other hygroscopic materials for desiccant wheels, MIL101@GO-6 (MILGO) and Campanian ignimbrite, in conjunction with an air-conditioning system driven by evacuated tube solar collectors equipped with a desiccant wheel. The numerical simulations were carried out by means of TRNSYS 17[®] (version 17, Thermal Energy System Specialists, Madison, WI, USA) to dynamically assess the energy flows in the considered plants and compared with that of a conventional system. The results demonstrate that primary energy savings of approximately 20%, 29%, and 15% can be reached with silica-gel, MILGO and zeolite-rich tuff desiccant wheel-based air handling units, respectively. Li et al. [28] investigated a two-stage rotary desiccant cooling/heating system driven by evacuated glass tube solar air collectors. The results show that the major advantage of the two-stage desiccant cooling system was that moisture removal reached 6.68–14.43 g/kg in hot and humid climate conditions. Solar heating with desiccant humidification can improve indoor comfort significantly. A solar hybrid SDCS is a good substitute for traditional vapor compression air conditioning systems, especially in hot and humid climates, since solar energy can result in energy savings in the range of 40–45% [29]. Rambhad [30] indicated regeneration temperatures of hot water in the range of 54.3 °C to 68.3 °C can be achieved in solar SDCS systems by simulation.

In previous studies, the solar-assisted hybrid SDCS system was considered a replacement of the refrigerant vapor compression air conditioning system due to its higher energy efficiency. However, most studies of the solar-assisted hybrid SDCS have been conducted using numerical analysis. Research into the system's long-term practical operations and analysis of its performance under the effects of different ambient temperatures and humidity ratios is less common, especially in hot and humid environments. The performance of the solar-assisted hybrid SDCS has not been investigated experimentally in detail. In this study, the effect of ambient humidity and temperature on the performance of a solar-assisted hybrid SDCS was investigated under the high humidity and high temperature ambient conditions of Taichung, Taiwan. The performance analysis and comparisons of solar-assisted hybrid SDCS, hybrid SDCS and solar SDCS systems were conducted through long-term experiments in order to understand their characteristics and operational ranges in terms of environmental conditions.

2. Experimental Configuration and Methods

The experiments were conducted in the typical hot and humid weather of Taichung. The city has four seasons with an average temperature of 23.3 °C. The monthly average temperature, average relative humidity (*RH*) and the solar irradiation of Taichung from April to October are listed in Table 1. In the solar hybrid SDCS system, the heat from solar collectors and the condenser of an integrated heat pump sub-system are used for regeneration.

Table 1. Monthly average temperature, relative humidity (*RH*) and solar irradiation of Taichung [31].

Month	Temperature (°C)	RH (%)	Solar Irradiation (W/m^2)
April	27.6	77.3	154.142
May	30.2	77.1	162.474
June	31.9	77.9	166.640
July	33.0	75.6	174.972
August	32.6	77.6	164.557
September	31.8	75.8	162.474
October	30.1	72.6	158.308

2.1. Proposed System Configuration

In general, the proposed solar-assisted hybrid SDCS is divided into two major sub-systems. The first is a hybrid SDCS and the second is a solar-heated water system. The solar collector is installed at a fixed tilt angle of 27° facing the south-east direction. Several configurations, including a hybrid SDCS system, solar-assisted SDCS systems and a solar-assisted hybrid SDCS system, were studied to investigate the effect of using a solar regeneration water system on system performance.

2.1.1. Hybrid SDCS Configuration

The hybrid SDCS configuration was the first configuration investigated. In this configuration, the only heat source for the regeneration process is the condenser of the integrated heat pump, as shown in Figure 1. In this configuration, chilled water is generated by the evaporator and stored in a chilled water tank. The chilled water is pumped to pre-cooling and cooling coils in order to cool process air from the environment. In the integrated heat pump sub-system, an additional condenser installed outside the air handling unit can eject excessive condensation heat to the surroundings and avoid the over-loading of the compressor due to higher condensation temperature in the refrigeration cycle. The return air stream gains heat from the condenser inside the air handling unit, then heats the upper part of the solid desiccant wheel and causes the vapor pressure in the solid desiccant to be higher than the vapor pressure in the air stream. The moisture in the solid desiccant is ejected to the surroundings with the air stream. The moisture removal from the desiccant wheel by additional heat in the return air stream is called the regeneration process.

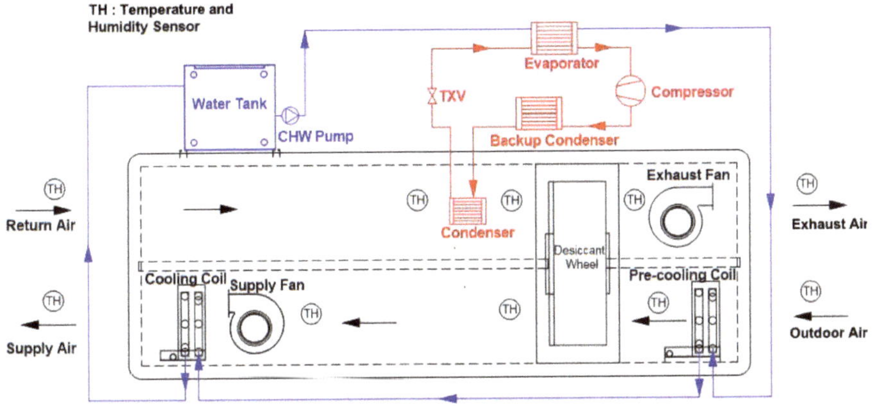

Figure 1. Hybrid solid desiccant cooling system (SDCS) configuration.

2.1.2. Solar-Assisted SDCS Configuration

The solar-assisted SDCS configuration is shown in Figure 2. The configuration only uses solar-heated water from a heat exchanger as the heat source for regeneration. Therefore, the regeneration temperature of the system will not be as high as the regeneration temperature of the hybrid SDCS configuration. In terms of process air, latent heat is handled by the desiccant wheel. However, since the heat pump is turned off, the temperature gradually increases with operation time due to adsorption heat in the dehumidification process. Therefore, the supply air temperature will be higher compared to other cases.

2.1.3. Solar-Assisted Hybrid SDCS Configuration

In the solar-assisted hybrid SDCS configuration, the condensing heat of the integrated heat pump and the heat of solar-heated water are used as heat sources for the regeneration process. In the regeneration air stream, the returning air is pre-heated by solar-heated water from a heat exchanger. Then, the returning air passes the condenser of the integrated heat pump to recover the dissipating heat from the condenser again in order to increase the regeneration temperature. In terms of process air, the air handling process concept is the same as in the hybrid SDCS configuration. The corresponding system configuration is shown in Figure 3.

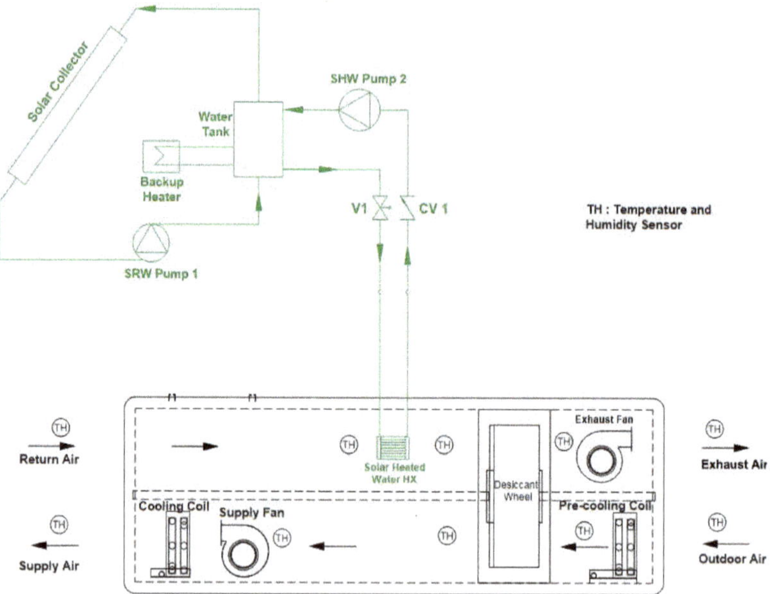

Figure 2. Solar-assisted SDCS configuration.

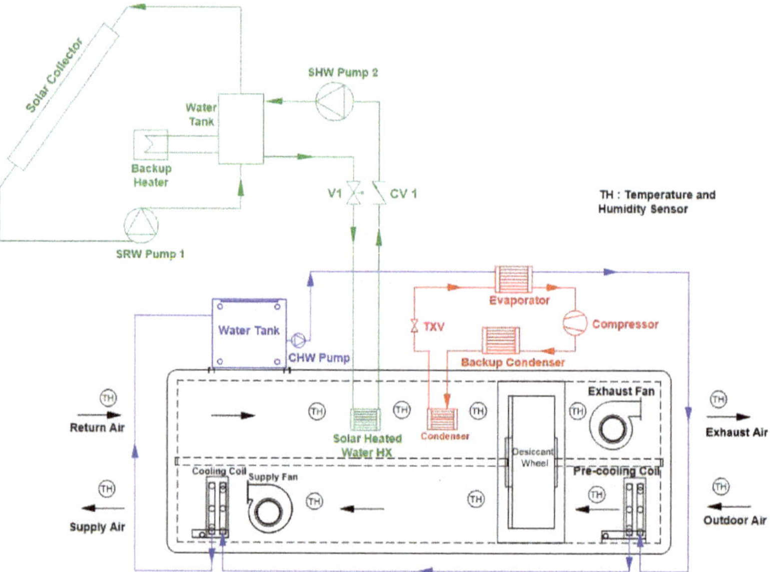

Figure 3. Solar-assisted hybrid SDCS configuration.

The system specifications are as follows:

(1)   Desiccant wheel: rotor depth 0.2 m, diameter 0.3 m, rotational speed 10 rph, thermal effectiveness 75%; silica gel, ProFlute (Stockholm, Sweden).

(2)   Compressor: capacity 6 kW, 220 volts (V)/7.3 amperes (A), locked rotor ampere (LRA), frequency 60 Hertz (Hz); R407c refrigerant, Tecumseh (Ann Arbor, MI, USA).

(3)   Condenser: finned and tube condenser, $730 \times 290 \times 200$ mm (main condenser), $530 \times 290 \times 100$ mm (auxiliary condenser).

(4)   Supply and exhaust fans: nominal power 0.7 kW, frequency 0–50 Hz.

(5)   Cooling water tank: capacity 71 L.

(6)   Chilled-water pump: nominal capacity 0.37 kW, rotational speed 3370 rpm, 220 V/1.5 A, flow rate 25 m^3/h, chilled water temperature: 13 °C.

(7)   Solar thermal collector (Sun Tech, Taichung, Taiwan): number of tubes 18, effective area 1.974 m^2, total gross area 11.699 m^2, effectiveness 94.5%.

(8)   Solar water tank: capacity 100 L.

(9)   Solar-heated water heat exchanger: finned and tube HX, $400 \times 360 \times 200$ mm.

(10)   Temperature and relative humidity (*RH*) sensors: ECOA EPRTH04101 (Ecoa Technologies, Taipei, Taiwan), 0–100 °C, 0–100% *RH*, temperature measurement accuracy ±0.3 °C at 25 °C, *RH* measurement accuracy ±4% *RH* (at 10–90%) and ±6% *RH* (at 0–10% and 90–100%).

(11)   Current sensors: CTT-CLS-CV clamp on type (U.R.D., Yokohama, Japan), current range 0–100 A, measurement accuracy ±2%.

(12)   Flow meter sensor: TESTO480 hot wire anemometer (Testo, West Chester, PA, USA), velocity measurement accuracy ±0.21 at 1.96 m/s, ±0.29 at 4.99 m/s, and ±0.41 at 10.06 m/s.

In terms of air flow rates, the flow rates of the process air and regeneration airstreams are not the same due to a slight leakage in the system. However, the gap between the two flow rates is small. The flow rate of the process stream is 465 m^3/hour, while the flow rate of the regeneration stream is 400 m^3/hour. The flow rates of both air streams are obtained by adjusting the frequency of the fans. In addition, the flow rate of the solar-heated water heat exchanger is 40 L/min. The thermal effectiveness of the solar-heated water heat exchanger is 62.03%.

2.2. Theoretical Analysis

The system's performance is analysed using theoretical analysis. The coefficient of performance, COP_{hvac} in Equation (1) is the ratio of the total cooling capacity, Q_c (kW), to the total power consumption of the system, E_{tot} (kW):

$$COP_{hvac} = Q_c/E_{tot} \tag{1}$$

The total cooling capacity of the system, Q_c (kW), is calculated by Equation (2):

$$Q_c = \dot{m}_p \left(h_{oa} - h_{sa}\right) \tag{2}$$

where h_{oa} (kJ/kg) is the specific enthalpy of outdoor air and h_{sa} (kJ/kg) is the specific enthalpy of supply air. The total power consumption in Equation (3) consists of the fan power, P_{fan} (kW), compressor power, P_{comp} (kW), water pump power, P_{pump} (kW) and other electrical components, P_{other} (kW):

$$E_{tot} = P_{fan} + P_{comp} + P_{pump} + P_{other} \tag{3}$$

In the supply air stream, the latent heat performance, COP_{lt}, of the system in Equation (4) is the ratio of latent heat capacity, Q_{lt} (kW), to the total power consumption of the system:

$$COP_{lt} = Q_{lt}/E_{tot} \tag{4}$$

The latent heat capacity, Q_{lt} (kW), of the system is calculated by Equation (5):

$$Q_{lt} = \dot{m}_p \, (h_{oa} - h')$$
(5)

where h' (kJ/kg) is the specific enthalpy of air with the temperature of the outside air and the humidity ratio of the supply air.

The specific moisture removal, SMR (kg/kg$_{da}$), is the humidity ratio difference between the outdoor air, ω_{oa} (kg/kg$_{da}$), and the supply air, ω_{sa} (kg/kg$_{da}$), along the process air stream. The SMR value can be calculated by Equation (6):

$$SMR = \omega_{oa} - \omega_{sa}$$
(6)

The effectiveness of the desiccant wheel, ε, is defined by Equation (7):

$$\varepsilon = SMR/(\omega_{oa} - \omega_i)$$
(7)

where the ideal specific moisture, ω_i, is assumed to be 0 kg/kg$_{da}$; thus, the denominator is the maximum moisture that can be removed by the desiccant wheel system. The moisture removal rate, MRR (kg/h), can be calculated by Equation (8):

$$MRR = \dot{m}_p \, (\omega_{oa} - \omega_{sa}) \, 3600$$
(8)

where $\dot{m}_p$ is the mass flow rate of the process air stream (kg/s). The solar fraction, SF, is determined by Equation (9):

$$SF = E_{sol}/E_{tot.m}$$
(9)

where E_{sol} (MJ) is the total useful solar thermal energy available in a month, which can be obtained from the solar collector capacity Q_{sol} (kW). Q_{sol} (kW) can be calculated by Equation (10).

$$Q_{sol} = \eta_{soc} \, A_{soc} \, R$$
(10)

where η_{soc} is the solar thermal collector overall efficiency, A_{soc} (m^2) is the gross area of the solar thermal collector, and R (kW/m^2) is the total incident solar radiation. In this study, the efficiency, η_{soc}, and the total area, A_{soc}, of the solar thermal collector are 94.5% and 11.699 m^2, respectively. The total energy input in a month, $E_{tot.m}$ (MJ), can be obtain from the total power P_{tot} (kW). P_{tot} (kW) can be calculated by Equation (11):

$$P_{tot} = I_{tot} \, V$$
(11)

where I_{tot} and V are the total current and voltage of the system, respectively.

3. Results and Discussion

3.1. Comparison of Average Temperature Declination and Specific Moisture Removal

A comparison of the temperature declination (T_d) and SMR of each system in different months in 2018 is shown in Table 2, where T_d is the temperature drop between the average outdoor air temperature (T_{oa}) and the average supply air temperature (T_{sa}).

According to the table, the temperature declination of the hybrid SDCS and solar-assisted hybrid SDCS are almost the same, while the solar-assisted SDCS does not have a cooling effect because the heat pump is turned off. The cooling and moisture removal of the process air by the solar-assisted hybrid SDCS are the highest among the three configurations. It can be seen that, for the solar-assisted SDCS, the average temperature declination is in the range of 3.65 °C to 6.79 °C, which gradually increases from April to June and then gradually decreases until October. Regarding moisture removal, the average *SMR* value is in the range of 0.0033 kg/kg$_{da}$ to 0.0075 kg/kg$_{da}$, which also gradually rises from April to June then declines until October.

Table 2. Comparison of T_d and specific moisture removal (*SMR*) for different system configurations.

Month	Average T_{oa}	Average W_{oa}	Hybrid SDCS		Solar-Assisted SDCS		Solar-assisted Hybrid SDCS	
			T_d (°C)	*SMR* (kg/kg$_{da}$)	T_d (°C)	*SMR* (kg/kg$_{da}$)	T_d (°C)	*SMR* (kg/kg$_{da}$)
April	24.75	0.0136	5.21	0.0037	N/A	0.0015	5.35	0.0043
May	28.36	0.0153	5.62	0.0041	N/A	0.0014	5.71	0.0054
June	28.27	0.0163	5.04	0.0061	N/A	0.0013	5.66	0.0055
July	29.93	0.0183	6.42	0.0066	N/A	0.0028	6.79	0.0075
August	27.06	0.0177	3.65	0.0052	N/A	0.0012	3.69	0.0053
September	28.26	0.0176	3.50	0.0040	N/A	0.0011	3.65	0.0046
October	24.65	0.0133	5.18	0.0031	N/A	0.0011	5.23	0.0033

3.2. Comparison of Regeneration Temperatures for Each System Configurations

A comparison of the regeneration temperature of each system during one day is shown in Figure 4. The regeneration temperature samples shown are for the system operating in better ambient conditions with enough sunlight intensity. As shown in Figure 4, the peak regeneration temperature value is generally reached between 14:00 and 16:00 each day. The regeneration temperature of each system is affected by the ambient temperature, especially for the solar-assisted SDCS and solar-assisted hybrid SDCS, in which the heat of solar-heated water is used as an additional heat source.

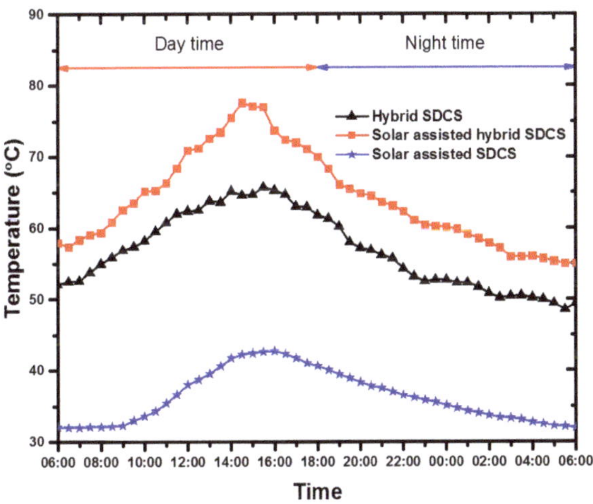

Figure 4. Comparison of regeneration temperature in each system configuration.

It can be observed from Figure 4 that, for a given time, the regeneration temperature of the solar-assisted hybrid SDCS is the highest among the three configurations; the temperature gradually rises from 57 °C at 06:00 to a peak value of 79 °C at 14:00, then gradually decreases to a lowest value of about 55 °C. In the case of the hybrid SDCS, the regeneration temperature distribution is similar to that of the solar-assisted hybrid SDCS, which is in a range of 49 °C to 65 °C. A peak value of 65 °C is attained at about 16:00. In the case of the solar-assisted SDCS, the regeneration temperature is in a range of 32 °C to 43 °C, and the peak value of 43 °C is attained at 16:00. These tendencies also occurred on other days. In general, the peak regeneration temperature is reached in the time range from 14:00 to 16:00.

Solar fraction (*SF*) is an important technical indicator to assess the feasibility of solar cooling systems. The higher the value of *SF*, the greater the contribution of solar energy to the system. *SF* is the ratio of the solar energy contribution to the total energy input needed to drive the solar cooling system. The total useful solar energy was described in the previous section and the total energy input of the system is the total energy input for the system's operation per month. The average *SF* value of the solar-assisted hybrid SDCS for each month is provided in Figure 5. As shown in Figure 5, it can be observed that the solar collector system shows a high contribution in July. This is because it is the peak summer season at this time, so the energy gained from the solar system is higher. On the other hand, the total energy required to operate the system is approximately constant in each month. The *SF* of the system in July is 0.82 and the lowest *SF* occurs in March with the value of 0.673.

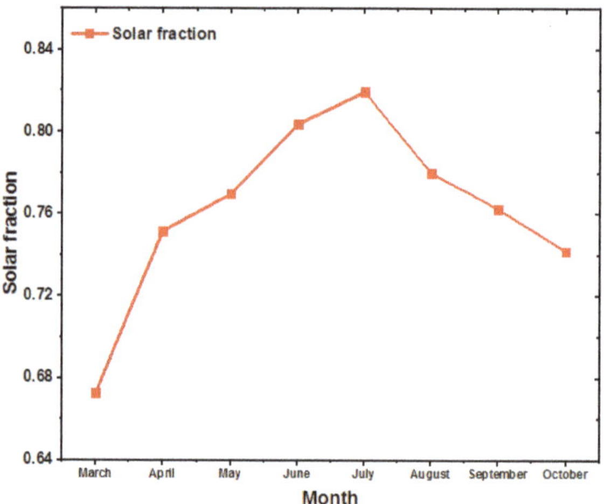

Figure 5. Solar fraction (*SF*) of solar-assisted hybrid SDCS.

3.3. Comparison of Specific Moisture Removal and Moisture Removal Rate

The effect of increasing relative humidity (*RH*) on *SMR* for each configuration is shown in Figure 6, where the outdoor air temperatures are 26 °C, 28 °C and 30 °C, respectively. However, the maximum *RH* attained by the three configurations at 28 °C and 30 °C is only 80% due to the effect of the ambient temperature of the seasons during the experiments. For a constant temperature, the value of relative humidity, *RH* (%), determines the humidity and specific vapor pressure of the process air. At the same process air temperature, higher *RH* values result in higher humidity ratios and vapor pressure. The higher vapor pressure value results in higher pressure differences between the process air and solid desiccant material, which can enhance the adsorption effect of the desiccant material. Thus, the *SMR* value of each configuration tends to increase with the increase in the *RH* value. At the same relative humidity value, the higher humidity ratio of the process air with higher temperature leads to a higher vapor pressure of the process air; thus, the *SMR* value of each configuration is also better for the higher outdoor temperature of 30 °C, as shown in Figure 6. In the solar-assisted SDCS, the *SMR* of the system is relatively low. The system only has a dehumidification effect if the *RH* value is greater than 65%. The *SMR* value of the configuration at an outdoor temperature of 26 °C and 85% *RH* is 0.00262 kg/kg$_{da}$.

The *SMR* of the hybrid SDCS is greater than the solar-assisted SDCS for any *RH* value, with the highest value of 0.0048 kg/kg$_{da}$ at an outdoor temperature of 26 °C and 85% *RH*. The *SMR* of the solar-assisted hybrid SDCS is the largest under all relative humidity conditions, with a maximum value of 0.0063 kg/kg$_{da}$ under the same operational conditions. The reason for this phenomena is that the solar-assisted hybrid SDCS has the highest regeneration temperature, which can enhance the adsorption effect of the desiccant material.

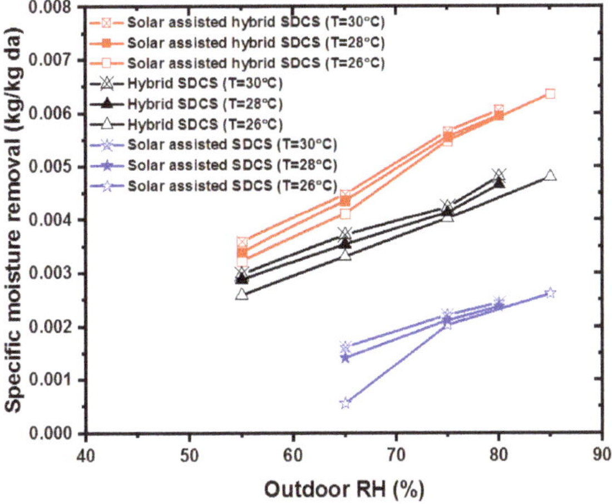

Figure 6. Effect of outdoor relative humidity (*RH*) on *SMR*.

Regarding the humidity ratio effect on *MRR* for each configuration, Figure 7 shows the distributions of moisture removal rate (*MRR*) of the three configurations with the increase in the humidity ratio at an outdoor operational temperature of 26 °C. However, the humidity ratio range of the solar-assisted SDCS and hybrid SDCS is narrower than the humidity ratio range of the solar-assisted hybrid SDCS due to the weather conditions during the experiments. As shown in Figure 7, larger humidity ratios lead to better *MRR* values. The solar-assisted SDCS has the lowest *MRR* value due to lower regeneration temperature. The *MRR* value of the hybrid SDCS and solar-assisted hybrid SDCS is better in comparison to the solar-assisted SDCS configuration. This indicates the regeneration temperature of both systems is sufficient to reactivate the desiccant material. Thus, the *MRR* values of the two configurations are also larger. When the humidity ratio of outdoor air reaches 0.018 kg/kg$_{da}$, the *MRR* can attain to a high value of 3.5 kg/h in the case of the solar-assisted hybrid SDCS.

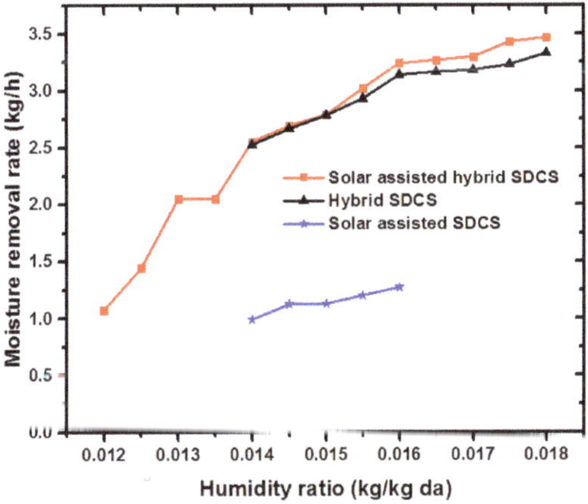

Figure 7. Effect of outdoor humidity ratio on moisture removal rate (*MRR*).

3.4. Effect of Relative Humidity to Desiccant Wheel Effectiveness

The specific moisture removal (*SMR*) value of a solid desiccant configuration determines the moisture removal capability of the system. It can represent the vapor mass of process air absorbed by the solid desiccant material to be ejected to the surroundings. The moisture removal effectiveness of solid desiccant is also determined by the *SMR* value. On the other hand, the *SMR* value is mostly affected by the relative humidity of the outdoor air. The effect of the relative humidity of outdoor air on solid desiccant effectiveness at constant air temperatures of 26 °C and 28 °C is shown in Figure 8. As shown in Figure 8, the effectiveness of solid desiccant increases with the rise in outdoor air relative humidity. At the outdoor temperature of 26 °C and outdoor relative humidity of 50%, the desiccant effectiveness is 0.274, and this value continues to increase as relative humidity increases. The effectiveness value at a relative humidity of 85% is 0.36. The increase in the relative humidity value at the same outdoor air temperature leads the humidity ratio of outdoor air, ω_{oa}, rises, together with the *SMR* value. In the case of this study, the solar-assisted hybrid SDCS increment value of *SMR* in all configurations is greater than the increment of the humidity ratio of outdoor air. Thus, from the definition of desiccant effectiveness, the effectiveness value gradually increases with the increase in relative humidity at the same outdoor air temperature. However, at the same relative humidity value, the effectiveness of solid desiccant wheel in the case with higher process air temperature is less than that with lower process air temperature. The effectiveness of the desiccant wheel is the ideal specific moisture, ω_i, which is assumed to be 0 kg/kg$_{da}$, so the denominator is the maximum moisture that can be removed by the desiccant wheel system. Increasing outdoor temperature with the same relative humidity causes the process air to have a higher humidity ratio. However, with the increase in outdoor air temperature, the increase in the *SMR* value is lower relative to the ideal moisture able to be removed by the system. Therefore, the effectiveness decreases with the temperature of the process air at the same relative humidity ratio.

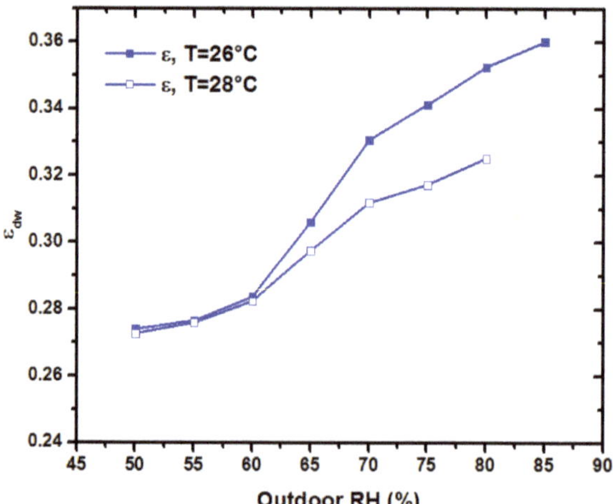

Figure 8. Effect of outdoor *RH* on desiccant effectiveness.

3.5. Effect of Regeneration Temperature to Specific Moisture Removal and Latent Heat Performance

It was mentioned in the previous section that the regeneration temperature impacts the moisture removal ability of the SDCS. The effect of regeneration temperature on specific moisture removal (*SMR*) of the SDCS is shown in Figure 9. The present investigation shows the regeneration temperature effect of each system. The regeneration temperature of the solar-assisted SDCS is in the range of 35–45 °C, while the regeneration temperature of the hybrid SDCS ranges from 50 to 65 °C and the regeneration temperature of the solar-assisted hybrid SDCS ranges from 55 to 70 °C. The regeneration temperatures can then be drawn as a single line as shown in Figure 9. The effect of regeneration temperature on specific moisture removal and COP_{lt} are thus investigated in this section. In addition to ambient temperature, the regeneration temperature also depends mostly on the capacity of the heat source. In Figure 9, the results are shown for a relative humidity ratio range of 58–62%. As shown in Figure 9, it can be concluded that a higher regeneration temperature leads to a higher *SMR* value. Consequently, the supply air condition will also be drier. If the regeneration temperature is in the range of 35–45 °C, the *SMR* is less than 0.002 kg/kg$_{da}$, which is not sufficient for the supply air condition. The average regeneration temperature of the solar-assisted SDCS in this study is about 40.4 °C. Even the dehumidification effect will not be apparent if the regeneration temperature is less than 35 °C. However, for the hybrid SDCS, the average regeneration temperature is about 53.9 °C. The higher average regeneration temperature of the hybrid SDCS provides a better *SMR* value than the solar-assisted SDCS. In addiition, the solar-assisted hybrid SDCS possesses the highest regeneration temperature, with an average value of 63.7 °C. Therefore, the *SMR* value of the solar-assisted hybrid SDCS is higher and the dehumidification performance is better among the three configurations.

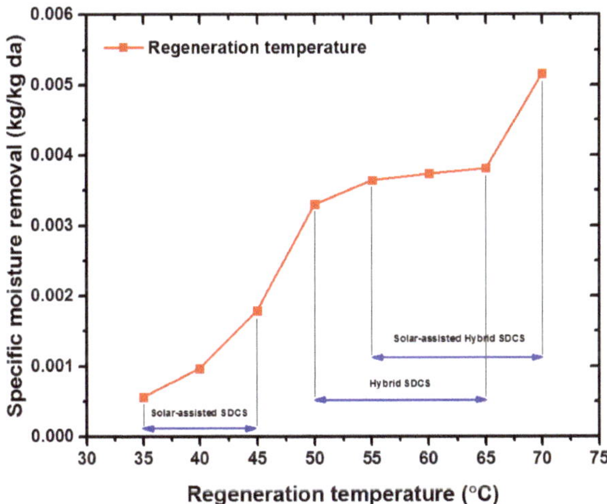

Figure 9. Effect of regeneration temperature on *SMR*.

The effect of regeneration temperature on latent heat performance (COP_{lt}) can be observed in Figure 10. In this case, the outside air relative humidity (*RH*) is also in the range of 58–62%. It can be seen that the COP_{lt} value increases as the regeneration temperature rises from 35 °C to 70 °C. The regeneration temperature depends on the heat generated by heat sources. In the solar-assisted SDCS with the regeneration temperature range of 30 °C to 45 °C, the COP_{lt} of the system is comparatively low. Therefore, the dehumidification effect is also low. The COP_{lt} of the system at a regeneration temperature of 45 °C is 0.556.

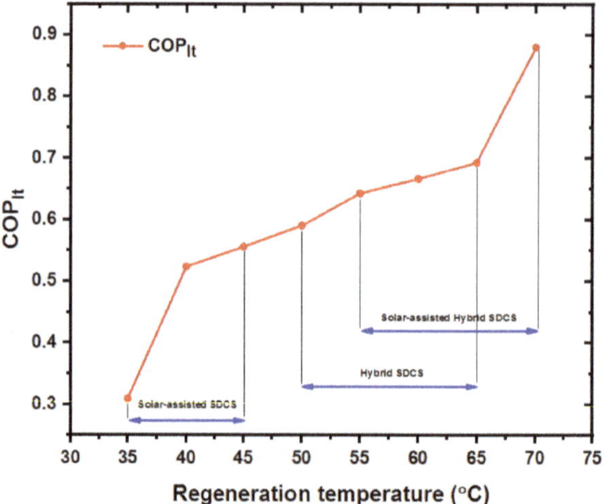

Figure 10. Effect of regeneration temperature on latent heat performance (COP_{lt}).

In the hybrid SDCS, in which the regeneration temperature is in the range of 50 °C to 65 °C, the dehumidification performance is better in comparison to the previous case. The increase in regeneration temperature shows a performance improvement in the system. The COP_{lt} of the hybrid SDCS under an average regeneration temperature of 55 °C is 0.643. The COP_{lt} value shows a massive inclination in the range from 55 °C to 70 °C. It indicates that the optimum regeneration temperature of the desiccant cooling system is 70 °C or higher. However, in this case, the maximum temperature that can be generated by the system is 70 °C. This condition occurs while the system utilizes the solar-assisted hybrid SDCS. The heat from the solar water heat exchanger can effectively increase the regeneration temperature of the system. The COP_{lt} of the solar-assisted hybrid SDCS at the average regeneration temperature of 65 °C is 0.694 and at the maximum regeneration temperature 70 °C is 0.88.

3.6. Comparison of System Total Performance for Different Outdoor Air Temperatures

Figure 11 shows a comparison of COP_{hvac} for each configuration at different outdoor temperatures.

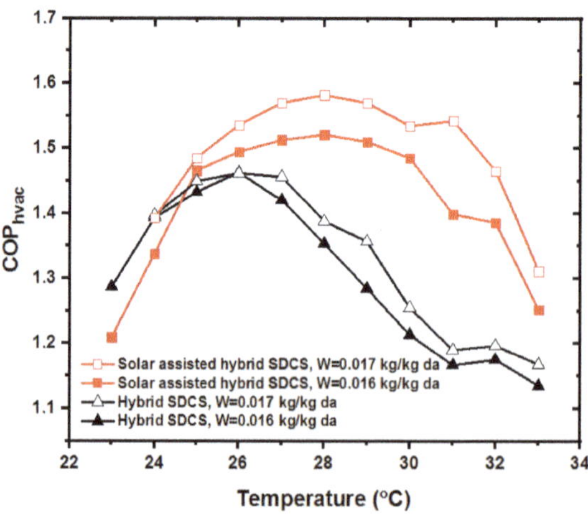

Figure 11. Comparison of the coefficient of performance (COP_{hvac}) at different outdoor temperatures.

The humidity ratio is maintained at 0.016 kg/kg$_{da}$ and 0.017 kg/kg$_{da}$. However, the COP_{hvac} of the solar-assisted SDCS is not investigated since it does not have a cooling effect. Figure 11 indicates the COP_{hvac} is greater for the solar-assisted hybrid SDCS than for the hybrid SDCS, especially at high outdoor temperatures. At low outdoor temperatures, the solar sub-system does not have a significant effect on the system, since the heat from solar-heated water and the regeneration temperature are smaller. At outdoor temperatures from 26 to 30 °C, the performance enhancement of the solar-assisted hybrid SDCS is obvious. At a humidity ratio of 0.016 kg/kg$_{da}$, the maximum COP_{hvac} of 1.52 can be reached by the solar-assisted hybrid SDCS at an outdoor temperature of 28 °C, and the maximum COP_{hvac} of 1.58 can be reached by the system at an outdoor humidity of 0.017 kg/kg$_{da}$.

The solar-assisted hybrid SDCS has better performance at high outdoor temperatures than the hybrid SDCS. However, if the outdoor temperature is too high, system performance is degraded because the condensation temperature rises with the increase in outdoor temperature, which leads to an increase in the power consumption of the compressor. The maximum outdoor temperature in the solar-assisted hybrid SDCS is 30 °C, and in the hybrid SDCS is 26 °C.

3.7. Comparison of System Total Performance for Different Outdoor Humidity

Figure 12 shows a comparison of COP_{hvac} for each system configuration at different outdoor humidities. The outdoor air temperatures in this figure are 26 °C and 32 °C, respectively. The COP_{hvac} of the configuration at an outdoor temperature of 32 °C and humidity less than 0.013 kg/kg$_{da}$ are not provided due to operational environment limitations. The result shows an increase in the outdoor humidity leads to better system performance. The COP_{hvac} inclination of solar assisted SDCS in every 1 g/kg$_{da}$ humidity ratio is about 6.17%, which is higher than that of the hybrid SDCS with the vaue of 5.4%. The COP_{hvac} of the solar-assisted hybrid SDCS is approximately 3.7% greater than that of the hybrid SDCS. The reason for this is that the solar-assisted hybrid SDCS has better *MRR* and dehumidification effects. The maximum COP_{hvac} at a humidity ratio of 0.018 kg/kg$_{da}$ and a temperature of 26 °C is 1.53 in the solar-assisted hybrid SDCS and 1.49 in the hybrid SDCS. The COP_{hvac} of systems at an outdoor temperature of 32 °C is approximately 18% lower than at 26 °C due to the temperature effect. At an outdoor temperature of 32 °C, the COP_{hvac} of the solar-assisted hybrid SDCS system is 7% greater than that of the hybrid SDCS. The maximum COP_{hvac} of the solar-assisted hybrid SDCS at an outdoor temperature of 32 °C is 1.369 for a humidity ratio of 0.0175 kg/kg$_{da}$, while the maximum COP_{hvac} of the hybrid SDCS is 1.218.

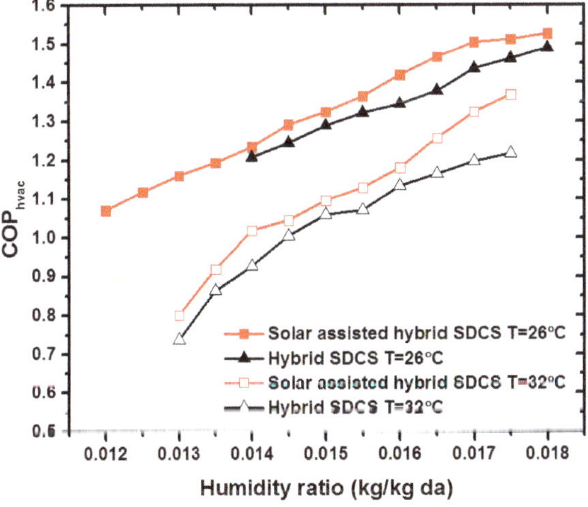

Figure 12. Comparison of COP_{hvac} at different outdoor humidity ratios.

4. Conclusions

In this study, a solar-assisted hybrid SDCS system was developed in which solar-heated water is used as a heat source for the regeneration process in addition to heat from the condenser of an integrated heat pump. A solar thermal collector sub-system is used to generate the solar regeneration water. Experiments were conducted in the typical hot and humid weather of Taichung, Taiwan, from the spring to fall seasons. According to the experiment results, several points can be concluded from the study, as follows:

(a) Solar-assisted hybrid SDCSs are feasible for use in hot and humid locations such as Taichung. Such systems can utilize solar energy as an additional heat source for the regeneration process. Therefore, the regeneration temperature of the system can be increased and the dehumidification effect can also be enhanced. Compared to the hybrid SDCS, the overall performance of the solar-assisted hybrid SDCS in terms of power consumption was found to be approximately 10% greater in the study. For a solar SDCS, a larger area and capacity of the solar thermal collector are required in high humidity environments.

(b) Higher humidity ratios lead to better *MRR* values for each solid desiccant cooling system configuration.

(c) The performance of a SDCS is very sensitive to changes in ambient conditions. The performance of each system configuration is better for higher outdoor humidity ratios. In terms of outdoor temperature, the COP_{hvac} of both systems increases with outdoor temperature. However, there are optimum values of the outdoor temperature for the COP_{hvac} of the system. When the ambient temperature is greater than the optimum value, the COP_{hvac} gradually decreases with the increase in ambient temperature. In this study, for the hybrid SDCS, the optimum outdoor temperature is between 26 and 27 °C, and for the solar-assisted hybrid SDCS, the optimum temperature range is 27–30 °C. Beyond these ranges, the overall performance of both systems will decline severely.

Author Contributions: Research concept were proposed by W.-J.L., D.F. and F.R.F.; data processing and the manuscript preparation were conducted by Y.-S.C., F.H.M. and U.N.A.; data analysis and interpretation were implemented by W.-J.L., D.F. and F.R.F.; manuscript editing was performed by W.-J.L., and F.H.M.

Funding: This research was funded by the Ministry of Science and Technology of Taiwan under grant number MOST 106-2221-E-167-026 and MOST 104-2221-E-167-026-MY2.

Conflicts of Interest: The authors declare no conflict of interest.

Nomenclature

A	area (m^2)	Q	heat capacity (kW)
COP	coefficient of performance	RH	relative humidity (%)
DCS	desiccant cooling system	rph	rotations per hour
E	power consumption (kW)	SDCS	solid desiccant cooling system
h	specific enthalpy (kJ/kg)	SF	solar fraction
I	solar irradiation (W/m^2)	SMR	specific moisture removal (kg/kg$_{da}$)
$\dot{m}$	mass flow rate (kg/s)	T	temperature (°C)
MRR	moisture removal rate (kg/h)	Td	temperature declination (°C)
P	power (kW)	ω	humidity ratio (kg/kg$_{da}$)

Subscripts

η	effectiveness	a	process air
c	cooling	ra	return air
soc	solar collector	reg	regeneration air
$hvac$	HVAC system	sa	supply air
lt	latent	sol	solar
m	month	tot	total
oa	outside air		

References

1. Wyon, D.P. The Effects of indoor air quality on performance and productivity. *Indoor Air* **2004**, *14*, 92–101. [CrossRef] [PubMed]
2. ANSI/ASHRAE Standard 55-2013. Available online: https://www.ashrae.org/technical-resources/bookstore/standard-55-thermal-environmental-conditions-for-human-occupancy (accessed on 15 May 2018).
3. Li, K.Y.; Luo, W.J.; Huang, J.Z.; Chan, Y.C.; Faridah, D. Operational Temperature Effect on Positioning Accuracy of a Single-Axial Moving Carrier. *Appl. Sci.* **2017**, *7*, 420. [CrossRef]
4. World Energy Outlook. Available online: https://www.iea.org/media/weowebsite/2008-1994/WEO2004.pdf (accessed on 15 May 2018).
5. Pérez-Lombard, L.; Ortiz, J.; Pout, C. A review on buildings energy consumption information. *Energy Build.* **2008**, *40*, 394–398. [CrossRef]
6. Inasawa, S.; Suzuki, R.; Qian, E.W.; Kitajima, T.; Yamashita, Y. Ozone layer depletion and its effects: A review. *Int. J. Environ. Sci. Dev.* **2011**, *2*, 30–37.
7. American Society of Heating, Refrigerating and Air Conditioning Engineers (ASHRAE). Desiccant dehumidification and pressure-drying equipment. In *2016 ASHRAE Handbook—Fundamentals (SI Edition)*; ASHRAE Inc.: Atlanta, GA, USA, 2016; p. 3.
8. Luo, W.J.; Kuo, H.C.; Wu, J.Y.; Faridah, D. Development and analysis of a new multi-function heat recovery split air conditioner with parallel refrigerant pipe. *Adv. Mech. Eng.* **2016**, *8*, 1–12. [CrossRef]
9. Mandegari, M.; Pahlavanzadeh, H. Introduction of a new definition for effectiveness of desiccant wheel. *Energy* **2009**, *34*, 797–803. [CrossRef]
10. Chang, C.C.; Luo, W.J.; Lu, C.W.; Cheng, Y.S.; Tsai, B.Y.; Lin, Z.H. Effects of process air conditions and switching cycle period on dehumidification performance of desiccant-coated heat exchangers. *Sci. Technol. Built Environ.* **2016**, *23*, 81–90. [CrossRef]
11. Narayanan, R.; Halawa, E.; Jain, S. Performance characteristic of solid-desiccant evaporative cooling system. *Energies* **2018**, *11*, 2574. [CrossRef]
12. Narayanan, R.; Halawa, E.; Jain, S. Dehumidification potential of a solid desiccant based evaporative cooling system with an enthalpy exchanger operating in subtropical and tropical climates. *Energies* **2019**, *12*, 2704. [CrossRef]
13. Guidara, Z.; Elleuch, M.; Bacha, H.B. New solid desiccant solar air conditioning unit in Tunisia: Design and simulation study. *Appl. Therm. Eng.* **2013**, *58*, 656–663. [CrossRef]
14. Enteria, N.; Yoshino, H.; Mochida, A.; Satake, A.; Yoshie, R.; Takaki, R.; Yonekura, H.; Mitamura, T.; Tanaka, Y. Performance of solar-desiccant cooling system with Silica-Gel (SiO_2) and Titanium Dioxide (TiO_2) desiccant wheel applied in East Asian climates. *Sol. Energy* **2012**, *86*, 1261–1279. [CrossRef]
15. Speerforck, A.; Ling, J.; Aute, V.; Radermacher, R.; Schmitz, G. Modeling and simulation of a desiccant assisted solar and geothermal air conditioning system. *Energy* **2017**, *141*, 2321–2336. [CrossRef]
16. White, S.D.; Kohlenbach, P.; Bongs, C. Indoor temperature variations resulting from solar desiccant cooling in a building without thermal backup. *Int. J. Refrig.* **2009**, *32*, 695–704. [CrossRef]
17. Jani, D.B.; Mishra, M.; Sahoo, P.K. Experimental investigation on solid desiccant–vapor compression hybrid air-conditioning system in hot and humid weather. *Appl. Therm. Eng.* **2016**, *104*, 556–564. [CrossRef]
18. Jani, D.B.; Mishra, M.; Sahoo, P.K. Performance analysis of a solid desiccant assisted hybrid space cooling system using TRNSYS. *J. Build. Eng.* **2018**, *19*, 26–35. [CrossRef]
19. Worek, W.M.; Moon, C.J. Simulation of an integrated hybrid desiccant vapor-compression cooling system. *Energy* **1986**, *11*, 1005–1021. [CrossRef]
20. Hwang, W.B.; Choi, S.; Lee, D.Y. In-depth analysis of the performance of hybrid desiccant cooling system incorporated with an electric heat pump. *Energy* **2017**, *118*, 324–332. [CrossRef]
21. Jani, D.B.; Mishra, M.; Sahoo, P.K. Performance studies of hybrid solid desiccant-vapor compression air-conditioning system for hot and humid climates. *Energy Build.* **2015**, *102*, 284–292. [CrossRef]
22. Merabtl, L.; Merzouk, M.; Merzouk, N.K.; Taane, W. Performance study of solar driven solid desiccant cooling system under Algerian coastal climate. *Int. J. Hydrogen Energy* **2017**, *42*, 28997–29005. [CrossRef]
23. Beccali, M.; Finocchiaro, P.; Nocke, B. Energy performance evaluation of a demo solar desiccant cooling system with heat recovery for the regeneration of the adsorption material. *Renew. Energy* **2012**, *44*, 40–52. [CrossRef]

24. Beccali, M.; Finocchiaro, P.; Nocke, B. Energy and economic assessment of desiccant cooling systems coupled with single glazed air and hybrid PV/thermal solar collectors for applications in hot and humid climate. *Sol. Energy* **2009**, *83*, 1828–1846. [CrossRef]

25. Fong, K.F.; Chow, T.T.; Lee, C.K.; Lin, Z.; Chan, L.S. Solar hybrid cooling system for high-tech offices in subtropical climate: Radiant cooling by absorption refrigeration and desiccant dehumidification. *Energy Convers. Manag.* **2011**, *52*, 2883–2894. [CrossRef]

26. Fong, K.F.; Chow, T.T.; Lee, C.K.; Lin, Z.; Chan, L.S. Advancement of solar desiccant cooling system for building use in subtropical Hong Kong. *Energy Build.* **2010**, *42*, 2386–2399. [CrossRef]

27. Bareschino, P.; Pepe, F.; Roselli, C.; Sasso, M.; Tariello, F. Desiccant-Based Air Handling Unit Alternatively Equipped with Three Hygroscopic Materials and Driven by Solar Energy. *Energy* **2019**, *12*, 1543. [CrossRef]

28. Li, H.; Dai, Y.J.; Li, Y.; La, D.; Wang, R.Z. Case study of a two-stage rotary desiccant cooling/heating system driven by evacuated glass tube solar air collectors. *Energy Build.* **2012**, *47*, 107–112. [CrossRef]

29. Jani, D.B.; Mishra, M.; Sahoo, P.K. A critical review on application of solar energy as renewable regeneration heat source in solid desiccant-vapor compression hybrid cooling system. *J. Build. Eng.* **2018**, *18*, 107–124. [CrossRef]

30. Rambhad, K.S.; Walke, P.V.; Tidke, D.J. Solid desiccant dehumidification and regeneration methods: A review. *Renew. Sustain. Energy Rev.* **2016**, *59*, 73–83. [CrossRef]

31. Relative Humidity Table of Taiwan. Available online: https://www.cwb.gov.tw/V7e/climate/monthlyMean/tx.htm (accessed on 14 April 2018).

Article

Condition Monitor System for Rotation Machine by CNN with Recurrence Plot

Yumin Hsueh, Veeresha Ramesha Ittangihala, Wei-Bin Wu, Hong-Chan Chang and Cheng-Chien Kuo *

Department of Electrical Engineering, National Taiwan University of Science and Technology, Taipei City 10607, Taiwan
* Correspondence: cckuo@mail.ntust.edu.tw

Received: 2 August 2019; Accepted: 19 August 2019; Published: 21 August 2019

Abstract: Induction motors face various stresses under operating conditions leading to some failure modes. Hence, health monitoring for motors becomes essential. In this paper, we introduce an effective framework for fault diagnosis of 3-phase induction motors. The proposed framework mainly consists of two parts. The first part explains the preprocessing method, in which the time-series data signals are converted into two-dimensional (2D) images. The preprocessing method generates recurrence plots (RP), which represent the transformation of time-series data such as 3-phase current signals into 2D texture images. The second part of the paper explains how the proposed convolutional neural network (CNN) extracts the robust features to diagnose the induction motor's fault conditions by classifying the images. The generated RP images are considered as input for the proposed CNN in the texture image recognition task. The proposed framework is tested on the dataset collected from different 3-phase induction motors working with different failure modes. The experimental results of the proposed framework show its competitive performance over traditional methodologies and other machine learning methods.

Keywords: induction motor; convolutional neural networks (CNN); recurrence plots (RP); time-series data (TSD)

1. Introduction

Induction motors are electromechanical devices used in most industrial applications. Due to their simple design and well-developed manufacturing technologies, induction motors are considered relatively reliable and robust [1]. However, the motors fall into failure mode and seriously affect industrial operations. Eventually, this leads to failure of the entire operating system if the failure condition is not identified or if it is neglected. Several types of faults related to winding, stator, rotor, and bearing can be observed in an induction motor [2,3]. There are mainly four types of fault diagnosis methods such as signal-based, model-based, knowledge-based, and active/hybrid methods. Signal-based methods use the measured signal to extract the features and make a diagnostic decision based on the prior knowledge of the diagnostic process. Signal-based methods can be classified into a time-domain signal-based method, frequency-domain signal-based method, and time-frequency signal-based method. Model-based methods can be categorized into deterministic fault diagnosis methods, fault diagnosis methods for discrete-events and hybrid systems; stochastic fault diagnosis; and fault diagnosis methods for distributed and network systems, which are categorized by the model type used [4]. Hybrid methods are studied as a combination of two or more fault diagnosis methods. For example, He et al. [5] diagnosed plastic bearing faults by combining the signal-based and data-driven methods. By combining signal-based and knowledge-based techniques, a fault diagnosis method was studied to detect the inter-turn faults in induction motors, in which wavelet transform is applied to extract the features from the collected vibration signals, and the principal component analysis (PCA) and neural network (NN) were used as classifiers to classify healthy from faulty motors [6]. Active methods are studied as a system in which a suitably designed input signal is injected into a dynamic process during a test period to distinguish accurately and quickly the faulty modes from the normal modes. Stochastic active fault diagnosis and deterministic active fault diagnosis are the two active methods studied [7–9]. For example, Campbell et al. [10], used two candidate models one as a normal system and another as a fault system in a multimodel system, and an auxiliary signal was designed to detect the correct model under a given interval of test time. As an extended study, an active fault detection method for multiple faults generated simultaneously or sequentially [11].

Knowledge-based methods, also considered as data-driven methods, are the most commonly used methods for analyzing signals such as vibration, temperature, electrical tension, and current. These methods require a huge amount of historical data to find the patterns in the given signal. However, data signals can be captured using sensors [12,13]. Signal-based features are extracted, and feature selection methods are applied to reduce feature dimensions and also to avoid the repeated information, which in turn improves the performance by holding the significant features [14]. The extracted features are used for fault diagnosis by various traditional machine learning methods [15–17]. Traditional machine learning methods have achieved prominent results. However, feature extraction depends significantly on diagnostic knowledge and signal processing expertise. Furthermore, traditional methods are incapable of extracting discriminative features from raw data and always require a process to extract the feature from the signal [16,18–21].

In spite of the advanced development in machine learning, deep learning (DP) has become the most effective study that can significantly overcome the drawbacks of traditional machine learning methods for fault diagnosis. DP can automatically extract and learn abstract features from raw data, and avoid manual feature extraction [18]. Many deep learning models, such as deep belief network (DBN) [22], stacked sparse auto-encoder [19], sparse auto-encoder [21], denoising auto-encoder [23] and sparse filtering [24] have been studied to diagnose the faults, and very significant results have been achieved. One of the most effective used deep learning model, called convolutional neural network (CNN), has been used to learn hierarchical feature representation from raw data and has delivered promising results [20–25].

Most pattern recognition tasks deal with time-series data signals. Financial data (stock and currency exchange rates), video processing, music mining, weather and forecasting, biometric data, and biomedical signal processing are few examples, in which time-series data have been studied [26–28]. Likewise, electrical industrial devices also often work with time-series data such as measurements of voltage, current, temperature, and vibration signals. One-dimensional CNN has been studied and applied to time-domain machinery data signals to diagnose the faults in induction motors [29]. In a few cases, machinery data can also be represented in two-dimensional (2D) model, such as the time-frequency domain using the wavelet transform technique [30]. In addition, time-series data can be represented in 2D texture images using the concept called recurrence plot (RP) [31]. Image representation of time-series data provides a different set of features that are not available for 1D signals. Therefore, 2D texture images can be used for classification [32].

In this study we use a time-series data signals converted as recurrence plots (RP) to feed the proposed deep CNN for fault diagnosis. Current signals with a phase difference were collected from 3-phase induction motors, and each phase current was used as an input data sample. Our results show that RP provides an effective method for viewing trajectory periodicity over a phase space, allowing us to discover specific elements of the m-dimensional phase space trajectory using a 2D figure. The main contributions of this paper are illustrated as follows. First, the raw current signals are represented as RP images. Second, an efficient deep CNN model is studied and applied on RP images to extract the multi-level features for fault classification. Lastly, the proposed deep CNN based framework achieves significant results compared to other deep learning methods.

The rest of the paper is arranged in the following way: Section 2 discusses the related works. Section 3 explains the proposed framework—time-series data to 2D texture image conversion—and proposed deep CNN models are discussed. Section 4 presents the experimental results followed by the conclusions and future work presented in Section 5.

2. Related Works

This section briefly reviews recent deep learning contributions on induction motors' fault diagnosis. Several types of signal processing methodologies have been studied in the time-domain, frequency-domain, and time-frequency domain to extract and learn the features in order to classify the working condition of the motor. Lee et al. [33] studied the convolutional deep belief network (CDBN) to classify audio signals. They converted time-domain into frequency-domain data to learn the features form audio signals. A multi-channel CNN has been studied to handle multi-variate time-series data [34]. A separate CNN is used to learn the features from individual time-series data, and result from all the CNNs are combined and classified using a fully connected multilayer perceptron (MLP) classifier. Audio signals are transformed into a time-frequency domain to feed into CNN for classification [35]. The Gramian Angular Field (GAF) and Markov Transition Field (MTF) are used to convert time-series signals as images. A tiled CNN is used to classify time-series images [36,37].

Ngaopitakkul et al. [15] explain a decision algorithm based on ANN for fault diagnosis. Pandya et al. [17] propose an efficient KNN (k-Nearest Neighbours) classifier using the asymmetric proximity function for fault diagnosis. Yang et al. [16] constructed an SVM (Support Vector Machine)-based method to diagnose the fault patterns of the roller bearings. Jia et al. [38] propose a fault diagnosis method based on deep neural networks using an auto-encoder. Deep learning models such as deep autoencoder (DAE), deep belief network (DBN), and CNN have been discussed for fault diagnosis [19,20,22]. Ince et al. [26] propose a one-dimensional (1D) CNN to diagnose faults using real-time motor data. Abdeljaber et al. [39] studied 1D CNN to detect real-time structural damages. A deep CNN was used to analyze multichannel time-series data signals for human activities [40]. However, these models only used a small amount of low-level features in hidden layers. However, in this paper, we study a deep CNN method to automatically learn the useful texture features in order to classify faults. 1D raw current signals were converted to 2D images, and the proposed CNN model was able to successfully capture the temporal and spatial dependencies in the images by applying relevant filters. Furthermore, the proposed model was able to extract and learn high-level features from these images along with the low-level features. The performance of fault classification improved by the combined implementation of feature extraction and the CNN classifier.

3. Proposed Study and Framework

This section explains the proposed framework based on RP images and deep CNN for fault diagnosis. It consists of two subsections: (1) the time-series data signals are converted into 2D texture images, and (2) a deep convolutional neural network model is discussed to learn features from texture images for fault classification.

3.1. Time-Series Data to 2D Texture Images

The time-series data can be categorized using a unique recurring behavior such as periodic and irregular cyclic aspects. Moreover, time-series data are generated as the repetition of states, which is a normal phenomenon for ever-changing irregular systems or random processes. RP [31,32] is a tool for visualizing and investigating the m-dimensional phase space trajectory using a 2D representation of its repetitive occurrences. The primary idea of RP is to disclose trajectory movements from the current state to the previous state and it can be formulated as:

$$R_{i,j} - \theta\left(\epsilon - \|\vec{s_l} - \vec{s_m}\|\right), \ \vec{s}(.) \in \mathfrak{R}^n, \ l,m = 1,2,\ldots\ldots,K$$

where K is the number of states $\vec{s}$, ϵ is a threshold value of distance, $\|.\ \|$ is the norm and $\theta(.)$ is the Heaviside function. The recurrence matrix (R) comprises two sets of values called texture and typology. The texture information belongs to individual dots, sloping lines, perpendicular lines, and horizontal lines, whereas the typology information categorized by uniform, regular, shift, and interrupted. Obviously, in RP, there are patterns and information that are not easily visually seen and interpreted. The detailed explanation can be found in [32].

Raw current signals are collected from 3-phase induction motors for the fault analysis every 5 s with a sampling rate of 10,000 samples per second. Data samples taken for different periods of time from 1 s to 5 s were investigated with recurrence plots. Five seconds of data samples gave the most distinguishable patterns in the recurrence plots. The collected raw current signals from the operating induction motor are represented as recurrence plot shown in Figure 1.

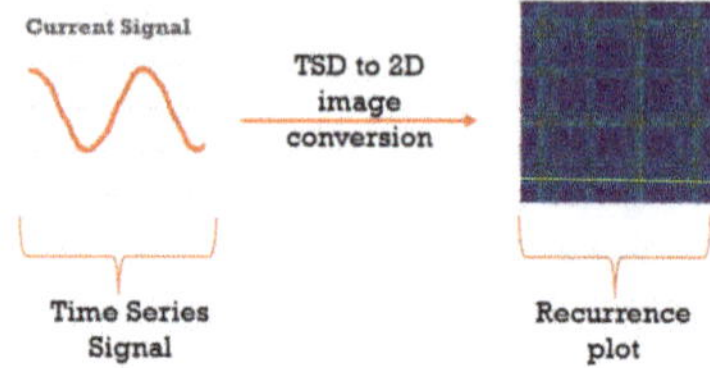

Figure 1. Feature representation. Time-series data into a 2D (Two dimensional) texture images (recurrence plot).

As shown in the Figure 2, nondistinguishable recurrence plots were generated for two different motors operating with different modes of failure, when the raw current signal values were used to generate recurrence plots. Even though, the motors working with different modes such as faulty or healthy, it can be clearly seen that it is almost impossible to find a distinguishable pattern in these recurrence plots—they look exactly the same with no difference in any color or pattern. However, to find distinguishable patterns in RP images, an effective preprocessing technique called Max–Min difference was used in this study, and it is implemented as follows:

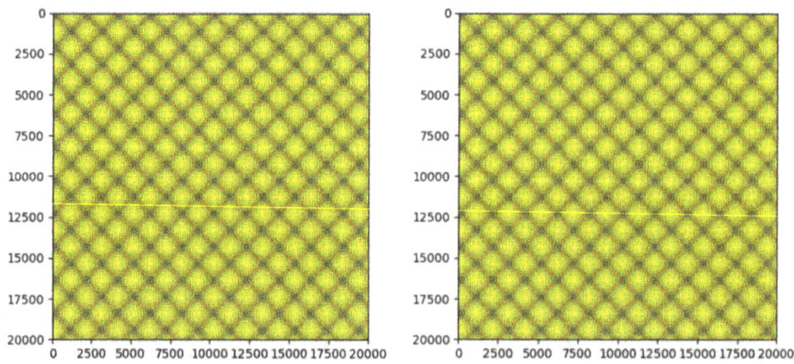

Figure 2. Recurrence plots before signal preprocessing. Left: recurrence plot for fault condition (bearing axis deviation) motor. Right: recurrence plot for healthy condition motor.

Step 1: maximum and minimum peaks of the current signal are collected for each one full cycle.

Step 2: difference between the maximum and minimum peak value is then used to generate the recurrence plot for the whole signal.

Step 3: the above two steps are repeated for all types of faults, and healthy motor signals generate distinguishable recurrence plot.

As shown in the Figure 3, after applying the preprocessing technique to raw current signal values, the generated recurrence plots are well distinguishable and can be considered for classification of different faults and healthy conditions of the motors by CNN. Initially, the 2D recurrence texture images were generated by raw one dimensional (1D) current signals and then classifier automatically learned the features from texture images to classify the motors' fault condition.

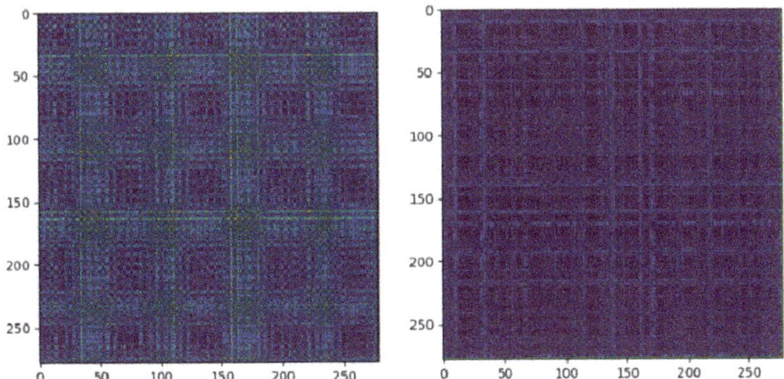

Figure 3. Recurrence plots with (Max–Min) preprocessing technique. Left: recurrence plot for fault condition (bearing axis deviation) motor. Right: recurrence plot for healthy condition motor.

3.2. The Methodology's Architecture

The methodology's architecture consists of two parts. Part one explains the architecture used to train the CNN model and part two belongs to motor testing using the trained model.

The relevant data were collected for a total of five conditions of the induction motor. A setup of four induction motors for the following four faults and one motor for the healthy condition were used. The four faults were: (1) bearing axis deviation, (2) stator and rotor friction, (3) rotor end ring break, and (4) poor insulation.

As shown in Figure 4, the training setup had two stages. The first stage setup was done in one of the lab servers (lab server). Data for all conditions of the motor were collected in this server in CSV format. The dataset comprised of 3-phase current signals. Data preprocessing was applied to the raw current signals to generate the recurrence plots. Recurrence plots' 2D texture images were stored into the S3 database hosted on the cloud platform (EI-PaaS).

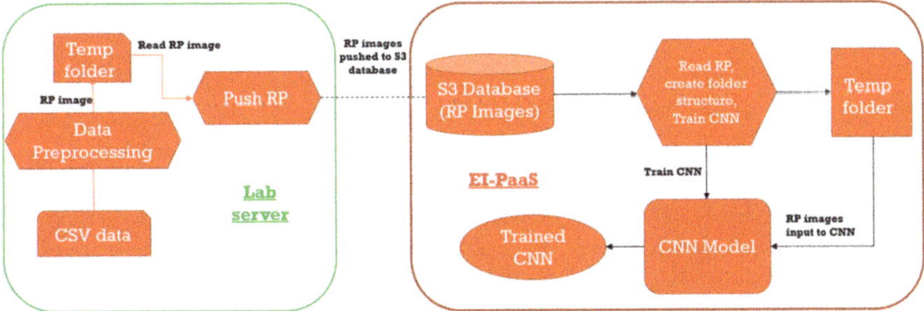

Figure 4. The architecture used while training the CNN (Convolutional Neural Network) model for motor fault diagnosis. Left: the setup used in one of the lab servers, where the historical motor data are collected and stored for different types of faults and healthy conditions. Right: this setup is on the cloud platform called EI-PaaS, where the S3 database and the CNN model are designed.

The second stage setup is in the cloud. The Edge Intelligent Platform as a Service (EI-PaaS) has the analytical framework service, where the CNN model was implemented along with the S3 database. While training the CNN model, the images were maintained in the temporary directory structure. Figure 5 illustrates the setup used for the deployment/testing phase of the application. The architecture has a 3-phase induction motor connected to a data acquisition system (DAQ), which sends motor-related data such as current signal values in binary format to an Edge device. The Edge device reads the binary data and converts the data into decimal format and stores them as a CSV file. The data stored in the CSV file are used to generate the relevant RP image. The generated RP texture image is fed to the well-trained CNN model to diagnose and classify the motor condition as one of the four faults or healthy condition.

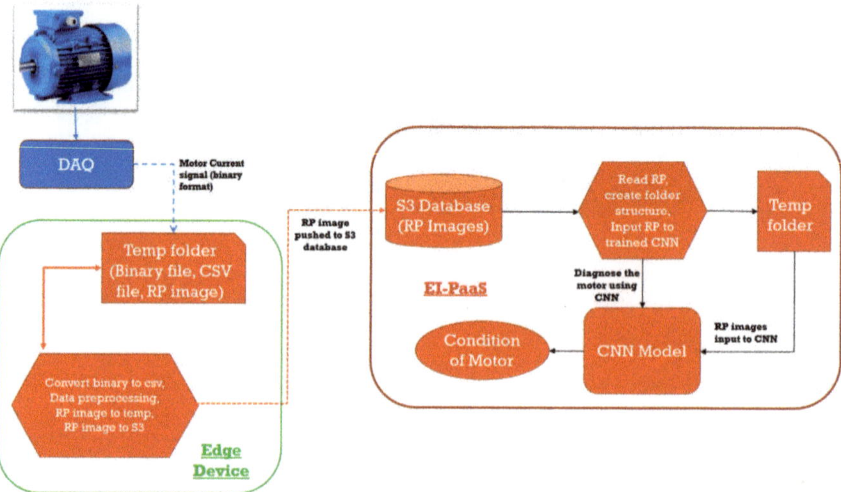

Figure 5. The architecture of the framework setup used in testing/deployment of the CNN model application for motor fault diagnosis. Left: this setup has an induction motor connected to a data acquisition system and Edge device for data preprocessing. Right: this setup has S3 database to store RP, the trained CNN model for motor fault diagnosis in the cloud platform (EI-PaaS).

3.3. Architecture of the Proposed CNN Model

The proposed deep CNN model has a three-stage structure. Each stage representing a feature learning stage with different feature-levels and it includes convolution, activation, and pooling layers.

As shown in Figure 6, the proposed deep CNN model has three convolutional layers with 32–3×3 filter, 64–3×3 filter, and 128–3×3 filter, respectively. In addition, three max-pooling layers of pooling size 2×2 were used. Type of layers, output shape of each layer, along with the number of trainable parameters are listed in Table 1.

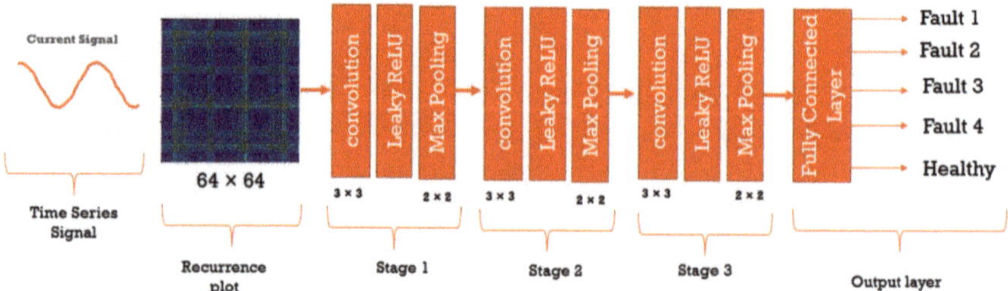

Figure 6. The proposed deep convolutional neural network architecture diagram. The RP images are resized to 64×64 and fed as input to the CNN model. The architecture consists of one input and two hidden layers followed by pooling layers and a fully connected layer.

Table 1. Proposed CNN model summary.

Layer (Type)	Output Shape	Param #
conv2d_4 (Conv2D)	(None, 1, 64, 32)	18,464
leaky_re_lu_5 (LeakyReLU)	(None, 1, 64, 32)	0
max_pooling2d_4 (MaxPooling2)	(None, 1, 32, 32)	0
conv2d_5 (Conv2D)	(None, 1, 32, 64)	18,496
leaky_re_lu_6 (LeakyReLU)	(None, 1, 32, 64)	0
max_pooling2d_5 (MaxPooling2)	(None, 1, 16, 64)	0
conv2d_6 (Conv2D)	(None, 1, 16, 128)	73,856
leaky_re_lu_7 (LeakyReLU)	(None, 1, 16, 128)	0
max_pooling2d_6 (MaxPooling2)	(None, 1, 8, 128)	0
flatten_2 (Flatten)	(None, 1024)	0
dense_2 (Dense)	(None, 128)	131,200
leaky_re_lu_8 (LeakyReLU)	(None, 128)	0
dense_3 (Dense)	(None, 5)	645

Total params: 242,661, Trainable params: 242,661, Non-trainable params: 0.

The activation function Leaky ReLU (Rectified Linear Units) was applied to introduce nonlinearity into each stage, allowing CNN to learn complex models. A specific reason for adding Leaky ReLU was to avoid and attempt to fix the problem of dying ReLUs. It has proven to be more effective than the logistic sigmoid function. However, during the training, ReLU units can die and this could occur when large gradient flows through a ReLU neuron. It causes the weights to update such that the neuron will never activate again on any data point. Leaky ReLU makes an attempt to solve this problem [41,42]. Pooling layers were introduced to reduce the resolution of the input image by the process of subsampling and the max-pooling was applied in the proposed model.

At the end of the three stages, the feature maps were flattened into a column vector. The flatted output vector supplied to a feed-forward neural network and backpropagation was employed to every iteration of training. During training, the proposed model was able to distinguish among the dominating and also low-level features in texture images and classify by a fully connected layer for five types of faults. To estimate the parameters of the proposed model, one of the gradient-based optimization (backpropagation algorithm) methods was used. Adam optimizer was used to update the parameters to achieve faster convergence [43].

4. Experimental Results and Discussion

To assess the performance of the proposed methodology, raw current signals from an experimental setup of a total of five induction motors with the same configuration were used. One healthy and four fault types of raw current data signals were collected from the experimental setup. The different five current signals were studied and analyzed for the healthy condition of the motor, as well as for the following four faulty conditions of the motors [44]. The raw signal to image conversion method and the CNN model implementation were written in python 3.6 with TensorFlow and run on the Windows 64 bit operating system.

4.1. Healthy and Fault Conditioned Motors

1. Bearing axis deviation: this condition is considered as class 'Fault 1'. This happens due to the offset of centers on both sides of coupling when the motor is connected to load.

2. Stator and rotor friction: this condition of the motor is considered as 'Fault 2'. Due to the friction and overheating, the stator or rotor coil is short-circuited and hence it will breakdown if it is not diagnosed and fixed.
3. Rotor end ring break: this condition is considered as 'Fault 3'. Due to the high frequency and overloaded operation of the motor, the excessive current may cause the breaking of the rotor bar.
4. Poor insulation: this condition is considered as 'Fault 4'. This occurs due to the rapid change of the current or voltage.
5. Healthy: this condition is considered as 'Fault 0' for the classification. Current signal values are collected from the motor working in normal condition.

The proposed framework uses induction motor raw current signal values to generate recurrence plots for respective conditions. Generated recurrence plots are used as input images to the CNN model for further classification on the fault conditions of motors.

4.2. Dataset

The dataset was collected from a lab setup for four fault scenarios and one healthy scenario. The setup included five motors each for four faults and one healthy condition operating at full load. The dataset consisted of 750 samples and 150 samples for each type of scenario as described in Table 2. At 5 s of sampling rate, 50,000 raw current signal data points were collected per sample.

Table 2. Dataset considered for evaluation.

Fault 0	Fault 1	Fault 2	Fault 3	Fault 4	Total
150	150	150	150	150	750

The dataset was divided mainly into three parts as described in the Table 3. Sixty percent of the dataset (450 data samples for training) and 15% of the dataset (112 data samples for validation) were used simultaneously to train the CNN model. The remaining 20% (188 data samples) were used to test the trained CNN model.

Table 3. Dataset samples used for evaluation.

Data Split Ratio		
Training	60%	450
Validation	15%	112
Testing	25%	188

4.3. Performance Results of the Proposed CNN

The proposed deep CNN model was trained over 100 epochs to learn the multi-level features for one healthy and for each type of the motors' faulty condition. The deep CNN model was trained to automatically train and learn the robust features from 450 samples of training data and simultaneously validated against 112 data samples during each iteration. Then, the trained deep CNN model was tested against the 188 samples of the testing dataset. The proposed deep CNN model was trained and verified with a batch of 16, 32, and 64. The deep CNN model achieved the best results with a batch of size 32.

The accuracies and losses were collected for each iteration while training the proposed deep CNN model, and plotted as shown in the Figure 7. The CNN model was able to learn the complex features efficiently with reaching almost 100% accuracy, and ~97–98% validation accuracy. Over the 100 epochs, the proposed deep CNN model was able to learn the generalized features from the recurrence plot texture images to diagnose the induction motor faults and classify the motors as faulty or healthy.

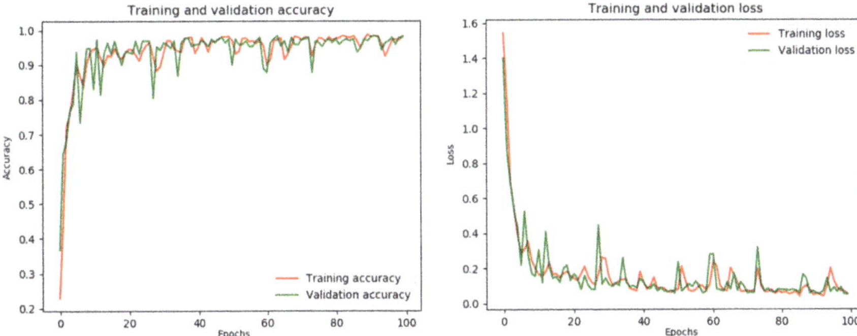

Figure 7. Accuracy and loss curves over 100 epochs of CNN training.

The performance of the proposed deep CNN was evaluated using a testing dataset consisting of 188 data samples. The performance results of the trained deep CNN model were very prominent with an average accuracy of 98% as described in Table 4. The classification report clearly explains how well the proposed deep CNN model was able to extract and learn the features from the testing data samples and classify the features into respective classes. In the classification report, the average values of the evaluation metrics such as precision, recall, and f1 score on test dataset were impressive, with 98% of accuracy. The proposed model was able to classify Fault 3 and Fault 0 (healthy) conditions accurately. However, the proposed model misclassified other faults, but with an acceptable margin. The confusion matrix for the test data was calculated using the trained deep CNN model (Figure 8). Almost all test samples were accurately classified with very few misclassifications.

In order to evaluate the performance of the proposed framework relative to traditional and other deep learning methods, the proposed framework was compared and the performance metrics were collected in terms of accuracy. The proposed framework was compared to support vector machine SVM [45], artificial neural network (ANN) [22], adaptive deep convolutional neural network (ADCNN) [46], sparse filter [24], and deep belief network (DBN) [22].

Table 4. Classification report on the test dataset.

Classification Report				
CLASS	Precision	Recall	F1-Score	Support
Fault 0	1.00	1.00	1.00	38
Fault 1	1.00	0.89	0.94	37
Fault 2	0.95	1.00	0.97	38
Fault 3	1.00	1.00	1.00	37
Fault 4	0.95	1.00	0.97	38
Accuracy			0.98	188
Macro avg	0.98	0.98	0.98	188
Weighted avg	0.98	0.98	0.98	188

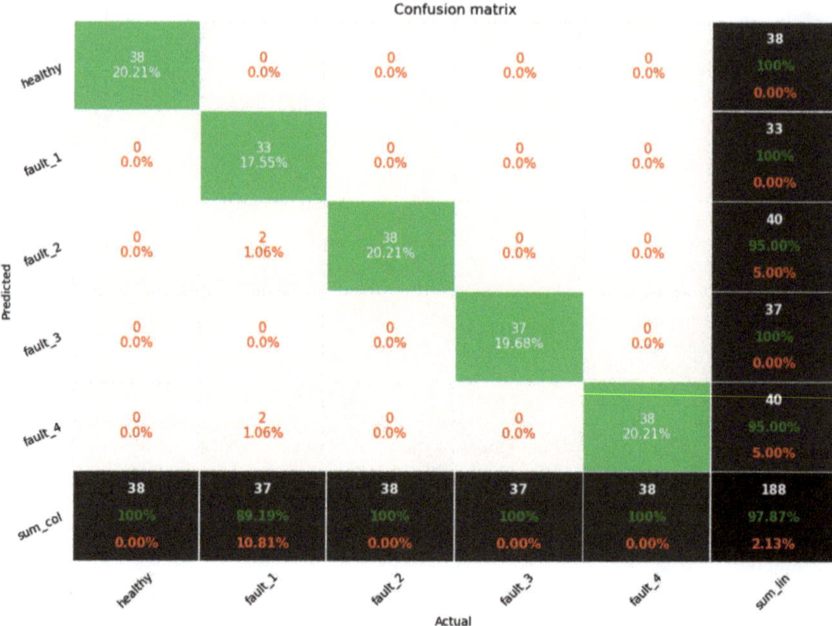

Figure 8. Confusion matrix for the testing dataset (Fault 0: healthy, Fault 1: bearing axis deviation, Fault 2: stator and rotor friction, Fault 3: rotor end ring break, and Fault 4: poor insulation).

The prediction accuracy of these methods was calculated (Table 5). Comparing the performance results explains the significant results achieved by the proposed deep CNN framework compared to the other methods listed above. A prediction accuracy of 99.81%, shows the significant performance from the proposed deep CNN model.

Table 5. Comparison results.

Methods	Accuracy (%)
SVM	88.2
ANN	87.8
ADCNN	98.4
Sparse filter	98.2
DBN	81.8
Proposed deep CNN	**99.81**

SVM: Support Vector Machine, DBN: Deep Belief Network, ANN: Artificial Neural Network, ADCNN: Adaptive Deep Convolutional Neural Network.

5. Conclusions and Future Work

In this study, we investigate and discuss a novel framework to diagnose the faults in 3-phase induction motors based on recurrence plots and the deep CNN model. The important contributions of this paper are: proposing a method to transform a time-series data signal to 2D texture images (recurrence plots) and applying the proposed deep CNN model to learn the features from the recurrence plots to classify the 2D texture images for the fault diagnosis. The proposed framework is implemented for four types of faults including bearing axis deviation, stator and rotor friction, rotor end ring break, and poor insulation, and achieved a very prominent accuracy of 99.81%. The proposed framework outperforms other traditional and deep learning models due to its ability to learn both high-level and low-level features. The proposed framework demonstrates promising results by considering a single variable as an input feature compared to the rule-based diagnosis methods that require multiple features for fault diagnosis.

The limitations of the proposed methodology are discussed as follows. First, the dataset collected for the experiment is comparatively small and a huge amount of data samples is needed for different load conditions such as no load, half load, or full load. Second, data from the motors with different specifications are needed to extract and learn more generalized features. Therefore, important future work should focus on motors working with different load conditions to collect more data samples, and investigating data to generate more generalized features for CNN model training. Furthermore, future work includes the review of transfer learning to avoid any unnecessary time required to train the model and utilize the model to learn other feature types.

Author Contributions: Y.H. has generated the data and analyzed the faults in Induction motors. W.B.W. validated the data for each kind of faults. V.R.I. performed data preprocessing to train the CNN model and evaluated the trained model for fault diagnosis. Y.H. and V.R.I. analyzed the experimental results with the guidance from C.C.K.; H.C.C. and C.C.K. revised the manuscript for submission.

Funding: This research received no external funding.

Conflicts of Interest: The authors declare no conflict of interest.

References

1. Albrecht, P.; Appiarius, J.; McCoy, R.; Owen, E.; Sharma, D. Assessment of the Reliability of Motors in Utility Applications—Updated. *IEEE Trans. Energy Convers.* **1986**, 39–46. [CrossRef]
2. Bonnett, A.; Soukup, G. Cause and analysis of stator and rotor failures in three-phase squirrel-cage induction motors. *IEEE Trans. Ind. Appl.* **1992**, *28*, 921–937. [CrossRef]
3. Dai, X.; Gao, Z. From model signal to knowledge: A data-driven perspective of fault detection and diagnosis. *IEEE Trans. Ind. Informat.* **2013**, *9*, 2226–2238. [CrossRef]
4. Gao, Z.; Cecati, C.; Ding, S.X. A Survey of Fault Diagnosis and Fault-Tolerant Techniques—Part I: Fault Diagnosis with Model-Based and Signal-Based Approaches. *IEEE Trans. Ind. Electron.* **2015**, *62*, 3757–3767. [CrossRef]
5. He, D.; Li, R.; Zhu, J. Plastic bearing fault diagnosis based on a twostep data mining approach. *IEEE Trans. Ind. Electron.* **2013**, *60*, 3429–3440.
6. Seshadrinath, J.; Singh, B.; Panigrahi, B.K. Vibration Analysis Based Interturn Fault Diagnosis in Induction Machines. *IEEE Trans. Ind. Inform.* **2014**, *10*, 340–350. [CrossRef]
7. Gao, Z.; Cecati, C.; Ding, S. A Survey of Fault Diagnosis and Fault-Tolerant Techniques Part II: Fault Diagnosis with Knowledge-Based and Hybrid/Active Approaches. *IEEE Trans. Ind. Electron.* **2015**, *62*, 3752–3756. [CrossRef]
8. Zhang, X. *Auxiliary Signal Design in Fault Detection and Diagnosis*; Springer: New York, NY, USA, 1989.
9. Kerestecioglu, F. *Change Detection and Input Design in Dynamic Systems*; Research Studies Press: Taunton, UK, 1993.
10. Campbell, S.L.; Nikoukhah, R. *Auxiliary Signal Design for Failure Detection*; Walter de Gruyter GmbH: Berlin, Germany, 2004.
11. Scott, J.K.; Findeisen, R.; Braatz, R.D.; Raimondo, D.M. Input design for guaranteed fault diagnosis using zonotopes. *Automatica* **2014**, *50*, 1580–1589. [CrossRef]
12. Lee, J.; Wu, F.; Zhao, W.; Ghaffari, M.; Liao, L.; Siegel, D. Prognostics and health management design for rotary machinery systems—Reviews, methodology and applications. *Mech. Syst. Signal Process.* **2014**, *42*, 314–334. [CrossRef]
13. Filho, P.L.; Pederiva, R.; Brito, J. Detection of stator winding faults in induction machines using flux and vibration analysis. *Mech. Syst. Signal Process.* **2014**, *42*, 377–387. [CrossRef]
14. Sun, W.; Chen, J.; Li, J. Decision tree and PCA-based fault diagnosis of rotating machinery. *Mech. Syst. Signal Process.* **2007**, *21*, 1300–1317. [CrossRef]

15. Ngaopitakkul, A.; Bunjongjit, S. An application of a discrete wavelet transform and a back-propagation neural network algorithm for fault diagnosis on single-circuit transmission line. *Int. J. Syst. Sci.* **2013**, *44*, 1745–1761. [CrossRef]
16. Yang, Y.; Yu, D.; Cheng, J. A fault diagnosis approach for roller bearing based on IMF envelope spectrum and SVM. *Measurements* **2007**, *40*, 943–950. [CrossRef]
17. Pandya, D.; Upadhyay, S.; Harsha, S. Fault diagnosis of rolling element bearing with intrinsic mode function of acoustic emission data using APF-KNN. *Expert Syst. Appl.* **2013**, *40*, 4137–4145. [CrossRef]
18. Lecun, Y.; Bengio, Y.; Hinton, G. Deep learning. *Nature* **2015**, *521*, 436–444. [CrossRef]
19. Qi, Y.; Shen, C.; Wang, D.; Shi, J.; Jiang, X.; Zhu, Z. Stacked Sparse Autoencoder-Based Deep Network for Fault Diagnosis of Rotating Machinery. *IEEE Access* **2017**, *5*, 15066–15079. [CrossRef]
20. Xia, M.; Li, T.; Xu, L.; Liu, L.; De Silva, C.W. Fault Diagnosis for Rotating Machinery Using Multiple Sensors and Convolutional Neural Networks. *IEEE/ASME Trans. Mechatron.* **2018**, *23*, 101–110. [CrossRef]
21. Wen, L.; Gao, L.; Li, X. A New Deep Transfer Learning Based on Sparse Auto-Encoder for Fault Diagnosis. *IEEE Trans. Syst. Man Cybern. Syst.* **2017**, *99*, 136–144. [CrossRef]
22. Shao, H.; Jiang, H.; Zhang, X.; Niu, M. Rolling bearing fault diagnosis using an optimization deep belief network. *Meas. Sci. Technol.* **2015**, *26*, 115002. [CrossRef]
23. Shao, H.; Jiang, H.; Wang, F.; Zhao, H. An enhancement deep feature fusion method for rotating machinery fault diagnosis. *Knowl. Based Syst.* **2017**, *119*, 200–220. [CrossRef]
24. Lei, Y.; Jia, F.; Lin, J.; Xing, S.; Ding, S.X. An Intelligent Fault Diagnosis Method Using Unsupervised Feature Learning Towards Mechanical Big Data. *IEEE Trans. Ind. Electron.* **2016**, *63*, 3137–3147. [CrossRef]
25. Lee, K.B.; Cheon, S.; Kim, C.O. A Convolutional Neural Network for Fault Classification and Diagnosis in Semiconductor Manufacturing Processes. *IEEE Trans. Semicond. Manuf.* **2017**, *30*, 135–142. [CrossRef]
26. Wang, J.; Liu, P.; She, M.F.; Nahavandi, S.; Kouzani, A. Bag-of-words representation for biomedical time series classification. *Biomed. Signal Process. Control* **2013**, *8*, 634–644. [CrossRef]
27. Hatami, N.; Chira, C.; Hatami, N. Classifiers with a Reject Option for Early Time-Series Classification. In Proceedings of the 2013 IEEE Symposium on Computational Intelligence and Ensemble Learning (CIEL), Singapore, 16–19 April 2013; pp. 9–16.
28. Wang, Z.; Oates, T. Pooling Sax-Bop Approaches with Boosting to Classify Multivariate Synchronous Physio-Logical Time-Series Data. In Proceedings of the FLAIRS Conference, Hollywood, FL, USA, 18–20 May 2015; pp. 335–341.
29. Ince, T.; Kiranyaz, S.; Eren, L.; Askar, M.; Gabbouj, M. Real-Time Motor Fault Detection by 1-D Convolutional Neural Networks. *IEEE Trans. Ind. Electron.* **2016**, *63*, 7067–7075. [CrossRef]
30. Gilles, J. Empirical Wavelet Transform. *IEEE Trans. Signal Process.* **2013**, *61*, 3999–4010. [CrossRef]
31. Eckmann, J.; Kamphorst, S.; Ruelle, D. Recurrence plots of dynamical systems. *EPL Euro Phys. Lett.* **1987**, *4*, 17.
32. Debayle, J.; Hatami, N.; Gavet, Y. Classification of Time-Series Images Using Deep Convolutional Neural Networks. In Proceedings of the Tenth International Conference on Machine Vision (ICMV 2017), Vienna, Austria, 13–15 November 2017. [CrossRef]
33. Lee, H.; Largman, Y.; Pham, P.; Ng, A. Unsupervised Feature Learning for Audio Classification Using Convolutional Deep Belief Networks. In Proceedings of the Conference on Neural Information Processing Systems (NIPS09), Vancouver, BC, Canada, 7–10 December 2009; pp. 1096–1104.
34. Zheng, Y.; Liu, Q.; Chen, E.; Ge, Y.; Zhao, J.L. Time Series Classification Using Multi-Channels Deep Convolutional Neural Networks. In *Web-Age Information Management*; WAIM 2014. Lecture Notes in Computer Science; Li, F., Li, G., Hwang, S., Yao, B., Zhang, Z., Eds.; Springer: Cham, Switzerland, 2014; Volume 8485.
35. Abdel-Hamid, O.; Deng, L.; Dong, Y. Exploring Convolutional Neural Network Structures and Optimization Techniques for Speech Recognition. In Proceedings of the 14th Annual Conference of the International Speech Communication Association (Interspeech 2013), Lyon, France, 25–29 August 2013.
36. Wang, Z.; Oates, T. Imaging Time-Series to Improve Classification and Imputation. In Proceedings of the International Joint Conference on Artificial Intelligence (IJCAI), Buenos Aires, Argentina, 25–31 July 2015; pp. 3939–3945.

37. Wang, Z.; Oates, T. Encoding Time-Series as Images for Visual Inspection and Classification Using Tiled Convolutional Neural Networks. In Proceedings of the Association for the Advancement of Artificial Intelligence (AAAI) Conference, Austin, TX, USA, 25–30 January 2015.
38. Jia, F.; Lei, Y.; Lin, J.; Zhou, X.; Lu, N. Deep neural networks: A promising tool for fault characteristic mining and intelligent diagnosis of rotating machinery with massive data. *Mech. Syst. Signal Process.* **2016**, *72*, 303–315. [CrossRef]
39. Abdeljaber, O.; Avci, O.; Kiranyaz, S.; Gabbouj, M.; Inman, D.J. Real-time vibration-based structural damage detection using one-dimensional convolutional neural networks. *J. Sound Vib.* **2017**, *388*, 154–170. [CrossRef]
40. Yang, J.; Nguyen, M.; San, P.; Li, X.; Krishnaswamy, S. Deep Convolutional Neural Networks on Multichannel Time-Series for Human Activity Recognition. In Proceedings of the International Joint Conference on Artificial Intelligence (IJCAI), Buenos Aires, Argentina, 25–31 July 2015; pp. 3995–4001.
41. Krizhevsky, A.; Sutskever, I.; Hinton, G.E. ImageNet Classification with Deep Convolutional Neural Networks. In Proceedings of the Conference on Neural Information Processing Systems (NIPS12), Tahoe, NV, USA, 3–6 December 2012; pp. 1097–1105.
42. Xu, B.; Wang, N.; Chen, T.; Li, M. Empirical Evaluation of Rectified Activations in Convolutional Network. *arXiv* **2015**, arXiv:150500853.
43. Kingma, D.P.; Ba, J. Adam: A method for stochastic optimization. *arXiv* **2014**, arXiv:14126980.
44. Chang, H.; Kuo, C.; Hsueh, Y.; Wang, Y.; Hsieh, C. Fuzzy-Based Fault Diagnosis System for Induction Motors on Smart Grid Structures. In Proceedings of the 2017 IEEE International Conference on Smart Energy Grid Engineering (SEGE), Oshawa, ON, Canada, 14–17 August 2017; pp. 103–109.
45. Zhang, X.; Liang, Y.; Zhou, J.; Zang, Y. A novel bearing fault diagnosis model integrated permutation entropy, ensemble empirical mode decomposition and optimized SVM. *Measurement* **2015**, *69*, 164–179. [CrossRef]
46. Guo, X.; Chen, L.; Shen, C. Hierarchical adaptive deep convolution neural network and its application to bearing fault diagnosis. *Measurement* **2016**, *93*, 490–502. [CrossRef]

Article

Design of a Logistics System with Privacy and Lightweight Verification

Chin-Ling Chen [1,2,3], Dong-Peng Lin [4], Hsing-Chung Chen [5,6,7,*], Yong-Yuan Deng [3,*] and Chin-Feng Lee [4,*]

1   School of Computer and Information Engineering, Xiamen University of Technology, Xiamen 361024, China
2   School of Information Engineering, Changchun Sci-Tech University, Changchun 130600, China
3   Department of Computer Science and Information Engineering, Chaoyang University of Technology, Taichung 41349, Taiwan
4   Department of Information Management, Chaoyang University of Technology, Taichung 41349, Taiwan
5   Department of Computer Science and Information Engineering, Asia University, Taichung 41354, Taiwan
6   Department of Medical Research, China Medical University Hospital, China Medical University, Taichung 40402, Taiwan
7   Department of Bioinformatics and Medical Engineering, Asia University, Taichung 41354, Taiwan
*   Correspondence: shin8409@ms6.hinet.net or cdma2000@asia.edu.tw (H.-C.C.); allen.nubi@gmail.com (Y.-Y.D.); lcf@cyut.edu.tw (C.-F.L.)

Received: 29 April 2019; Accepted: 27 June 2019; Published: 8 August 2019

Abstract: Presently, E-commerce has developed rapidly as a result of many services and applications integrating e-commerce technologies offered online. Buyers can buy goods online and sellers can then deliver the goods to them. Logistics therefore plays an important role in online e-commerce applications, with a focus on rapid delivery, the integrity of goods, and the privacy of personal information. Previous studies have proposed secure mechanisms for the transfer of electronic cash and digital content, in which only the sender and the receiver know the secret information hidden in the signature. However, they did not consider requirements such as the anonymous and lightweight verification in the logistics architecture. Therefore, this study designs a secure logistics system, with anonymous and lightweight verification, in order to meet the following requirements: Mutual authentication, non-repudiation, anonymity, integrity and a low overhead for the logistics environment. A buyer could check the goods and know if the parcel has been exchanged by a malicious person. Moreover, the proposed scheme not only presents a solution to meet the logistics system's requirements, but also to reduce both computational and communication costs.

Keywords: mutual authentication; privacy; logistics system; ECC; ban logic

---

1. Introduction

Background

In recent years, with the rapid development of e-commerce, online shopping has become a current trend and many shopping and financial transactions can now be completed via online shopping. These activities include online orders and online payments [1]. As buyers and sellers interact online, the purchase of goods is divided into digital and physical products. If a product is physical, the seller will entrust their logistics to deliver the goods to the buyer. As these logistics requirements grow, greater focus is required, not only on rapid delivery, but also on ensuring the integrity of goods and the privacy of personal information [2–7].

Unfortunately, the current process of goods delivery and online shopping does not entail an immediate physical exchange of goods. There is therefore a risk of counterfeiting and fraud, in addition to a risk that goods may be lost due to human error, and this may be compounded by information errors, which could mean that it cannot be determined where the goods were lost. In 2016, Liu and Wang [6] noted that the means of preventing the loss of goods has become a very important issue in this field.

However, during the goods transportation process, the logistics provider will copy both the buyer and seller's personal information on an order detail and paste the order detail on the packages. That is, the goods can only be accurately delivered to the buyer, but this process includes the risk of private personal information being leaked, which may result in improper use or theft of that personal information. The delivery verification can also include the risks of identity impersonation, parcel exchange, and the loss of packages. Since there is no reliable mechanism for the buyer and seller to identify each other, it is impossible to know who has the goods or when goods are lost.

In 2006, Aijaz et al. [1] classified various attacker behaviors as active and passive attackers, internal and external personnel, and malicious and rational attackers. Active attackers tamper with shopping information, while passive attackers do not actively participate in tampering with information, but rather eavesdrop on shopping information. The stolen information may then be forwarded to other attackers. Internal attackers are very dangerous in the transmission process. As a consequence of their good understanding of items and personal information, internal personnel can cause various kinds of complex attacks. External personnel are not members of the transaction process, so they are much less harmful than internal personnel. The main goal of malicious attackers is to steal or tamper with information and cause the loss of property.

This paper proposes a logistics method using the public key crypto system to protect the personal privacy and the shopping information of buyers, sellers, and logistics companies during the transmission process to prevent information from being stolen. In addition, lightweight encryption technology is used to protect personal tag information to prevent personal information from being leaked during the delivery process.

In 2016, Liu and Wang [6] published papers on an NFC-based security-enhanced express delivery systems, in which the individuals' personal information was hidden in tags and only authorized people could get permission to access that information, thus protecting personal information from being stolen and achieving fast identity authentication. Digital signatures are then used by buyers, sellers, and logistics companies to achieve non-repudiation. The proposed system thus achieves mutual authentication, lightweight and fast verification, cost savings, the anonymity of personal information, non-repudiation in the transaction, and the completeness of the product.

The remainder of this paper is arranged as follows. Section 2 presents the system architecture. Section 3 presents the proposed secure package logistics system, based on protecting personal information anonymously by tag. Section 4 presents a security analysis and then illustrates the computation cost, communication cost, and performance analysis of the proposed scheme. Section 5 offers conclusions.

2. Related Works and Requirement

2.1. Related Works

In 2016, Liu and Wang noted that digital tags may not be able to perform complex encryption and decryption operations due to computation limitations [6]. In general, current logistics schemes lack face to face package checking procedures and rely on buyers to ensure that packages are intact upon receipt. In addition, the security issues of RFID systems are not completely suitable for the scheme proposed in this study [8]. Instead, this study uses ECC (elliptic curve cryptography) to generate session keys that are used to secure data transmissions and the BAN logic model [9] to prove the correctness of the proposed scheme with mutual authentication. Recently, many authentication schemes have applied BAN logic to prove the correctness of authentication and key establishment. The following is the notation of BAN logic.

$P \mid\equiv X$	P believes X, or P would be entitled to believe X.
$P \lhd X$	P sees X. Someone has sent a message containing X to P, who can read and repeat X.
$P \mid\sim X$	P once said X. P at some time sent a message including X.
$P \mid\Rightarrow X$	P has jurisdiction over X. P is an authority on X and should be trusted on this matter.
$< X >_Y$	This represents X combined with Y.
$\#(X)$	The formula X is fresh, that is, X has not been sent in a message at any time before the current run of the protocol.
$P \overset{K}{\leftrightarrow} Q$	P and Q may use the shared key K to communicate.
$P \overset{S}{\rightleftharpoons} Q$	The formula S is a secret known only to P and Q and possibly to principals trusted by them.

2.2. Requirements

In order to achieve a good logistics system, the following security requirements must be met and known attacks must be prevented:

(1) Mutual authentication: The basic requirement for good system communication is the identity authentication during the transmission process. The message must guarantee the validity of a sender and receiver [10,11].

(2) Non-repudiation: In the information transmission process, if each identity is not authenticated, the sender and the receiver are vulnerable to being sent false information by an impersonation attack. Therefore, the non-repudiation of information is crucial to effectively prevent impersonation [4,12].

(3) Anonymity: It is easy for buyer and seller to disclose information in the goods delivery process. Therefore, the contents should not disclose any information about the buyer and seller [13].

(4) Integrity: In an unencrypted environment, information is easily tampered with in the transmission process, resulting in the receiver being vulnerable to the information received not being that sent by the original sender's information. Therefore, the integrity of the information must be ensured during transmission [14].

(5) Low overhead: Identity verification in the information transmission process must ensure information integrity and maintain transmission speed, so reduced computation is necessary for a faster system [15,16].

There are several common malicious attack patterns that can target package logistic systems [15,17–19], as follows:

(1) Modification attack: The attacker intercepts the information of the transmitting party and the receiving party, and modifies the contents of the shopping information, resulting in the loss of the transmitting party and the receiving party. Therefore, the transmitted information must be secure against modification in order to prevent such attacks.

(2) Impersonation attack: The attacker uses a fake identity to disguise themselves as a sender and sends a fake message to the receiver, causing the receiver to receive a false message.

(3) Man-in-the-middle-attack: The attacker establishes independent contact with both ends of the communication and exchanges the information so that both sender and receiver of the communication think that they are talking directly with each other through a private connection. In fact, the entire conversation is completely controlled by the attacker. In a man-in-the-middle attack, an attacker can intercept calls from both parties and insert new content.

(4) Clone attack: An attacker steals items by copying a label and impersonating a deliverer to deliver a non-original item.

3. The Proposed Scheme

3.1. System Architecture

The system consists of the following parties: Seller (S), logistics (L), buyer (B), and deliverer (D). The architecture and information flow are shown in Figure 1. The four parties in the scheme, in detail, are the following:

(1) Seller: An online shopping store. People can shop there and the seller sends the goods to the buyer, who will sign for the delivered package.
(2) Logistics: A company entrusted to deliver the packages to the buyer.
(3) Deliverer: The logistics employee. They assist the logistics company to deliver the package to the buyer.
(4) Buyer: Someone who buys something from the seller and who signs for the delivered package.

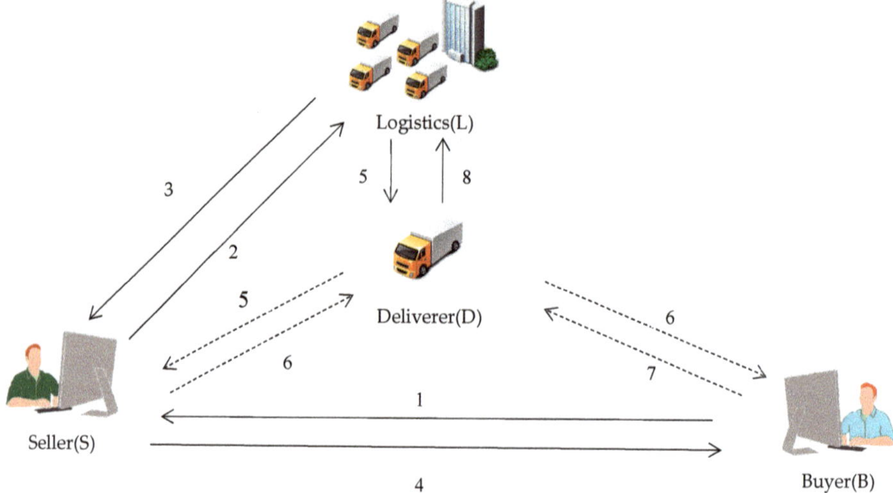

Figure 1. Logistics system architecture.

The eight steps in the scheme, in detail are as follows:

(1) The buyer requests a product from the seller.
(2) The seller provides the buyer and seller's information to the logistics.
(3) The logistics generates the transaction number and sends the buyer's tag to the seller.
(4) The seller sends the transaction number and the buyer's tag to the buyer.
(5) The logistics gives the tag of the seller's information to the deliverer. The deliverer goes to the seller's home and sends his/her identity to the seller.
(6) The seller verifies the deliverer's identity. The seller then transmits their signature and gives the goods and the buyer's tag to the deliverer. The deliverer goes to the buyer's house and sends his/her identity to the buyer.
(7) The buyer transmits their signature to the deliverer.
(8) The deliverer brings the buyer and the seller's signature back to the logistics to verify the signature and complete the transaction.

3.2. Notations

The notations used in this paper are listed below:

ID_X	Identification of X
M_X	X's address and telephone information
$M_{Product}$	Product information
TID	Transaction number
$Tag_{X,Y}$	The tag used for X to Y
Puk_X, Prk_X	The public key and private key of X, respectively
$Sig_{X,Y}$	The signature for X to Y
C_i	The ith ciphertext
P	Base point
SK_{XY}	Session key between X and Y
$E_{SK_{XY}}(M)$	Encrypt message M with session key SK_{XY}
$D_{prkX}(M)$	Decrypt message M with session key SK_{XY}
$S_{prkX}(M)$	Sign message M with X's private key prk_X
$V_{pukX}(M)$	Verify message M with X's public key puk_X
$E_{pukX}(M)$	Encrypt message M with X's public key puk_X
$D_{prkX}(M)$	Decrypt message M with X's private key prk_X
$h(M)$	The message M calculated by one-way hash function
$\oplus$	Exclusive-or operation for any two operands with same binary size
$\parallel$	Concatenation operator
$A \overset{?}{=} B$	Determine if A is equal to B
$\dashrightarrow$	A secure channel
$\longrightarrow$	An insecure channel

3.3. Initialization Phase

During the initialization phase, the Certificate Authority (CA) issues the public key and private key, and selects a large prime, P, and elliptic curve, E, over a finite field for each party.

3.4. Session Key Generation Phase and Order Request Phase

In the session key generation phase and order request phase, the buyer provides shopping information to the seller. The seller sends the buyer and the seller's information to the logistics and asks for the goods to be delivered. The logistics generates the transaction number and the tag for the seller and sends the transaction numbers to the buyer. Figure 2 presents the session key generation and order request phase of the proposed scheme.

Step 1: The buyer selects a random r_B and computes R_B as follows:

$$R_B = r_B{}^*P, \tag{1}$$

The buyer signs the (R_B, ID_B) with the private key Prk_B, as follows:

$$Sig_{BS} = S_{prkB}(R_B, ID_B), \tag{2}$$

The buyer then sends (R_B, ID_B, Sig_{BS}) to the seller.

Step 2: The seller selects a random number r_S and then computes R_S and signs the (R_S, ID_S) with the private key Prk_S, as follows:

$$R_S = r_S * P, \tag{3}$$

$$Sig_{SL} = S_{prkS}(R_S, ID_S). \tag{4}$$

The seller then sends (R_S, ID_S, Sig_{SL}) to logistics. The seller then verifies the Sig_{BS} with the public key Puk_B to determine whether the signagture is legal or not, as follows:

$$V_{pukB}(Sig_{BS}) \overset{?}{=} (R_B, ID_B). \tag{5}$$

If it passes the verification, the seller computes session key SK_{BS}, as follows:

$$SK_{BS} = h((r_s * R_B)\|ID_B\|ID_S), \tag{6}$$

uses the SK_{BS} to encrypt (R_B, ID_B) with SK_{BS}, as follows:

$$C_1 = E_{SK_{BS}}(R_B, ID_B), \tag{7}$$

and signs the (R_S, ID_S) with the private key Prk_S, as follows:

$$Sig_{SB} = S_{prkS}(R_S, ID_S), \tag{8}$$

The seller then sends $(R_S, ID_S, C_1, Sig_{SB})$ to the buyer.

Step 3: The logistics selects a random number r_L, and computes R_L, as follows:

$$R_L = r_L * P. \tag{9}$$

Logistics then verifies the Sig_{SL} with the public key Puk_S to determine whether the signagture is legal or not, as follows:

$$V_{pukS}(Sig_{SL}) \overset{?}{=} (R_S, ID_S), \tag{10}$$

If it holds, logistics computes session key SK_{SL}, as follows:

$$SK_{SL} = h((r_L * R_S)\|ID_S\|ID_L). \tag{11}$$

Then the logistics encrypts (R_S, ID_S) with SK_{SL}, as follows:

$$C_3 = E_{SK_{SL}}(R_S, ID_S). \tag{12}$$

Next, logistics signs the (R_L, ID_L) with the private key Prk_L, as follows:

$$Sig_{LS} = S_{prkL}(R_L, ID_L), \tag{13}$$

and sends $(R_L, ID_L, C_3, Sig_{LS})$ to the seller.

Step 4: The buyer verifies the Sig_{SB} with the public key Puk_S to determine whether the signagture is legal or not, as follows:

$$V_{pukS}(Sig_{SB}) \overset{?}{=} (R_S, ID_S). \tag{14}$$

The buyer then computes session key SK_{BS}, as follows:

$$SK_{BS} = h((r_B * R_S)\|ID_B\|ID_S), \tag{15}$$

and uses the SK_{BS} to decrypt C_1, as follows:

$$(R_B*, ID_B*) = D_{SK_{BS}}(C_1), \tag{16}$$

and determines whether (R_B, ID_B) is equal or not, as follows:

$$(R_B, ID_B) \stackrel{?}{=} (R_B, ID_B)* . \tag{17}$$

The seller then encrypts $(R_S, ID_S, ID_B, M_B, M_{product})$ with SK_{BS}, as follows:

$$C_2 = E_{SK_{BS}}\left(R_S, ID_S, ID_B, M_B, M_{product}\right), \tag{18}$$

Then buyer then sends (ID_B, C_2) to the seller.

Step 5: The seller decrypts C_2 with SK_{BS}, as follows:

$$\left(R_S*, ID_S*, ID_B, M_B, M_{product}\right) = D_{SK_{BS}}(C_2), \tag{19}$$

and then gets (R_S*, ID_S*), and determines whether (R_S, ID_S) is equal or not, as follows:

$$(R_S, ID_S) \stackrel{?}{=} (R_S*, ID_S*). \tag{20}$$

The seller verifies the Sig_{LS} with the public key Puk_L to determine whether the signagture is legal or not, as follows

$$V_{pukL}(Sig_{LS}) \stackrel{?}{=} (R_L, ID_L). \tag{21}$$

If it passes the verification, the seller computes SK_{SL}, as follows:

$$SK_{SL} = h((r_S * R_L)\|ID_S\|ID_L), \tag{22}$$

and decrypts C_3 with SK_{SL}, as follows:

$$(R_S*, ID_S) = D_{SK_{SL}}(C_3). \tag{23}$$

The seller gets (R_S*, ID_S*), determines whether (R_S, ID_S) is equal or not, as follows:

$$(R_S, ID_S) \stackrel{?}{=} (R_S*, ID_S*), \tag{24}$$

If it holds, the seller encrypts $(R_L, ID_L, ID_S, M_S, ID_B, M_B)$ with SK_{SL}, as follows:

$$C_4 = E_{SK_{SL}}(R_L, ID_L, ID_S, M_S, ID_B, M_B), \tag{25}$$

and then sends (ID_S, C_4) to logistics.

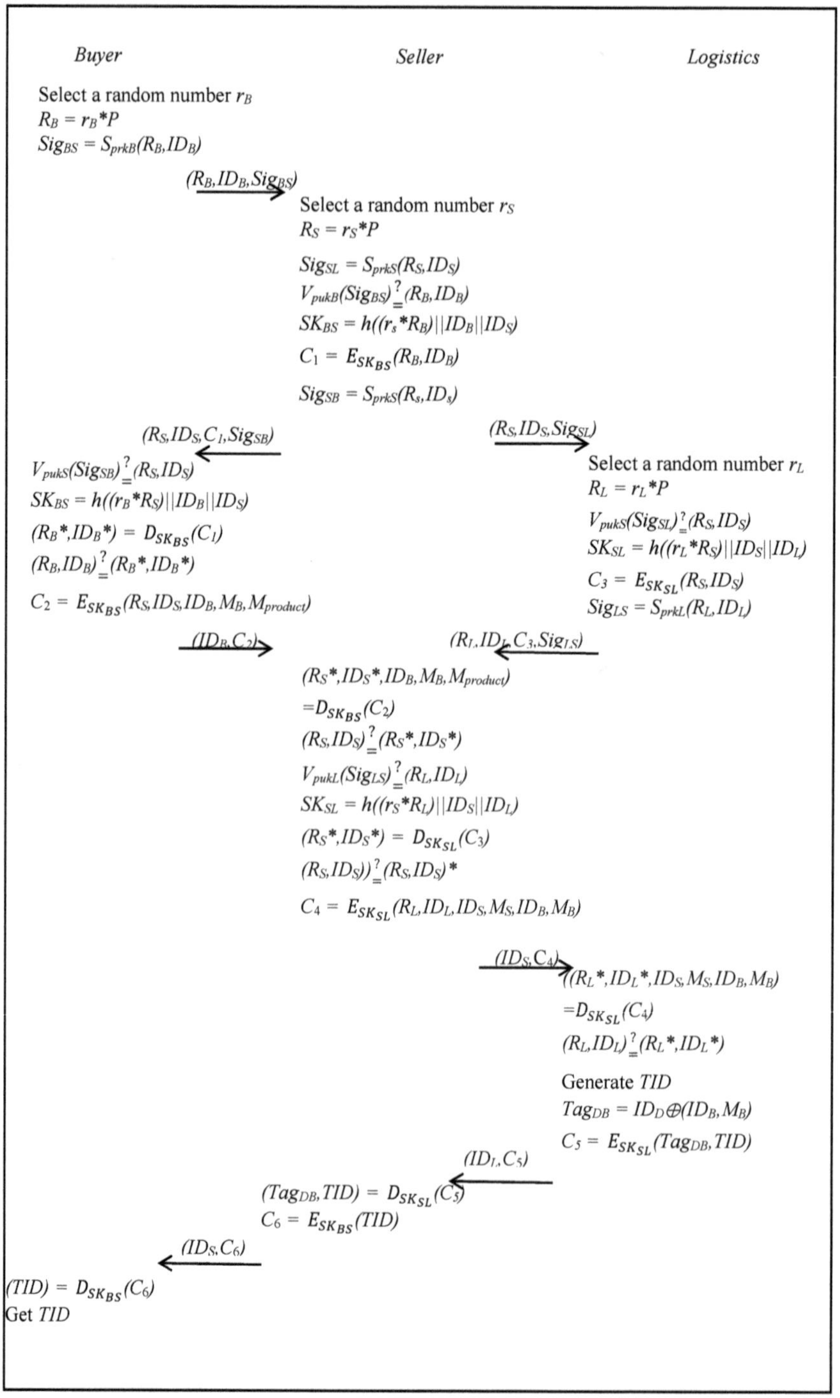

Buyer	Seller	Logistics

Select a random number r_B
$R_B = r_B*P$
$Sig_{BS} = S_{prkB}(R_B, ID_B)$

$\xrightarrow{(R_B, ID_B, Sig_{BS})}$

Select a random number r_S
$R_S = r_S*P$
$Sig_{SL} = S_{prkS}(R_S, ID_S)$
$V_{pukB}(Sig_{BS}) \stackrel{?}{=} (R_B, ID_B)$
$SK_{BS} = h((r_s*R_B)||ID_B||ID_S)$
$C_1 = E_{SK_{BS}}(R_B, ID_B)$
$Sig_{SB} = S_{prkS}(R_s, ID_s)$

$\xleftarrow{(R_S, ID_S, C_1, Sig_{SB})}$ $\xrightarrow{(R_S, ID_S, Sig_{SL})}$

$V_{pukS}(Sig_{SB}) \stackrel{?}{=} (R_S, ID_S)$
$SK_{BS} = h((r_B*R_S)||ID_B||ID_S)$
$(R_B*, ID_B*) = D_{SK_{BS}}(C_1)$
$(R_B, ID_B) \stackrel{?}{=} (R_B*, ID_B*)$
$C_2 = E_{SK_{BS}}(R_S, ID_S, ID_B, M_B, M_{product})$

Select a random number r_L
$R_L = r_L*P$
$V_{pukS}(Sig_{SL}) \stackrel{?}{=} (R_S, ID_S)$
$SK_{SL} = h((r_L*R_S)||ID_S||ID_L)$
$C_3 = E_{SK_{SL}}(R_S, ID_S)$
$Sig_{LS} = S_{prkL}(R_L, ID_L)$

$\xrightarrow{(ID_B, C_2)}$ $\xleftarrow{(R_L, ID_L, C_3, Sig_{LS})}$

$(R_S*, ID_S*, ID_B, M_B, M_{product})$
$= D_{SK_{BS}}(C_2)$
$(R_S, ID_S) \stackrel{?}{=} (R_S*, ID_S*)$
$V_{pukL}(Sig_{LS}) \stackrel{?}{=} (R_L, ID_L)$
$SK_{SL} = h((r_S*R_L)||ID_S||ID_L)$
$(R_S*, ID_S*) = D_{SK_{SL}}(C_3)$
$(R_S, ID_S)) \stackrel{?}{=} (R_S, ID_S)*$
$C_4 = E_{SK_{SL}}(R_L, ID_L, ID_S, M_S, ID_B, M_B)$

$\xrightarrow{(ID_S, C_4)}$

$(R_L*, ID_L*, ID_S, M_S, ID_B, M_B)$
$= D_{SK_{SL}}(C_4)$
$(R_L, ID_L) \stackrel{?}{=} (R_L*, ID_L*)$
Generate TID
$Tag_{DB} = ID_D \oplus (ID_B, M_B)$
$C_5 = E_{SK_{SL}}(Tag_{DB}, TID)$

$\xleftarrow{(ID_L, C_5)}$

$(Tag_{DB}, TID) = D_{SK_{SL}}(C_5)$
$C_6 = E_{SK_{BS}}(TID)$

$\xleftarrow{(ID_S, C_6)}$

$(TID) = D_{SK_{BS}}(C_6)$
Get TID

Figure 2. Session key generation and order request phase.

Step 6: The logistics decrypts C_4 with SK_{SL}, as follows:

$$(R_L*, ID_L*, ID_S, M_S, ID_B, M_B) = D_{SK_{SL}}(C_4),\tag{26}$$

and then gets (R_L*,ID_L*) and determines whether (R_L,ID_L) is equal or not, as follows:

$$(R_L, ID_L) \overset{?}{=} (R_L*, ID_L*),\tag{27}$$

Logistics generates *TID* and $Tag_{DB,}$ and computes the following:

$$Tag_{DB} = ID_D \oplus (ID_B,M_B).\tag{28}$$

and then uses the SK_{SL} to encrypt (Tag_{DB},TID), as follows:

$$C_5 = E_{SK_{SL}}(Tag_{DB}, TID),\tag{29}$$

then sends (ID_L,C_5) to the seller.

Step 7: The seller decrypts C_5 with SK_{SL}, as follows:

$$(Tag_{DB}, TID) = D_{SK_{SL}}(C_5).\tag{30}$$

The seller encrypts *(TID)* with SK_{BS}, as follows:

$$C_6 = E_{SK_{BS}}(TID),\tag{31}$$

then sends (IDS,C6) to the buyer.

Step 8: The buyer decrypts C6 with SKBS, as follows:

$$TID = D_{SK_{BS}}(C_6),\tag{32}$$

and then gets *TID*.

3.5. Package Collection Phase

The logistics sends the tag containing the seller information to the deliverer. The deliverer decrypts the tag and goes to the seller's house. After verifying the delivery identity, the seller transmits their signature to give the goods and the buyer's tag to the deliverer. The package collection phase is illustrated in Figure 3.

Step 1: The logistics signs (ID_D,ID_L,TID) with private key Prk_L, as follows:

$$Sig_{LS} = S_{prkL}(ID_D,ID_L,TID),\tag{33}$$

The logistics uses SK_{SL} to encrypt (ID_D,ID_L,TID), as follows:

$$C_7 = E_{SK_{SL}}(ID_D, ID_L, TID),\tag{34}$$

then generates Tag_{DS}, as follows:

$$Tag_{DS} = ID_D \oplus (ID_S,M_S),\tag{35}$$

and sends $(ID_L,Tag_{DS},C_7,Sig_{LS})$ to the deliverer.

Step 2: The deliverer computes the following formula:

$$(ID_S, M_S) = Tag_{DS} \oplus ID_D, \tag{36}$$

and the deliverer can then get (ID_S, M_S).

Step 3: The deliverer sends $(IDD, IDL, TagDS, C7, SigLS)$ to the seller for verification and the seller computes IDD^* as follows:

$$ID_D^* = Tag_{DS} \oplus (ID_S, M_S), \tag{37}$$

and verifies whether ID_D is equal or not, as follows:

$$ID_D \overset{?}{=} ID_D. \tag{38}$$

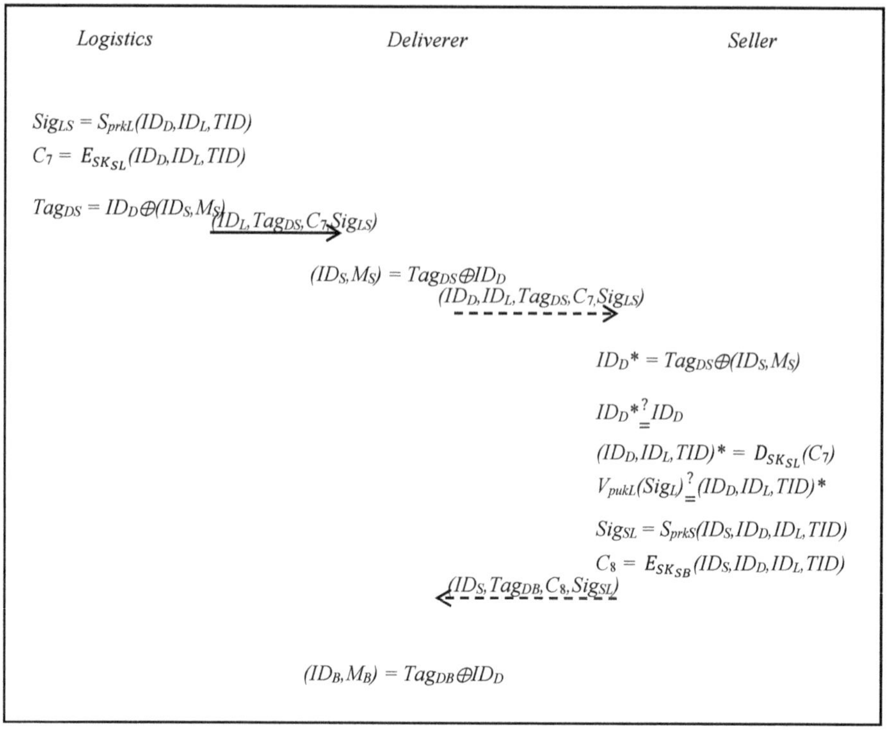

Figure 3. Package collection phase of the proposed scheme.

The seller decrypts C_7 with SK_{SL}, as follows:

$$(ID_D, ID_L TID)* = D_{SK_{SL}}(C_7). \tag{39}$$

The seller verifies the Sig_{LS} with the public key Puk_L to determine whether the signagture is legal or not, as follows:

$$V_{pukL}(Sig_{LS}) \overset{?}{=} (ID_D, ID_L, TID)*. \tag{40}$$

If it passes the verification, the seller signs the (ID_S, ID_D, ID_L, TID) with the private key Prk_S, as follows:

$$Sig_{SL} = S_{prkS}(ID_S, ID_D, ID_L, TID),\tag{41}$$

and uses SK_{SB} to encrypt (ID_S, D_D, ID_L, TID), as follows:

$$C_8 = E_{SK_{SB}}(ID_S, ID_D, ID_L, TID).\tag{42}$$

The seller then gives the goods and $(ID_S, Tag_{DB}, C_8, Sig_{SL})$ to the deliverer.

Step 4: The deliverer computes as following formula:

$$(ID_B, M_B) = Tag_{DB} \oplus ID_D,\tag{43}$$

and gets (ID_B, M_B).

3.6. Product Transfer Phase

The deliverer decrypts the tag and sends the goods to the buyer's address. After verifying the deliverer's identity, the buyer obtains the goods and sends a signature to the deliverer. The deliverer takes the signatures of the buyer and the seller. The deliverer then returns to the logistics for confirmation and completes the transaction. The product transfer phase is illustrated in Figure 4.

Step 1: The deliverer sends the goods and (ID_D, Tag_{DB}, C_8) to the buyer to verify the identity, using the following formula:

$$ID_D{}^* = Tag_{DB} \oplus (ID_B, M_B),\tag{44}$$

Once the deliverer has $ID_D{}^*$, they determine whether the ID_D is equal or not, as follows:

$$ID_D{}^* \stackrel{?}{=} ID_D.\tag{45}$$

The buyer decrypts C_8 with SK_{SB}, as follows:

$$(ID_S, ID_D, ID_L, TID) = D_{SK_{SB}}(C_8),\tag{46}$$

and then gets TID^* and determines whether the TID, which is stored in the session key generation and order request phase, is equal or not, as follows:

$$TID{}^* \stackrel{?}{=} TID.\tag{47}$$

The deliverer uses Prk_B to sign $(ID_B, ID_S, ID_D, ID_L, TID)$, as follows:

$$Sig_{BL} = S_{prkB}(ID_B, ID_S, ID_D, ID_L, TID),\tag{48}$$

The buyer sends (Sig_{BL}, ID_B) to the deliverer.

Step 2: The deliverer returns to the logistics. Logistics verifies Sig_S with public key Puk_S, as follows:

$$V_{pukS}(Sig_{SL}) \stackrel{?}{=} (ID_S, ID_D, ID_L, TID),\tag{49}$$

and then determines whether the signagture is legal or not. Logistics verifies Sig_B with public key Puk_B, as follows:

$$V_{pukB}(Sig_{BL}) \stackrel{?}{=} (ID_B, ID_S, ID_D, ID_L, TID),\tag{50}$$

and determines whether the signagture is legal or not.

If it passes the verification, the transaction is completed.

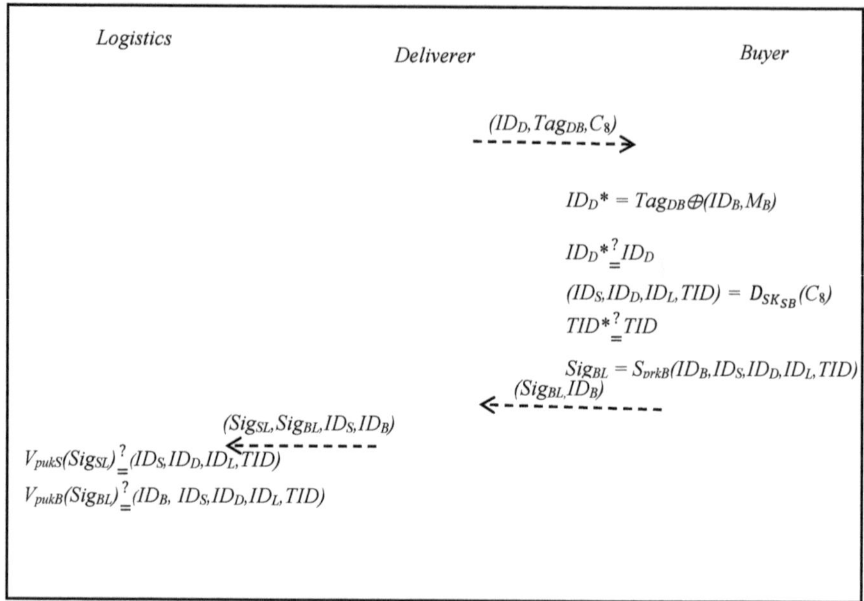

Figure 4. Product transfer phase of the proposed scheme.

4. Security Analysis and Discussion

4.1. Mutual Authentication Issue

This study uses BAN logic to prove that the proposed scheme achieves mutual authentication in each phase. In the session key generation and order request phases of the proposed scheme, the main goal is to determine whether the data has been modified between the buyer and seller, or the seller and the logistics provider.

The notation of BAN logic is described below:

$P\|\equiv X$	*P believes X, or P would be entitled to believe X.*
$P\triangleleft X$	*P sees X, someone has sent a message containing X to P, who can read and repeat X.*
$P\|\sim X$	*P once said X. P at some time sent a message including X.*
$P\|\stackrel{K}{\rightarrow}X$	*P has X as a public key.*
$P\stackrel{k}{\leftrightarrow}X$	*P and X may use the session key K to communicate*
$P\|\Rightarrow X$	*P has jurisdiction over x.*
$\#(X)$	*The formula X is fresh.*
$\{X\}_K$	*The formula X encrypted by K.*

The main goals of the scheme must be achieved in order to verify that the transmitted data has not been modified between buyer and seller, or between the seller and the logistics provider. These goals are listed below:

G1 $\quad S|\equiv S \overset{SK_{BS}}{\leftrightarrow} B$

G2 $\quad S|\equiv B|\equiv S \overset{SK_{BS}}{\leftrightarrow} B$

G3 $\quad B|\equiv S \overset{SK_{BS}}{\leftrightarrow} B$

G4 $\quad B|\equiv B|\equiv S \overset{SK_{BS}}{\leftrightarrow} B$

G5 $\quad S|\equiv S \overset{SK_{SL}}{\leftrightarrow} L$

G6 $\quad S|\equiv L|\equiv S \overset{SK_{SL}}{\leftrightarrow} L$

G7 $\quad L|\equiv S \overset{SK_{SL}}{\leftrightarrow} L$

G8 $\quad L|\equiv S|\equiv S \overset{SK_{SL}}{\leftrightarrow} L$

G9 $\quad S|\equiv ID_B$

G10 $\quad S|\equiv B|\equiv ID_B$

G11 $\quad B|\equiv ID_S$

G12 $\quad B|\equiv S|\equiv ID_S$

G11 $\quad L|\equiv ID_S$

G12 $\quad L|\equiv S|\equiv ID_S$

G11 $\quad S|\equiv ID_L$

G12 $\quad S|\equiv L|\equiv ID_L$

According to the purchase phase, BAN logic is used to produce an idealized form, as follows:

M1 $\quad (\{R_B,ID_B\}prk_B,\{R_S,ID_S,ID_B,M_B,M_{product}\}SK_{BS}),$

M2 $\quad (\{R_S,ID_S\}prk_S,\{R_B,ID_B\}SK_{BS}),$

M3 $\quad (\{R_S,ID_S\}prk_S,\{R_L,ID_L,ID_S,M_S,ID_B,M_B\}SK_{SL}),$

M4 $\quad (\{R_L,ID_L\}prk_L,\{R_S,ID_S\}SK_{BS}).$

In order to analyze the proposed improved scheme, this study makes the following assumptions:

A1 $\quad S|\equiv \#(R_B)$

A2 $\quad B|\equiv \#(R_B)$

A3 $\quad S|\equiv \#(R_S)$

A4 $\quad B|\equiv \#(R_S)$

A5 $\quad L|\equiv \#(R_S)$

A6 $\quad S|\equiv \#(R_S)$

A7 $\quad L|\equiv \#(R_L)$

A8 $\quad S|\equiv \#(R_L)$

A9 $\quad S|\equiv \#SK_{BS}$

A10 $\quad B|\equiv \#SK_{BS}$

A11 $\quad L|\equiv \#SK_{SL}$

A12 $\quad S|\equiv \#SK_{SL}$

A13 $\quad B|\equiv |\overset{pubB}{\rightarrow} S$

A14 $\quad S|\equiv |\overset{pubB}{\rightarrow} S$

A15 $\quad B|\equiv |\overset{pubS}{\rightarrow} B$

A16 $\quad S|\equiv |\overset{pubS}{\rightarrow} B$

A17 $\quad L|\equiv |\overset{pubS}{\rightarrow} L$

A18 $\quad S|\equiv |\overset{pubS}{\rightarrow} L$

A19 $\quad L|\equiv |\overset{pubL}{\rightarrow} S$

A20 $\quad S|\equiv |\overset{pubL}{\rightarrow} S$

A21 $\quad S|\equiv B|\Rightarrow S \overset{SK_{BS}}{\leftrightarrow} B$

A22 $\quad B|\equiv S|\Rightarrow S \overset{SK_{BS}}{\leftrightarrow} B$

$A23$	$L\|\equiv S\|\Rightarrow S\overset{SK_{SL}}{\leftrightarrow}L$
$A24$	$S\|\equiv L\|\Rightarrow S\overset{SK_{SL}}{\leftrightarrow}L$
$A25$	$S\|\equiv B\|\Rightarrow ID_B$
$A26$	$B\|\equiv S\|\Rightarrow ID_S$
$A27$	$L\|\equiv S\|\Rightarrow ID_S$
$A28$	$S\|\equiv L\|\Rightarrow ID_L$

According to these assumptions and the rules of BAN logic, this study shows the session key generation and order request phases of the proposed scheme as follows:

a. Seller S authenticates Buyer B By $M1$ and the seeing rule, derive the following:

$$S\triangleleft (\{R_B,ID_B\}prk_B,\{R_S,ID_S,ID_B,M_B,M_{product} \}SK_{BS}). \tag{Statement 1}$$

By $A1$ and $A2$ and the freshness rule, derive the following:

$$S\|\equiv\#(\{R_B,ID_B\}prk_B,\{R_S,ID_S,ID_B,M_B,M_{product} \}SK_{BS}). \tag{Statement 2}$$

By (Statement 1), $A9$, $A13$, and $A14$ and the message meaning rule, derive the following:

$$S\|\equiv B\|\sim\#(\{R_B,ID_B\}puk_B,\{R_S,ID_S,ID_B,M_B,M_{product} \}SK_{BS}). \tag{Statement 3}$$

By (Statement 2), (Statement 3), and the verification rule, derive the following:

$$S\|\equiv B\|\equiv(\{R_B,ID_B\}puk_B,\{R_S,ID_S,ID_B,M_B,M_{product} \}SK_{BS}. \tag{Statement 4}$$

By (Statement 4) and the belief rule, derive the following:

$$S\| \equiv B\| \equiv S \overset{SK_{BS}}{\leftrightarrow} B. \tag{Statement 5}$$

By (Statement 5), $A21$, and the jurisdiction rule, derive the following:

$$S\| \equiv S \overset{SK_{BS}}{\leftrightarrow} B. \tag{Statement 6}$$

By (Statement 6) and the belief rule, derive the following:

$$S\|\equiv B\|\equiv ID_B. \tag{Statement 7}$$

By (Statement 7), $A25$, and the belief rule, derive the following:

$$S\|\equiv ID_B. \tag{Statement 8}$$

b. Buyer B authenticates Seller S By $M2$ and the seeing rule, derive the following:

$$B\triangleleft(\{R_S,ID_S\}prk_S,\{R_B,ID_B\}SK_{BS}). \tag{Statement 9}$$

By $A3$, $A4$, and the freshness rule, derive the following:

$$B\|\equiv\#(\{R_S,ID_S\}prk_S,\{R_B,ID_B\}SK_{BS}). \tag{Statement 10}$$

By (Statement 9), $A10$, $A15$, $A16$, and the message meaning rule, derive the following:

$$B\|\equiv S\|\sim\#(\{R_S,ID_S\}puk_S,\{R_B,ID_B\}SK_{BS}). \tag{Statement 11}$$

By (Statement 10), (Statement 11) and the verification rule, derive the following:

$$B|{\equiv}S|{\equiv}(\{R_S,ID_S\}prk_S,\{R_B,ID_B\}SK_{BS}).$$ (Statement 12)

By (Statement 12) and the belief rule, derive the following:

$$B| \equiv S| \equiv S \xleftrightarrow{SK_{BS}} B.$$ (Statement 13)

By (Statement 13), $A22$ and the jurisdiction rule, derive the following:

$$B| \equiv S \xleftrightarrow{SK_{BS}} B.$$ (Statement 14)

By (Statement 14) and the belief rule, derive the following:

$$B|{\equiv}S|{\equiv}IDS$$ (Statement 15)

By (Statement 15), $A26$ and the belief rule, derive the following:

$$B|{\equiv}ID_S.$$ (Statement 16)

By (Statement 6), (Statement 8), (Statement 14), and (Statement 16), it is proved that buyer B and seller S authenticate each other in the proposed scheme. The seller authenticates the buyer by (5). If it passes the verification, the seller authenticates the legality of the buyer and then the buyer authenticates the seller by (14).

c.   Logistics L authenticates Seller S By $M3$ and the seeing rule, derive the following:

$$L{\vartriangleleft}(\{R_S,ID_S\}prk_S, \{R_L,ID_L,ID_S,M_S,ID_B,M_B\}SK_{SL}).$$ (Statement 17)

By $A5$, $A6$, and the freshness rule, derive:

$$L|{\equiv}\#(\{R_S,ID_S\}prk_S, \{R_L,ID_L,ID_S,M_S,ID_B,M_B\}SK_{SL}).$$ (Statement 18)

By (Statement 17), $A11$, $A17$, $A18$, and the message meaning rule, derive the following:

$$L|{\equiv}S|{\sim}\#(\{R_S,ID_S\}puk_S, \{R_L,ID_L,ID_S,M_S,ID_B,M_B\}SK_{SL}).$$ (Statement 19)

By (Statement 18), (Statement 19), and the verification rule, derive the following:

$$L|{\equiv}S|{\equiv}(\{R_S,ID_S\}puk_S, \{R_L,ID_L,ID_S,M_S,ID_B,M_B\}SK_{SL}).$$ (Statement 20)

By (Statement 20) and the belief rule, derive the following:

$$L| \equiv S| \equiv S \xleftrightarrow{SK_{BS}} B.$$ (Statement 21)

By (Statement 21), $A23$, and the jurisdiction rule, derive the following:

$$L| \equiv S \xleftrightarrow{SK_{BS}} B.$$ (Statement 22)

By (Statement 22) and the belief rule, derive the following:

$$L|{\equiv}S|{\equiv}ID_S.$$ (Statement 23)

By (Statement 23), $A27$, and the belief rule, derive the following:

$$L|\equiv ID_S. \tag{Statement 24}$$

d. Logistics L authenticates Seller S By $M4$ and the seeing rule, derive the following:

$$S\triangleleft(\{R_L,ID_L\}prk_L, \{R_S,ID_S\}SK_{BS}). \tag{Statement 25}$$

By $A7$, $A8$, and the freshness rule, derive the following:

$$S|\equiv\#(\{R_L,ID_L\}prk_L, \{R_S,ID_S\}SK_{BS}). \tag{Statement 26}$$

By (Statement 25), $A12$, $A19$, $A20$ and the message meaning rule, derive the following:

$$S|\equiv L|\sim\#(\{R_L,ID_L\}prk_L, \{R_S,ID_S\}SK_{BS}). \tag{Statement 27}$$

By (Statement 26), (Statement 27), and the verification rule, derive the following:

$$S|\equiv L|\equiv(\{R_L,ID_L\}prk_L, \{R_S,ID_S\}SK_{BS}). \tag{Statement 28}$$

By (Statement 28) and the belief rule, derive the following:

$$S| \equiv L| \equiv S \overset{SK_{SL}}{\leftrightarrow} L. \tag{Statement 29}$$

By (Statement 29), $A24$, and the jurisdiction rule, derive the following:

$$S| \equiv S \overset{SK_{SL}}{\leftrightarrow} L. \tag{Statement 30}$$

By (Statement 30) and the belief rule, derive the following:

$$S|\equiv L|\equiv ID_L. \tag{Statement 31}$$

By (Statement 31), $A28$, and the belief rule, derive the following:

$$S|\equiv ID_L. \tag{Statement 32}$$

By (Statement 22), (Statement 24), (Statement 30), and (Statement 32), it is proved that logistics L and seller S authenticate each other in the proposed scheme. The logistics authenticates the seller by (14): If it passes the verification, the logistics provider authenticates the legality of the seller and then the buyer authenticates the logistics as (21).

4.2. Non-Repudiation Issue

The proposed scheme uses digital signatures to achieve non-repudiation between the parties in each phase. The sender uses their private key to sign the transmitted message and then the receiver verifies the received message. The receiver uses their private key to sign the response message. Table 1 shows the non-repudiation of the proposed scheme.

Table 1. Non-repudiation of the proposed scheme.

Phase \ Party	Proof	Issuer	Holder	Verification
Session key generation and order request phase	(R_B,ID_B)	Buyer	Seller	$V_{pukB}(Sig_{BS}) \overset{?}{=} (R_B,ID_B)$
	(R_S,ID_S)	Seller	Buyer	$V_{pukS}(Sig_{SB}) \overset{?}{=} (R_S,ID_S)$
	(R_S,ID_S)	Seller	Logistics	$V_{pukS}(Sig_{SL}) \overset{?}{=} (R_S,ID_S)$
	(R_L,ID_L)	Logistics	Seller	$V_{pukL}(Sig_{LS}) \overset{?}{=} (R_L,ID_L)$
Package collection phase	(ID_D,ID_L,TID,Sig_L)	Logistics	Seller	$V_{pukL}(Sig_L) \overset{?}{=} (ID_D,ID_L,TID)$
Product transfer phase	$(ID_S,ID_D,ID_L,TID,Sig_S)$	Seller	Buyer	$V_{pukS}(Sig_S) \overset{?}{=} (ID_S,ID_D,ID_L,TID)$
	$(ID_B,ID_S,ID_D,ID_L,TID,Sig_B)$	Buyer	Logistics	$V_{pukB}(Sig_B) \overset{?}{=} (ID_B,ID_S,ID_D,ID_L,TID)$

4.3. Anonymity Issue

All personal information, $Tag_{DS} = (ID_S,M_S)\oplus ID_D$ and $Tag_{DB} = (ID_B,M_B)\oplus ID_D$, is protected so that only the legal identities ID_D, ID_S, and ID_B can read the content. Therefore, the contents will not disclose any information about buyer or seller.

4.4. Low Overhead Issue

In the package collection phase and the product transfer phase, this study uses exclusive operation or encryption to quickly verify and reduce the verification cost. This study also uses session keys to substitute public key encryption to enhance calculation speed, thus meeting the low overhead requirement.

4.5. Data Integrity Issue

This study uses digital signatures to ensure data integrity. A malicious attack can be detected using digital signatures to verify the integrity of the data, even if an attacker has tampered with the transmitted data. Thus, attackers cannot tamper with the transmitted data without being detected. Therefore, the proposed scheme achieves data integrity.

4.6. Security Against Known Attacks

4.6.1. Modification Attack

In the information transmission process, encryption is performed using session keys, preventing the modification of transmitted data:

(1) The session key generation and order request phase is as follows:

$$C_1 = E_{SK_{BS}}(R_B,ID_B), \tag{7}$$
$$C_3 = E_{SK_{SL}}(R_S,ID_S), \tag{12}$$
$$C_2 = E_{SK_{BS}}(R_S,ID_S,ID_B,M_B,M_{product}), \tag{18}$$
$$C_4 = E_{SK_{SL}}(R_L,ID_L,ID_S,M_S,ID_B,M_B), \tag{25}$$
$$C_5 = E_{SK_{SL}}(Tag_{DB},TID), \tag{29}$$
$$C_6 = E_{SK_{BS}}(TID), \tag{31}$$

(2) Package collection phase:

$$C_7 = E_{SK_{SL}}(ID_D,ID_L,TID), \tag{34}$$
$$C_8 = E_{SK_{SB}}(ID_S,ID_D,ID_L,TID). \tag{42}$$

4.6.2. Impersonation Attack

In the session key generation and order request phase, package collection phase, and product transfer phase of information transmission, digital signatures cannot be disguised.

(1)   The session key generation and order request phase is as follows:

$$Sig_{BS} = S_{prkB}(R_B,ID_B), \qquad (2)$$
$$Sig_{SL} = S_{prkS}(R_S,ID_S), \qquad (4)$$
$$Sig_{SB} = S_{prkS}(R_S,ID_S), \qquad (8)$$
$$Sig_{LS} = S_{prkL}(R_L,ID_L). \qquad (13)$$

(2)   Package collection phase:

$$Sig_{SL} = S_{prkS}\,(ID_S,ID_D,ID_L,TID). \qquad (41)$$

(3)   Product transfer phase:

$$Sig_{BL} = S_{prkB}(ID_B,ID_S,ID_D,ID_L,TID), \qquad (48)$$

4.6.3. Man-in-the-Middle Attack

The proposed scheme uses signature mechanisms $Sig_{BS} = S_{prkB}(R_B,ID_B)$, $Sig_{SL} = S_{prkS}(R_S,ID_S)$, and $Sig_{LS} = S_{prkL}(R_L,ID_L)$ to prevent modification of the R_B, R_S, and R_L, and uses those variables to generate session keys $SK_{BS} = h((r_s{*}R_B)\|ID_B\|ID_S)$ and $SK_{SL} = h((r_L{*}R_S)\|ID_S\|ID_L)$. The session key encryption/decryption offers security against man-in-the-middle attacks.

4.6.4. Clone Attack

In the package collection phase and the product transfer phase, the deliverer must give their own information, ID_D and Tag_{DS}, and the seller can then execute the exclusive-or operation or encrypt the Tag_{DS} and verify the identity of the deliverer $ID_D{}^* = Tag_{DS}{\oplus}(ID_S,M_S)$. In the product transfer phase of the proposed scheme, the deliverer must give their own information, ID_D and Tag_{DB}, and the buyer can then execute the exclusive-or operation or encrypt the Tag_{DB} and verify the identity of the deliverer $ID_D{}^* = Tag_{DB}{\oplus}(ID_B,M_B)$, thus preventing a clone attack.

4.7. Computation Cost

Table 2 shows the computation costs of the proposed scheme.

Table 2. Computation costs of the proposed scheme.

Party \ Phase	Buyer	Seller	Logistics	Deliverer
Session key generation and order request phase	$2T_{asy} + 3T_{sys} + 1T_h + 1T_{mul}$	$4T_{asy} + 6T_{sys} + 2T_h + 2T_{mul}$	$2T_{asy} + 3T_{sys} + 1T_h + 1T_{xor} + 1T_{mul}$	N/A
Package collection phase	N/A	$2T_{asy} + 2T_{sys} + 1T_{xor}$	$1T_{asy} + 1T_{sys} + 1T_{xor}$	$2T_{xor}$
Product transfer phase	$1T_{asy} + 1T_{sys} + 1T_{xor}$	N/A	$2T_{asy}$	N/A

Notes:

T_{asy}	The time required for an asymmetrical signature/verifying a signature.
T_{sys}	The time required for a symmetrical encryption/decryption operation.
T_h	The time required for a one-way hash function.
T_{xor}	The time required for an exclusive-or operation.
T_{mul}	The time required for a multiplication operation.

In Table 2, the proposed scheme's computation costs are analyzed for the buyer, seller, logistics, and deliverer in each phase. Due to the insignificant comparison operation impacts, they are not considered. For the highest computation cost reduction in the session key generation and order request phase, a buyer needs three asymmetrical signatures/verifying a signature, three symmetrical encryption/decryption operations, one hash function operation, and one multiplication operation. A seller needs four asymmetrical signatures/verifying a signature, six symmetrical encryption/decryption operations, two hash function operations, and two multiplication operations. The logistics provider needs two asymmetrical signatures/verifying a signature, three symmetrical encryption/decryption operations, one hash function operation, one exclusive-or operation, and one multiplication operation.

4.8. Communication Cost

Table 3 shows the communication cost of the proposed scheme.

Table 3. Communication cost of the proposed scheme.

Phase \ Party	Message Length	Round	3.5G (14 Mbps)	4G (100 Mbps)
Session key generation and order request phase	$4T_{sig} + 6T_{sys} = 4 \times 1024 + 6 \times 256 = 5632$ bits	8	5632/14000 = 0.402 ms	5632/100000 = 0.056 ms
Package collection phase	$3T_{sig} + 3T_{sys} + 3T_{xor} = 3 \times 1024 + 3 \times 256 + 3 \times 80 = 4080$ bits	3	4080/14000 = 0.291 ms	4080/100000 = 0.041 ms
Product transfer phase	$3T_{Sig} + 1T_{sys} + 1T_{xor} = 3 \times 1024 + 1 \times 256 + 1 \times 80 = 3408$ bits	3	4432/14000 = 0.243 ms	4432/100000 = 0.044 ms

Notes:

T_{sig}	The time required to transmit a signature (1024 bits).
T_{sys}	The time required to transmit a symmetric encryption/decryption ciphertext (256 bits).
T_{xor}	The time required to transmit an exclusive-or operation (80 bits).

From Table 3, the communication cost of the proposed scheme during the transaction process of each phase was analyzed and, since other operations have little impact, they were not considered in the communication cost. For the highest communication cost reduction in the session key generation and order request phase, four signature operations and six symmetric encryption/decryption operations must be transmitted. It thus requires $1024 \times 4 + 256 \times 6 = 5632$ bits. In a 3.5G environment, the maximum transmission speed is 14 Mbps, which only takes 0.402 ms to transfer all messages. In a 4G environment, the maximum transmission speed is 100 Mbps and the transmission time is reduced to 0.056 ms (ITU 2016).

4.9. Storage Cost

Table 4 shows the storage cost of the proposed scheme.

Table 4. Storage cost of the proposed scheme.

Phase / Party	Buyer	Seller	Logistics	Deliverer
Session key generation and order request phase	$1T_{asy} + 1T_{sys} + 1T_h + 2T_{mul} + 4T_{other} = 2176$ bits	$2T_{asy} + 3T_{sys} + 2T_h + 3T_{mul} + 5T_{other} = 4208$ bits	$1T_{asy} + 2T_{sys} + 1T_h + 2T_{mul} + 7T_{other} = 2672$ bits	N/A
Package collection phase	N/A	$1T_{asy} + 1T_{sys} + 5T_{other} = 1680$ bits	$1T_{asy} + 1T_{sys} + 6T_{other} = 1760$ bits	$1T_{other} = 80$ bits
Product transfer phase	$5T_{other} = 400$ bits	N/A	$1T_{asy} + 5T_{other} = 1424$ bits	N/A

Notes:

T_{asy}	The space required to storage an asymmetrical signature (1024 bits).
T_{sys}	The space required to storage a symmetrical encryption/decryption ciphertext (256 bits).
T_h	The space required to storage a one-way hash function calculated message (256 bits).
T_{mul}	The space required to storage a multiplication calculated message (160 bits).
T_{other}	The space required to storage other messages (80 bits).

In Table 4, the storage cost of the proposed scheme was analyzed for the buyer, seller, logistics and deliverer in each phase. For the highest storage cost in the session key generation and order request phase, a seller needs two asymmetrical signatures storage space, three symmetrical encryption/decryption ciphertexts storage space, two one-way hash function calculated messages storage space, three multiplication calculated messages storage space, and five other messages storage space. It thus requires $1024 \times 2 + 256 \times 3 + 256 \times 2 + 160 \times 3 + 80 \times 5 = 4208$ bits storage space.

5. Conclusions

In recent years, e-commerce services have prospered and online shopping has become a current trend. The security of personal information exchanged when purchasing a product online has thus become an important issue. This paper proposes a tag-based protection of personal information and a non-repudiable logistics system. The proposed scheme can effectively provide the secure transmission of personal information transmitted by items.

In the session key generation and order request phases, digital signatures are used to transmit data from the sender to the receiver, which ensures that the data cannot be modified. In the package collection phase and product transfer phase, tags containing hidden personal information are used to prevent personal information being leaked and to speed up the verification of the deliverer for buyers and sellers. The proposed scheme offers a reduction of computation costs, compared to other related works. The logistics can use the proposed system to achieve non-repudiation and to complete transactions by examining the digital signatures of the buyer and seller.

(1) The process of communication between buyers and sellers is mutual authentication.
(2) The non-repudiation of the goods delivery process is achieved through the signature mechanism.
(3) Personal information protection is achieved through exclusive-or operations.
(4) Tags use lightweight authentication technology to reduce the computation cost, compared to related works.

Future work will include the payment flow and applying block-chain technology to track the stream of and to prevent the loss of goods.

Author Contributions: Supervision and methodology, C.-L.C.; writing—original draft, D.-P.L.; validation, Y.-Y.D.; surveyed related work, H.-C.C. and C.-F.L.

Funding: This research was funded by the Ministry of Science and Technology, Taiwan, ROC, under contract number MOST 108-2221-E-324-013.

Conflicts of Interest: The authors declare no conflict of interest.

References

1. Aijaz, A.; Bochow, B.; Dotzer, F.; Festag, A.; Gerlach, M.; Kroh, R.; Leinmuller, T. Attacks on inter vehicle communication systems—An analysis. In Proceedings of the 3rd International Workshop on Intelligent Transportation, Hamburg, Germany, 14–15 March 2006; pp. 189–194.

2. Burrows, M.; Abadi, M.; Needham, R. A logic of authentication. *ACM Trans. Comput. Syst.* **1990**, *8*, 18–36. [CrossRef]

3. Chen, C.L.; Chiang, M.L.; Peng, C.C.; Chang, C.H.; Sui, Q.R. A Secure Mutual Authentication Scheme with Non-repudiation for Vehicular Ad Hoc Networks. *Int. J. Commun. Syst.* **2015**, *30*, e3081. [CrossRef]

4. Cui, J.; She, D.; Ma, J.; Wu, Q.; Liu, J. A New Logistics Distribution Scheme Based on NFC. In Proceedings of the 2015 International Conference on Network and Information Systems for Computers, Wuhan, China, 23–25 January 2015; pp. 492–495.

5. Cho, J.-S.; Jeong, Y.-S.; Park, S. Consideration on the brute-force attack cost and retrieval cost: A hash-based radio-frequency identification (RFID) tag mutual authentication protocol. *Comput. Math. Appl.* **2015**, *69*, 58–65. [CrossRef]

6. Liu, S.; Wang, J. A Security-Enhanced Express Delivery System Based on NFC. In Proceedings of the 2016 13th IEEE International Conference on Solid-State and Integrated Circuit Technology, Hangzhou, China, 25–28 October 2016; pp. 1534–1536.

7. Speranza, M.G. Trends in transportation and logistics. *Eur. J. Oper. Res.* **2018**, *264*, 830–836. [CrossRef]

8. Gope, P.; Amin, R.; Hafizul Islam, S.K.; Kumar, N.; Bhalla, V.K. Lightweight and privacy-preserving RFID authentication scheme for distributed IoT infrastructure with secure localization services for smart city environment. *Future Gener. Comput. Syst.* **2018**, *83*, 629–637. [CrossRef]

9. Liang, K.; Susilo, W. Searchable attribute-based mechanism with efficient data sharing for secure cloud storage. *IEEE Trans. Inf. Forensics Secur.* **2015**, *10*, 1981–1992. [CrossRef]

10. Das, A.K.; Goswami, A. A robust anonymous biometric-based remote user authentication scheme using smart cards. *J. King Saud Univ. Comput. Inf. Sci.* **2015**, *27*, 193–210. [CrossRef]

11. Madhusudhan, R.; Hegde, M. Security bound enhancement of remote user authentication using smart card. *J. Inf. Secur. Appl.* **2017**, *36*, 59–68. [CrossRef]

12. Qi, M.; Chen, J. A fresh Two-party Authentication Key Exchange Protocol for Mobile Environment. In Proceedings of the International Conference on Industrial Technology and Management Science, Tianjin, China, 27–28 March 2015; Volume 30, pp. 933–936.

13. Ray, B.R.; Abawajy, J.; Chowdhury, M.; Alelaiwi, A. Universal and secure object ownership transfer protocol for the Internet of Things. *Future Gener. Comput. Syst.* **2017**, *78*, 838–849. [CrossRef]

14. Rajput, U.; Abbas, F.; Eun, H.; Oh, H. A Hybrid approach for Efficient Privacy-Preserving Authentication in VANET. *IEEE Access* **2017**, *5*, 12014–12030. [CrossRef]

15. Sharma, V.; Vithalkar, A.; Hashmi, M. Lightweight security protocol for chipless RFID in Internet of Things (IoT) applications. In Proceedings of the 2018 10th International Conference on Communication Systems & Networks (COMSNETS), Bengaluru, India, 3–7 January 2018; pp. 468–471.

16. Tu, Y.; Piramuthu, S. Lightweight non-distance-bounding means to address RFID relay attacks. *Decis. Support Syst.* **2017**, *102*, 12–21. [CrossRef]

17. Whitmore, A.; Agarwal, A.; Da, X.L. The internet of things: A survey of topics and trends. *Inf. Syst. Front.* **2015**, *17*, 261–274. [CrossRef]

18. Wang, J.; Floerkemeier, C.; Sarma, S.E. Session-based security enhancement of RFID systems for emerging open-loop applications. *Pers. Ubiquitous Comput.* **2014**, *18*, 1881–1891. [CrossRef]
19. Zhao, S.; Aggarwal, A.; Frost, R.; Bai, X. A survey of applications of identity-based cryptography in mobile ad-hoc networks. *IEEE Commun. Surv. Tutor.* **2012**, *14*, 380–400. [CrossRef]

Article

Current Control of the Permanent-Magnet Synchronous Generator Using Interval Type-2 T-S Fuzzy Systems

Yuan-Chih Chang *, Chi-Ting Tsai and Yong-Lin Lu

Department of Electrical Engineering and Advanced Institute of Manufacturing with High-tech Innovations, National Chung Cheng University, Chiayi 62102, Taiwan
* Correspondence: ycchang@ccu.edu.tw; Tel.: +886-5-272-9108

Received: 10 July 2019; Accepted: 29 July 2019; Published: 31 July 2019

Abstract: The current control of the permanent-magnet synchronous generator (PMSG) using an interval type-2 (IT2) Takagi-Sugeno (T-S) fuzzy systems is designed and implemented. PMSG is an energy conversion unit widely used in wind energy generation systems and energy storage systems. Its performance is determined by the current control approach. IT2 T-S fuzzy systems are implemented to deal with the nonlinearity of a PMSG system in this paper. First, the IT2 T-S fuzzy model of a PMSG is obtained. Second, the IT2 T-S fuzzy controller is designed based on the concept of parallel distributed compensation (PDC). Next, the stability analysis can be conducted through the Lyapunov theorem. Accordingly, the stability conditions of the closed-loop system are expressed in Linear Matrix Inequality (LMI) form. The AC power from a PMSG is converted to DC power via a three-phase six-switch full bridge converter. The six-switch full bridge converter is controlled by the proposed IT2 T-S fuzzy controller. The analog-to-digital (ADC) conversion, rotor position calculation and duty ratio determination are digitally accomplished by the microcontroller. Finally, simulation and experimental results verify the performance of the proposed current control.

Keywords: permanent-magnet synchronous generator; T-S fuzzy; current control; interval type-2

1. Introduction

A permanent-magnet synchronous generator (PMSG) [1] is an essential unit implemented to convert energy. Its power density and power conversion efficiency is high and its maintenance cost is low. Consequently, PMSG can be implemented in various utilizations like electric vehicles (EV) and hybrid electric vehicles (HEV) [2,3], home appliances [4], wind generation systems [5], flywheel energy storage systems (FESS) [6] and ultrahigh-speed elevators [7]. The embedded type of the permanent magnet will affect the characteristics of a PMSG. The interior-type [8–10] has high inductance saliency and generates higher torque. However, its torque ripple is also higher. On the contrary, the torque ripple of the surface-mounted type [10,11] is lower and its reluctance torque is nearly absent [12]. In motor design concept, magnet design methods [13,14] are presented to optimize the performance and minimize the torque ripple.

If the d-axis current is controlled at zero, then the electromagnetic torque of a PMSG is proportional to the q-axis current. This characteristic enhances the importance of current controls [15–20] for the PMSG. Since the power generating performance of a PMSG is affected by its winding current, the torque ripple can be eliminated by reducing winding current harmonics [16]. Moreover, the generating capability of a PMSG can be improved via adopting proper and good current control algorithms. The traditionally implemented current control schemes include fixed-frequency control [17,18], hysteresis control [19] and predictive control [20]. The state space equation of a PMSG is nonlinear. Therefore, the Takagi-Sugeno fuzzy [21,22] (T-S fuzzy) system is implemented to design a speed controller for the permanent-magnet synchronous motor (PMSM) and a current controller for the PMSG. System uncertainty [23] generally exists in nonlinear systems. Therefore, the interval type-2 (IT2) fuzzy logic system [24,25] is proposed to deal with this problem. In this paper, IT2 T-S fuzzy models are implemented in the design of a current controller for the PMSG.

In this study, system configuration of the PMSG is first introduced. Then the dynamic model of a PMSG is conducted. Next, IT2 T-S fuzzy models of a PMSG are established for the design of an IT2 T-S fuzzy current controller. The stability of the IT2 T-S fuzzy control system for PMSG is analyzed using the Lyapunov theorem. The stability conditions of the proposed current controller are expressed in Linear Matrix Inequality (LMI) form. Experimental results, including constant current command tracking, variable current command tracking and computation time of the microcontroller, are demonstrated to verify the performance of the designed IT2 T-S fuzzy current controller.

2. System Configuration and Dynamic Model

Figure 1 introduces the system configuration of the PMSG based on the IT2 T-S fuzzy control systems. The prime mover of the PMSG consists of a PMSM with a PMSM driver. The input torque of PMSG is provided by the prime mover. The encoder signals on the shaft are utilized to calculate rotor speed and rotor position. Three-phase winding currents are sensed through the analog-to-digital converter (ADC). Then the abc-frame currents are transformed to dq-frame currents. The current commands and dq-frame currents are implemented in the IT2 T-S fuzzy current controller to calculate the dq-frame voltage commands. The corresponding duty ratio of six switches is determined via space-vector pulse-width modulation (SVPWM). All the control and transformation schemes are digitally realized using the microcontroller Renesas RX62T (Renesas Electronics Corp., Tokyo, Japan).

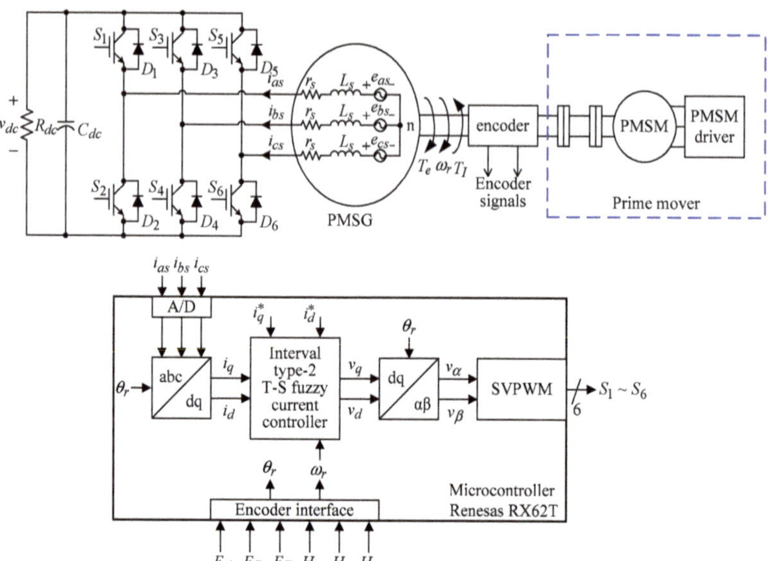

Figure 1. System configuration of the PMSG based on theIT2 T-S fuzzy control systems.

For convenience, voltage equations of the PMSG are expressed in dq-frame [1]:

$$v_q = -r_s i_q - L_q \dot{i}_q - \omega_r L_d i_d + \omega_r \lambda_m$$
$$v_d = -r_s i_d - L_d \dot{i}_d + \omega_r L_q i_q \tag{1}$$

where v_d and v_q are dq-frame voltages, i_d and i_q are dq-frame currents, r_s is winding resistance, L_d and L_q are dq-frame inductances, ω_r is electrical rotor speed and λ_m is the flux linkage established by the permanent magnet.

A PMSG can produce electromagnetic torque as follows:

$$T_e = \left(\frac{3}{2}\right)\left(\frac{P}{2}\right)\left[\lambda_m i_q + (L_q - L_d)i_q i_d\right]. \tag{2}$$

The electromagnetic torque is related to electrical rotor speed in the mechanical equation:

$$T_e = -J\left(\frac{2}{P}\right)\dot{\omega}_r - B\left(\frac{2}{P}\right)\omega_r + T_I \tag{3}$$

where J is the inertia of a PMSG, B is the damping coefficient of a PMSG, T_I is the input torque and P is magnetic pole number.

If $i_d = 0$ is satisfied in Equation (2), then the state equations of a PMSG are obtained:

$$\begin{aligned}
\dot{i}_q &= \tfrac{1}{L_s}\left(-v_q - r_s i_q - \omega_r L_s i_d + \omega_r \lambda_m\right) \\
\dot{i}_d &= \tfrac{1}{L_s}\left(-v_d - r_s i_d + \omega_r L_s i_q\right) \\
\dot{\omega}_r &= \tfrac{1}{J}\left[\left(\tfrac{P}{2}\right)(T_I - T_e) - B\omega_r\right] = \left(-\tfrac{3P^2}{8J}\lambda_m i_q - \tfrac{B}{J}\omega_r + \tfrac{P}{2J}T_I\right)
\end{aligned} \tag{4}$$

3. Design of the IT2 T-S Fuzzy Current Controller

Two new state variables are defined to guarantee current tracking capability:

$$\begin{aligned}
s_1 &= \int [r_1 - i_q]dt \\
s_2 &= \int [r_2 - i_d]dt
\end{aligned} \Rightarrow \begin{aligned}
\dot{s}_1 &= r_1 - i_q \\
\dot{s}_2 &= r_2 - i_d
\end{aligned} \tag{5}$$

where r_1 is the target value of i_q and r_2 is target value of i_d.

Extended state equations of the PMSG are obtained by combing Equations (4) and (5):

$$\begin{bmatrix} \dot{\omega}_r \\ \dot{i}_q \\ \dot{i}_d \\ \dot{s}_1 \\ \dot{s}_1 \end{bmatrix} = \begin{bmatrix} -\frac{B}{J} & -\frac{3P^2\lambda_m}{8J} & 0 & 0 & 0 \\ -\frac{\lambda_m}{L_q} & -\frac{r_s}{L_q} & -\frac{L_d}{L_q}\omega_r & 0 & 0 \\ \frac{L_q}{L_d}i_q & 0 & -\frac{r_s}{L_d} & 0 & 0 \\ 0 & -1 & 0 & 0 & 0 \\ 0 & 0 & -1 & 0 & 0 \end{bmatrix} \begin{bmatrix} \omega_r \\ i_q \\ i_d \\ s_1 \\ s_2 \end{bmatrix} + \begin{bmatrix} 0 & 0 \\ -\frac{1}{L_q} & 0 \\ 0 & -\frac{1}{L_d} \\ 0 & 0 \\ 0 & 0 \end{bmatrix}\begin{bmatrix} v_q \\ v_d \end{bmatrix} + \begin{bmatrix} \frac{PT_I}{2J} \\ 0 \\ 0 \\ r_1 \\ r_2 \end{bmatrix} \tag{6}$$

$$\Rightarrow \dot{x}(t) = Ax(t) + Bu(t) + Ev(t) \tag{7}$$

where $v(t)$ represent disturbances in PMSG systems.

The required output function is:

$$y(t) = \begin{bmatrix} 0 & 1 & 0 & 0 & 0 \\ 0 & 0 & 1 & 0 & 0 \end{bmatrix} \begin{bmatrix} \omega_r \\ i_q \\ i_d \\ s_1 \\ s_2 \end{bmatrix} = Cx(t). \tag{8}$$

The IT2 T-S fuzzy models and IT2 T-S fuzzy controller are combined to form the IT2 T-S fuzzy control systems. The characteristics of IT2 T-S fuzzy control systems employ the upper membership function and lower membership function to represent the model uncertainty of a nonlinear system. In the nonlinear system matrix of the PMSG, state variables i_q and ω_r are found. Therefore, i_q and ω_r are selected as Antecedent z_1 and z_2, respectively. The membership function of the IT2 T-S fuzzy is shown in Figure 2. z_p is the Antecedent variable. a_p, b_p, c_p and d_p are the boundaries of the upper and lower membership functions. In the developed PMSG system, the rating of i_q is 18A and the rating of ω_r is 754 rad/sec. In order to represent the model uncertainty of the PMSG system, 5% variation of the Antecedent variable is selected. Therefore, the Antecedent variables and boundaries of the PMSG are summarized in Table 1.

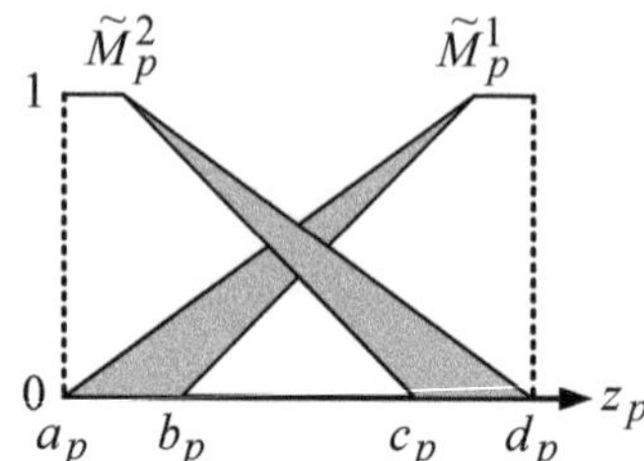

Figure 2. Membership function of the IT2 T-S fuzzy control systems.

Table 1. Antecedent variables and boundaries of the PMSG.

Antecedent \ Boundaries	a_p	b_p	c_p	d_p
z_1	0 A	1.8 A	17.1 A	18.9 A
z_2	0 rad/sec	76 rad/sec	716 rad/sec	792 rad/sec

In the design of the IT2 T-S fuzzy current controller, the nonlinear PMSG system is represented via linear sub-systems according to the model rules of the IT2 T-S fuzzy models:

$$
\begin{aligned}
&\textit{Model rules } i: \\
&\textit{If } z_1(t) \textit{ is } \widetilde{M}_1^i \textit{ and } \cdots \textit{ and } z_p(t) \textit{ is } \widetilde{M}_p^i, \\
&\textit{then } \dot{x}(t) = A_i x(t) + B_i u(t) + E_i v(t), \\
&y = C_i x(t),\ i = 1, 2, \cdots, r
\end{aligned}
\tag{9}
$$

where $\widetilde{M}_p^i$ are IT2 fuzzy sets, $x(t)$ are state variables, $u(t)$ are control inputs, A_i, B_i are state and input matrices of the sub-system, E_i is a constant matrix and $r = 4$ is the number of rules. The firing strength of the i-th rule is represented as follows:

$$
\widetilde{w}^i(z(t)) = \left[\underline{w}_i(z(t)), \overline{w}_i(z(t)) \right],\ i = 1, 2, \cdots, r
\tag{10}
$$

and

$$
\overline{w}_i(z(t)) = \prod_{j=1}^{p} \overline{\mu}_{\widetilde{M}_j^i}(z_j(t))
\tag{11}
$$

$$
\underline{w}_i(z(t)) = \prod_{j=1}^{p} \underline{\mu}_{\widetilde{M}_j^i}(z_j(t))
\tag{12}
$$

$$
\overline{w}_i(z(t)) \geq \underline{w}_i(z(t)) \geq 0,\ \forall i
\tag{13}
$$

where $\overline{w}_i(z(t))$ is the upper grade of membership, $\underline{w}_i(z(t)) \geq 0$ is the lower grade of membership, $\overline{\mu}_{\widetilde{M}_j^i}(z_j(t))$ is the upper membership function and $\underline{\mu}_{\widetilde{M}_j^i}(z_j(t))$ is the lower membership function. Then, the inferred IT2 T-S fuzzy model can be described as:

$$\dot{x}(t) = m\frac{\sum\limits_{i=1}^{r} \overline{w}_i(z(t))(A_i x(t) + B_i u(t) + E_i v(t))}{\sum\limits_{i=1}^{r} \overline{w}_i(z(t))} + n\frac{\sum\limits_{i=1}^{r} \underline{w}_i(z(t))(A_i x(t) + B_i u(t) + E_i v(t))}{\sum\limits_{i=1}^{r} \underline{w}_i(z(t))} \tag{14}$$

$$y(t) = m\frac{\sum\limits_{i=1}^{r} \overline{w}_i(z(t))(C_i x(t))}{\sum\limits_{i=1}^{r} \overline{w}_i(z(t))} + n\frac{\sum\limits_{i=1}^{r} \underline{w}_i(z(t))(C_i x(t))}{\sum\limits_{i=1}^{r} \underline{w}_i(z(t))} \tag{15}$$

where m and n are tuning parameters.

The parallel distributed compensation (PDC) of the IT2 T-S fuzzy controllers corresponding to the model rules are:

$$\text{Control rules } i: If z_1(t) \text{ is } \widetilde{M}_1^i \text{ and...and } z_p(t) \text{ is } \widetilde{M}_p^i, \text{then } u(t) = K_i x(t), \ i = 1, 2, \cdots, r \tag{16}$$

where K_i is the controller gain. The inferred IT2 T-S fuzzy controller can be expressed as:

$$u(t) = m\frac{\sum\limits_{i=1}^{r} \overline{w}_i(z(t))(K_i x(t))}{\sum\limits_{i=1}^{r} \overline{w}_i(z(t))} + n\frac{\sum\limits_{i=1}^{r} \underline{w}_i(z(t))(K_i x(t))}{\sum\limits_{i=1}^{r} \underline{w}_i(z(t))}. \tag{17}$$

The close loop IT2 T-S fuzzy control system can be obtained by substituting Equation (17) into Equation (14).

4. Stability Analysis

Define the H_∞ performance index:

$$\sup_{\|v(t)\|_2 \neq 0} \frac{\|y(t)\|_2}{\|v(t)\|_2} \leq \gamma, \ 0 \leq \gamma \leq 1 \tag{18}$$

where γ represents disturbance suppression ability of the IT2 T-S fuzzy control system.

Lemma [26]: Assume there exists a positive definite matrix $X \in \mathcal{R}^{n \times n}$ and matrix $M_i \in \mathcal{R}^{m \times n}$, which makes the following LMI condition feasible:

$$\begin{bmatrix} \Phi_{iii} & * & * \\ -(mE_i + nE_i)^T & \gamma^2 I & 0 \\ (m+n)C_i X & 0 & I \end{bmatrix} \geq 0, \tag{19}$$

$$\begin{bmatrix} \Phi_{ijj} + \Phi_{jii} & * & * \\ -((m+n)(E_i + E_j))^T & 2\gamma^2 I & 0 \\ ((m+n)(C_i + C_j))X & 0 & 2I \end{bmatrix} > 0, i < j \tag{20}$$

$$\begin{bmatrix} \Phi_{ijk} + \Phi_{ikj} & * & * \\ -(2mE_i + n(E_j + E_k))^T & 2\gamma^2 I & 0 \\ (2mC_i + n(C_j + C_k))X & 0 & 2I \end{bmatrix} \geq 0, j < k \tag{21}$$

where

$$X = P^{-1}, \; M_i = K_i X \tag{22}$$

$$\Phi_{ijk} = -m(A_i X + XA_i^T) - n(A_j X + XA_j^T) + m(B_i M_k + M_k^T B_i^T) + n(B_j M_k + M_k^T B_j^T). \tag{23}$$

Then, the designed current controller in Equation (17) will guarantee the current tracking capability, which means the current tracking error will converge to zero. The detailed proof procedure can be referred in [26]. The controller gain can be found through using $K_i = M_i X^{-1}$.

Let $\gamma = 0.86$, $m = 0.4$ and $n = 0.5$; the following controller gains are obtained from the LMI conditions listed in Equations (19) to (21):

$$
\begin{aligned}
K_1 &= \begin{bmatrix} 1.47332 & -0.28632 & -0.7274 & -893.16 & -0.18312 \\ 0.0222 & -0.7851 & -0.315 & 0.002 & -72.6 \end{bmatrix} \\
K_2 &= \begin{bmatrix} 1.3759 & -0.28656 & -0.2424 & -796.401 & -0.0618 \\ 0.0208 & -0.2617 & -0.31524 & 0.001 & -72.6 \end{bmatrix} \\
K_3 &= \begin{bmatrix} 1.5066 & -0.28668 & -0.7274 & -1414.17 & -0.18312 \\ 0.0021 & -0.7851 & -0.31536 & 0.002 & -72.6 \end{bmatrix} \\
K_4 &= \begin{bmatrix} 1.3759 & -0.2391 & -0.2425 & -796.401 & -0.0618 \\ 0.0007 & -0.2617 & -0.3156 & 0.001 & -69.3 \end{bmatrix}
\end{aligned}
\tag{24}
$$

5. Results and Discussions

The parameters of the PMSG are listed in Table 2. Table 3 shows the specifications of the PMSG drive. Figure 3 demonstrates the experimental equipment of the PMSG system based on the IT2 T-S fuzzy systems. The adopted PMSG is the model YBL17B-200L manufactured by YELI electric & machinery Co., LTD, Taiwan. The oscilloscope is KEYSIGHT DSO-X 3014T (Keysight Technologies, Santa Rosa, CA, USA). The current measurement system includes current probe Tektronix TCP303 and amplifier Tektronix TCPA300. Constant and variable current commend experimental results are given as follows.

Table 2. Parameters of the PMSG.

Poles	r_s	L_d	L_q
8	0.24 Ω	1.896 mH	2.131 mH
Rated Speed	**Rated Torque**	**Rated Current**	**Rated Power**
1800 rpm	23 N·m	11.8 Arms	4.5 kW

Table 3. Specifications of the PMSG Drive.

Rated power	5 kW	DC-link voltage	380 Vdc
Rated voltage	220 Vrms	DC-link capacitance	5600 μF
Rated current	13.1 Arms	Switching frequency	20 kHz

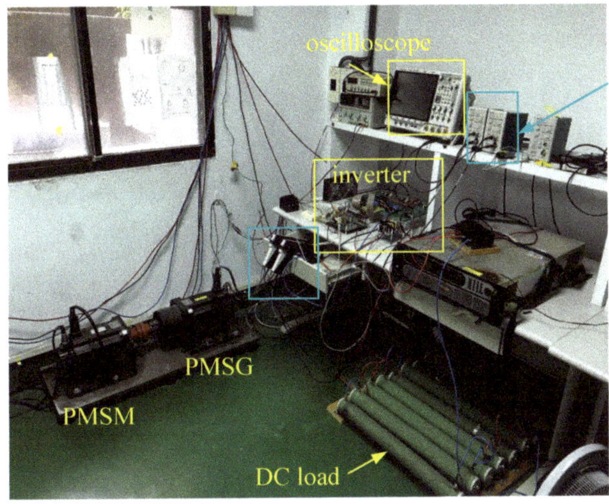

Figure 3. Experimental equipment for the PMSG system based on the IT2 T-S fuzzy systems.

5.1. Constant Current Command

Let the current command be $i_q^* = 15\,\text{A}$, $i_d^* = 0\,\text{A}$. In this case, the three-phase winding current will be balanced with peak value 15 A. Figure 4a–c show the winding current waveforms measured at different generator speeds. Table 4 summarizes the error and total harmonic distortion (THD) of the measured waveforms. It can be found that the current command tracking error is less than 2.5% in all conditions. The winding current can exactly track the current command. Moreover, the THD is less than 2%, and this means winding currents are nearly sinusoidal.

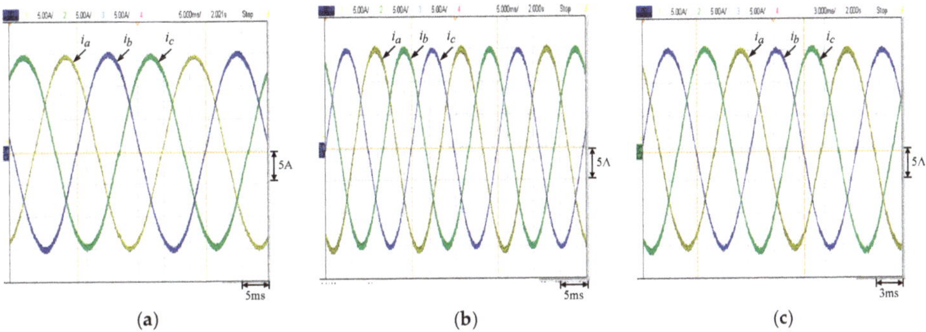

Figure 4. Winding current waveforms by letting $i_q^* = 15$ A and $i_d^* = 0$ A at different generator speeds: (**a**) 600 rpm, (**b**) 900 rpm and (**c**) 1200 rpm.

Table 4. Error and THD of the measured waveforms.

Generator Speed	Measured Current (A)	Error (%)	THD (%)
600 rpm	14.849	1.01	1.35
900 rpm	15.179	1.19	1.20
1200 rpm	15.368	2.45	1.50

Note: $\text{error}(\%) = \left| \frac{\text{measured current} - \text{current command}}{\text{current command}} \right| \times 100\%$, $\text{THD}(\%) = \left(\frac{\text{THD}_a + \text{THD}_b + \text{THD}_c}{3} \right)$.

To improve the scientific merit of this paper, the simulation is performed by PSIM Ver. 10. The same conditions with Figure 4a–c are simulated and shown in Figure 5a–c, respectively. From the simulation results, it can be found that the three-phase winding currents are balanced with peak value 15 A. Moreover, the THD is less than 2%; this means winding currents are nearly sinusoidal.

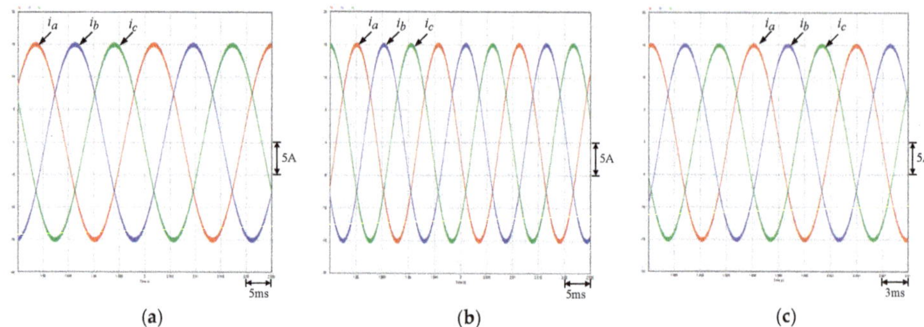

Figure 5. Simulation winding current waveforms by letting $i_q^* = 15$ A and $i_d^* = 0$ A at different generator speeds: (**a**) 600 rpm, (**b**) 900 rpm and (**c**) 1200 rpm.

5.2. Variable Current Command

First, the current command i_q^* is stepped from 5 A to 10 A ($i_d^* = 0$) at 900 rpm. Figure 6a is the step occurring at 60° and Figure 6b is the step occurring at 120°. The results using the IT1 T-S fuzzy system are compared in Figure 6c,d, respectively. The overshoot and settling time of the IT1 and IT2 T-S fuzzy systems are summarized in Table 5. It can be found that the winding currents can achieve the step current command faster by using the proposed IT2 T-S fuzzy control. The overshoot of the IT2 T-S fuzzy system is also less than the IT1 T-S fuzzy system, which makes the developed system more reliable and flexible.

Table 5. Overshoot and settling time of the IT1 and IT2 T-S fuzzy systems.

Step Position	Control System	Overshoot	Settling Time
Step occurs at 60°	IT1 T-S fuzzy	2.125 A	1.12 ms
	IT2 T-S fuzzy	2.03 A	1.03 ms

Note: overshoot = (maximum current − desired current) settling time: the required time when current tracking error is less than 5%.

Second, the speed is 1200 rpm and the current command is going from zero to a constant value. Figure 7a,b show the winding current waveforms of the current command that goes from zero to a constant value. The current tracking capability is verified.

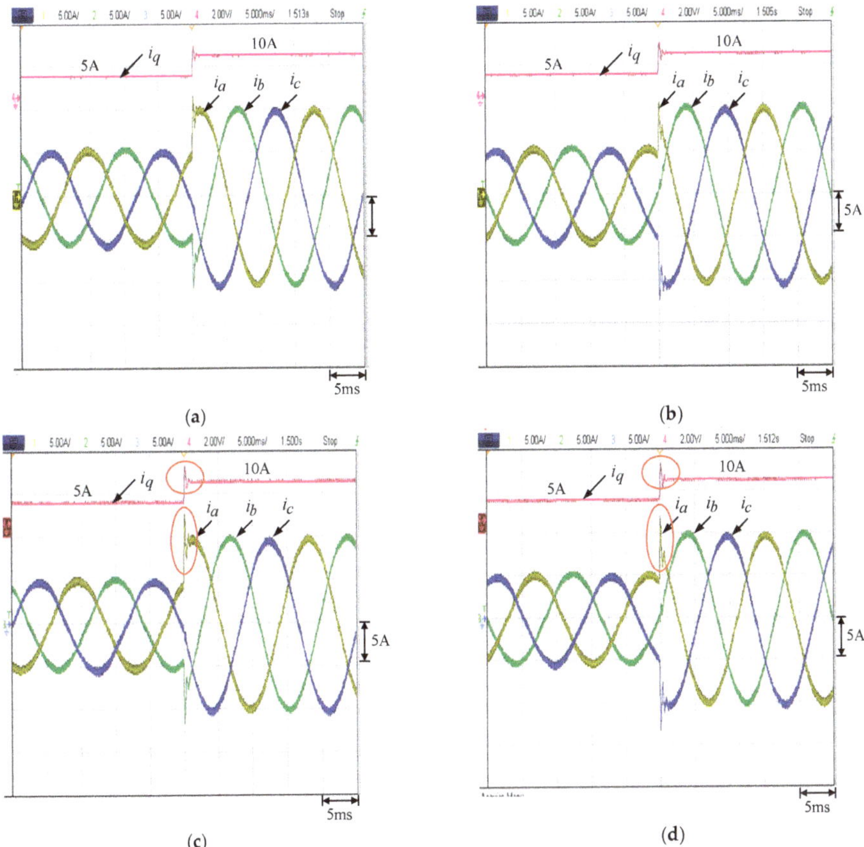

(a)

(b)

(c)

(d)

Figure 6. Winding current waveforms of the current command stepped from 5 A to 10 A at 900 rpm: (**a**) step occurs at 60° with IT2 T-S fuzzy, (**b**) step occurs at 120° with IT2 T-S fuzzy, (**c**) step occurs at 60° with IT1 T-S fuzzy and (**d**) step occurs at 120° with IT1 T-S fuzzy.

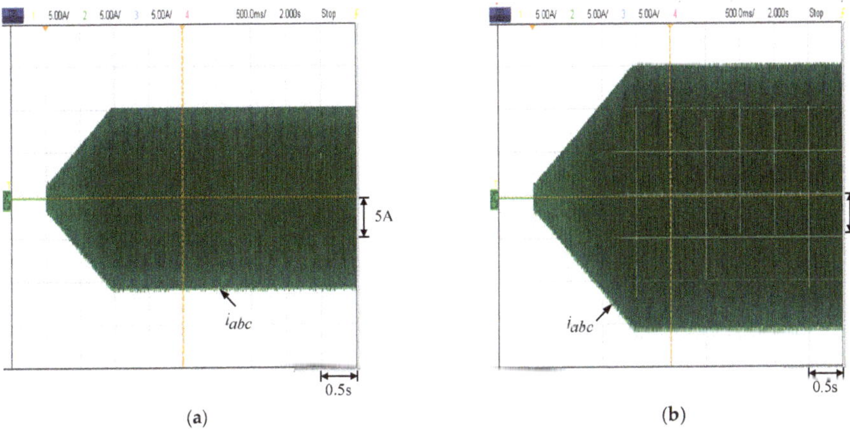

(a)

(b)

Figure 7. Winding current waveforms of the current command goes from zero to a constant value at 1200 rpm: (**a**) 10 A; (**b**) 15 A.

Finally, the speed is 900 rpm and 1500 rpm, respectively, and the current command is varying as 0 A→5 A→10 A→15 A→10 A→5 A. Figure 8a,b show the winding current waveforms, respectively. It is obvious that the winding current can track the variable current command very well. The output power of the PMSG can be adjusted by changing the current command in this situation.

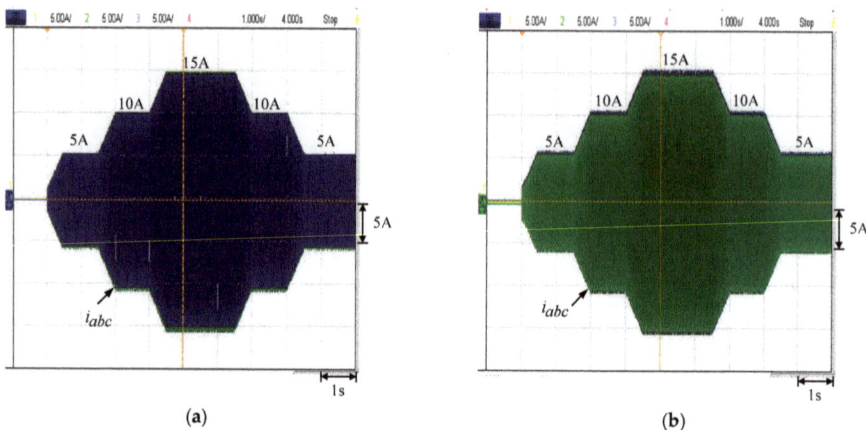

(a)   (b)

Figure 8. Winding current waveforms of the current command is varying as 0 A→5 A→10 A→15 A→10 A→5 A: (**a**) 900 rpm; (**b**) 1500 rpm.

5.3. Calculation Time

The calculation time of the IT2 T-S fuzzy and the IT1 T-S fuzzy systems is compared in Figure 9a,b. The calculation time of the IT2 T-S fuzzy is 32.6 μs and the calculation time of the IT1 T-S fuzzy is 30 μs. The IT2 T-S fuzzy system requires 2.6 μs more than the IT1 T-S fuzzy system to process the control algorithm. The switching period of the developed system is 50 μs. Therefore, the proposed algorithm is acceptable in implementation.

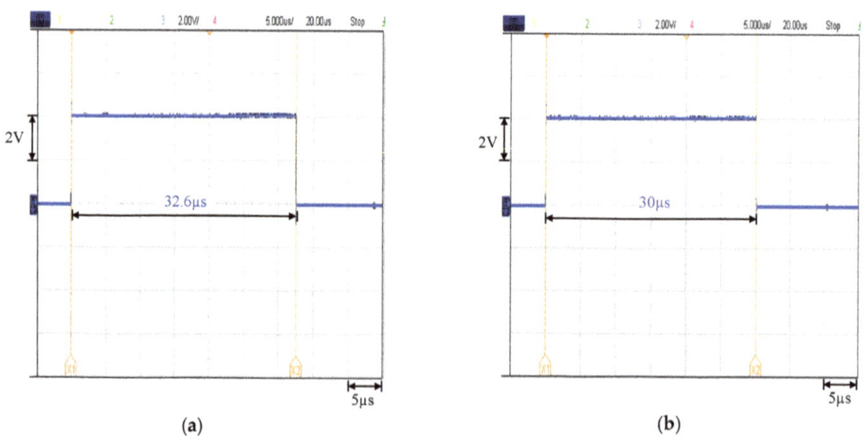

(a)   (b)

Figure 9. Calculation time of the IT2 T-S fuzzy and the IT1 T-S fuzzy systems: (**a**) IT2 T-S fuzzy; (**b**) IT1 T-S fuzzy.

6. Conclusions

The current control of the PMSG was designed and implemented based on the IT2 T-S fuzzy systems. First, the system configuration and the dynamic model were introduced. Next, the current controller was designed based on the IT2 T-S fuzzy models. The IT2 T-S fuzzy control system was implemented to consider the uncertainty of nonlinear systems. The stability analysis and detailed design process were also demonstrated. The controller gain could be found by using the LMI conditions. Furthermore, the experimental equipment of the PMSG was illustrated. Experimental results, including constant current command, variable current command and calculation time, were demonstrated to verify the performance of the proposed current control. Simulation results were also performed in constant current command to improve the scientific merit of this paper. The IT2 T-S fuzzy system is more complex than the IT1 T-S fuzzy system. However, its calculation time is acceptable. Furthermore, its overshoot and settling time under the current command variation is better than the IT1 T-S fuzzy system. The output power of a PMSG can be adjusted by changing the peak value of three-phase balanced winding currents. In the future, the sensorless control combined with the IT2 T-S fuzzy systems may be adopted to increase the practicability of the PMSG system.

Author Contributions: The authors contributed equally to this work.

Funding: This research was supported by the Ministry of Science and Technology, Taiwan, under the Grant of MOST 108-2623-E-194-001-D.

Conflicts of Interest: The authors declare no conflict of interest.

References

1. Krause, P.; Wasynczuk, O.; Sudhoff, S.; Pekarek, S. *Analysis of Electric Machinery and Drive Systems*, 3rd ed.; IEEE Press: Piscataway, NJ, USA, 2013; pp. 121–141. ISBN 9781118024294.
2. Hong, J.; Lee, H.; Nam, K. Charging method for the secondary battery in dual-inverter drive systems for electric vehicles. *IEEE Trans. Power Electron.* **2015**, *30*, 909–921. [CrossRef]
3. Lai, Y.S.; Lee, W.T.; Lin, Y.K.; Tsai, J.F. Integrated inverter/converter circuit and control technique of motor drives with dual-mode control for EV/HEV Applications. *IEEE Trans. Power Electron.* **2014**, *29*, 1358–1365. [CrossRef]
4. Lee, K.W.; Park, S.; Jeong, S. A seamless transition control of sensorless PMSM compressor drives for improving efficiency based on a dual-mode operation. *IEEE Trans. Power Electron.* **2015**, *30*, 1446–1456. [CrossRef]
5. Yaramasu, V.; Wu, B. Predictive control of a three-level boost converter and an NPC inverter for high-power PMSG-based medium voltage wind energy conversion systems. *IEEE Trans. Power Electron.* **2014**, *29*, 5308–5322. [CrossRef]
6. García-Gracia, M.; Cova, M.A.; Villen, M.T.; Uson, A. Novel modular and retractable permanent magnet motor/generator for flywheel applications with reduced iron losses in stand-by mode. *IET Renew. Power Gener.* **2014**, *8*, 551–557. [CrossRef]
7. Jung, E.; Yoo, H.; Sul, S.K.; Choi, H.S.; Choi, Y.Y. A nine-phase permanent-magnet motor drive system for an ultrahigh-speed elevator. *IEEE Trans. Ind. Appl.* **2016**, *48*, 987–995. [CrossRef]
8. Reddy, P.B.; El-Refaie, A.M.; Huh, K.K. Effect of number of layers on performance of fractional-slot concentrated-windings interior permanent magnet machines. *IEEE Trans. Power Electron.* **2015**, *30*, 2205–2218. [CrossRef]
9. Do, T.D.; Kwak, S.; Choi, H.H.; Jung, J.-W. Suboptimal control scheme design for interior permanent-magnet synchronous motors: An SDRE-Based approach. *IEEE Trans. Power Electron.* **2014**, *29*, 3020–3031. [CrossRef]
10. Pellegrino, G.; Vagati, A.; Guglielmi, P.; Boazzo, B. Performance comparison between surface-mounted and interior PM motor drives for electric vehicle applications. *IEEE Trans. Ind. Electron.* **2012**, *59*, 803–811. [CrossRef]
11. Wang, Z.; Lu, K.; Blaabjerg, F. A simple startup strategy based on current regulation for back-EMF-based sensorless control of PMSM. *IEEE Trans. Power Electron.* **2012**, *27*, 3817–3825. [CrossRef]
12. Jahns, T.M.; Soong, W.L. Pulsating torque minimization techniques for permanent magnet AC motor drives—A review. *IEEE Trans. Ind. Electron.* **1996**, *43*, 321–330. [CrossRef]
13. Kim, K.C.; Lee, J.; Kim, H.J.; Koo, D.H. Multiobjective optimal design for interior permanent magnet synchronous motor. *IEEE Trans. Magn.* **2009**, *45*, 1780–1783. [CrossRef]

14. Islam, R.; Husain, I.; Fardoun, A.; McLaughlin, K. Permanent-magnet synchronous motor magnet designs with skewing for torque tipple and cogging torque reduction. *IEEE Trans. Ind. Appl.* **2009**, *45*, 152–160. [CrossRef]

15. Chang, Y.C.; Wang, S.Y.; Dai, W.F.; Chang, H.F. Division-summation current control and one-cycle voltage regulation of the surface-mounted permanent-magnet synchronous generator. *IEEE Trans. Power Electron.* **2016**, *31*, 1391–1400. [CrossRef]

16. Hwang, J.-C.; Wei, H.-T. The current harmonics elimination control strategy for six-leg three-phase permanent magnet synchronous motor drives. *IEEE Trans. Power Electron.* **2014**, *29*, 3032–3040. [CrossRef]

17. Chou, M.C.; Liaw, C.M. Development of robust current 2-DOF controllers for a permanent magnet synchronous motor drive with reaction wheel load. *IEEE Trans. Power Electron.* **2009**, *24*, 1304–1320. [CrossRef]

18. Uddin, M.N.; Radwan, T.S.; George, G.H.; Rahman, M.A. Performance of current controllers for VSI-fed IPMSM drive. *IEEE Trans. Ind. Appl.* **2000**, *36*, 1531–1538.

19. Liaw, C.M.; Kang, B.J. A robust hysteresis current-controlled PWM inverter for linear PMSM driven magnetic suspended positioning system. *IEEE Trans. Ind. Electron.* **2001**, *48*, 956–967. [CrossRef]

20. Weigold, J.; Braun, M. Predictive current control using identification of current ripple. *IEEE Trans. Ind. Electron.* **2008**, *55*, 4346–4353. [CrossRef]

21. Chang, Y.C.; Chen, C.H.; Zhu, Z.C.; Huang, Y.W. Speed control of the surface-mounted permanent-magnet synchronous motor based on Takagi-Sugeno fuzzy models. *IEEE Trans. Power Electron.* **2016**, *31*, 6504–6510. [CrossRef]

22. Chang, Y.C.; Chang, H.C.; Huang, C.Y. Design and implementation of the permanent-magnet synchronous generator drive in wind generation systems. *Energies* **2018**, *11*, 1634. [CrossRef]

23. Kamiński, M.; Corigliano, A. Numerical solution of the Duffing equation with random coefficients. *Meccanica* **2015**, *50*, 1841–1853. [CrossRef]

24. Karnik, N.N.; Mendel, J.M.; Liang, Q. Type-2 fuzzy logic systems. *IEEE Trans. Fuzzy Syst.* **1999**, *7*, 643–658. [CrossRef]

25. Mendel, J.M.; John, R.I.; Liu, F. Interval type-2 fuzzy logic systems made simple. *IEEE Trans. Fuzzy Syst.* **2006**, *14*, 808–821. [CrossRef]

26. Lai, C.C. Design and Implementation of a Single-Phase Bidirectional Inverter Using Interval Type-2 T-S Fuzzy Control Systems. Master's Thesis, National Chung Cheng University, Chiayi, Taiwan, 2016.

Article

The Optimal Energy Dispatch of Cogeneration Systems in a Liberty Market

Whei-Min Lin [1], Chung-Yuen Yang [1], Chia-Sheng Tu [2], Hsi-Shan Huang [2] and Ming-Tang Tsai [3,*]

[1]   Department of Electrical Engineering, National Sun Yat-Sen University, Kaohsiung 807, Taiwan
[2]   College of Intelligence Robot, Fuzhou Polytechnic, Fujian 350108, China
[3]   Department of Electrical Engineering, Cheng-Shiu University, Kaohsiung 833, Taiwan
*   Correspondence: k0217@gcloud.csu.edu.tw; Tel.: +886-7-7310606

Received: 20 June 2019; Accepted: 20 July 2019; Published: 25 July 2019

Abstract: This paper proposes a novel approach toward solving the optimal energy dispatch of cogeneration systems under a liberty market in consideration of power transfer, cost of exhausted carbon, and the operation condition restrictions required to attain maximal profit. This paper investigates the cogeneration systems of industrial users and collects fuel consumption data and data concerning the steam output of boilers. On the basis of the relation between the fuel enthalpy and steam output, the Least Squares Support Vector Machine (LSSVM) is used to derive boiler and turbine Input/Output (I/O) operation models to provide fuel cost functions. The CO_2 emission of pollutants generated by various types of units is also calculated. The objective function is formulated as a maximal profit model that includes profit from steam sold, profit from electricity sold, fuel costs, costs of exhausting carbon, wheeling costs, and water costs. By considering Time-of-Use (TOU) and carbon trading prices, the profits of a cogeneration system in different scenarios are evaluated. By integrating the Ant Colony Optimization (ACO) and Genetic Algorithm (GA), an Enhanced ACO (EACO) is proposed to come up with the most efficient model. The EACO uses a crossover and mutation mechanism to alleviate the local optimal solution problem, and to seek a system that offers an overall global solution using competition and selection procedures. Results show that these mechanisms provide a good direction for the energy trading operations of a cogeneration system. This approach also provides a better guide for operation dispatch to use in determining the benefits accounting for both cost and the environment in a liberty market.

Keywords: cogeneration systems; Time-of-Use (TOU); CO_2 emission; Ant Colony Optimization

---

1. Introduction

Cogeneration systems, which combine heat and power (CHP) systems, have previously been extensively applied in industry. They offer an economic strategy providing both heat and power, which can then be passed on to buyers. Cogeneration systems offer a significant advantage when it comes to consideration of environmental issues. They are used as a distributed energy source, which can simultaneously sell both thermal steam and electricity to other industries. They can also be constructed in urban areas and used as distributed energy resources in microgrids [1–3]. In recent decades, consolidated cogeneration solutions have been used in industrial applications [4], while cogeneration system applications continue to grow. However, more experience is required in order to achieve the most efficient and energy-saving operation of these systems. To improve the competiveness of cogeneration systems in a liberalized market, an efficient tool for achieving the optimal operation of these systems must be developed.

To date, several efficiency strategies have been developed to achieve this optimal operation [5–19]. Ref. [5] presented a generalized network programming (GNP) to perform the economic dispatch of electricity and steam in a cogeneration plant. Ref. [6] presented an dispatching scheme which economically transfers energy between facilities and utilities. An optimal operation of the cogeneration system is proposed, which will integrate energy into the electricity grid by using the decision-making technique [7]. Ref. [8] is used to assess the potential process of using micro-cogeneration systems based on Stirling engines. The results demonstrate that a numerical analysis of the Stirling engine can accurately indicate the operation of the actual machine. Ref. [9] used a non-linear programming with Time-of-Use (TOU) rates considered during operation. Ref. [10] presented the operation of steam turbines experiencing blades failures during peak load times of the summer months at a cogeneration plant. The author of [11] applied some possible technologies to integrate pulp and paper production within the context of a high-efficiency cogeneration system. Grey wolf optimization [12] and cuckoo search algorithms [13] are proposed to simultaneously solve the economic logistics of using a combined heat and power system. Ref. [14,15] presented the suggested economical operation of a cogeneration system under control, with resultant multi-pollutants from a fossil-fuels-based thermal energy generation. The author of [16] introduced an original framework based on identifying the characteristics of small-scale and large-scale uncertainties, whereby a comprehensive approach based on multiple time frames was formulated. Ref. [17] developed a tool for long-term optimization of cogeneration systems based on mixed integer linear-programming and a Lagrangian relation. Ref. [18] proposed an enhanced immune algorithm to solve the scheduling of cogeneration plants in a deregulated market. Ref. [19] addressed an optimal strategy for the daily energy exchange of a 22-MW combined-cycle cogeneration plant in a liberty market.

One of the key issues of a cogeneration operation is heat and power modeling. In the papers described above, pure power dispatch was a major objective. Inevitably, though, more design objectives coupled with higher constraints will have to be incorporated. The energy trading dispatch of cogeneration systems is a complicated process, especially when the solution is being sought in a world of uncertainty. Conventional methods have thus become more difficult to solve. Recently, artificial intelligence (AI) has been applied in the economic dispatch of cogeneration systems [20–23]. The strategies proposed by AI algorithms must consider computer execution efficiency and a large computing space. Conventional algorithms may be faster, but are very often limited by the problem structure, and may diverge, or lead to a local minimum. This paper therefore proposes an Enhanced Ant Colony Optimization (EACO) to solve the energy trading dispatch of cogeneration systems.

Ant Colony Optimization (ACO) applies the activity characteristics of biotic populations to search optimization problems [24,25]. When ants are foraging, they not only refer to their own information but also learn from the most efficient ants in order to correct their route. They learn and exchange their information to search for the shortest route between their colony and food sources, and pass this information on until the whole ant colony reaches optimal status. The advantages of the ACO algorithm are that individual solutions within a range of possible solutions can converge to discover the optimal solution through a small number of evolution iterations. ACO has previously been applied to the economic dispatch of power systems [26–30]; however, while ACO is good at global searches, the populations produced are still a dilemma. In this paper, an EACO algorithm is proposed to improve this search ability. In the EACO, the crossover and mutation mechanisms [31] are used to generate offspring equipped to escape the local optimum. This paper proposes the use of EACO to solve the energy trading dispatch of the cogeneration systems by considering the TOU rate [32]. The different carbon prices of CO_2 emissions are also simulated and analyzed in the energy trading dispatch of these cogeneration systems. It can show the performance of the energy trading dispatch of the cogeneration systems to obtain the maximal profit.

2. Problem Formulation

Figure 1 shows M back pressure steam turbines, N extraction condensing steam turbines, and K high pressure steam boilers, where the high and medium pressure steam systems are connected by a common pipeline. Part of the generated electricity is supplied to the service power, and the excess electricity is sold to the power company. Input/Output (I/O) models of the boiler and turbine are described as follows.

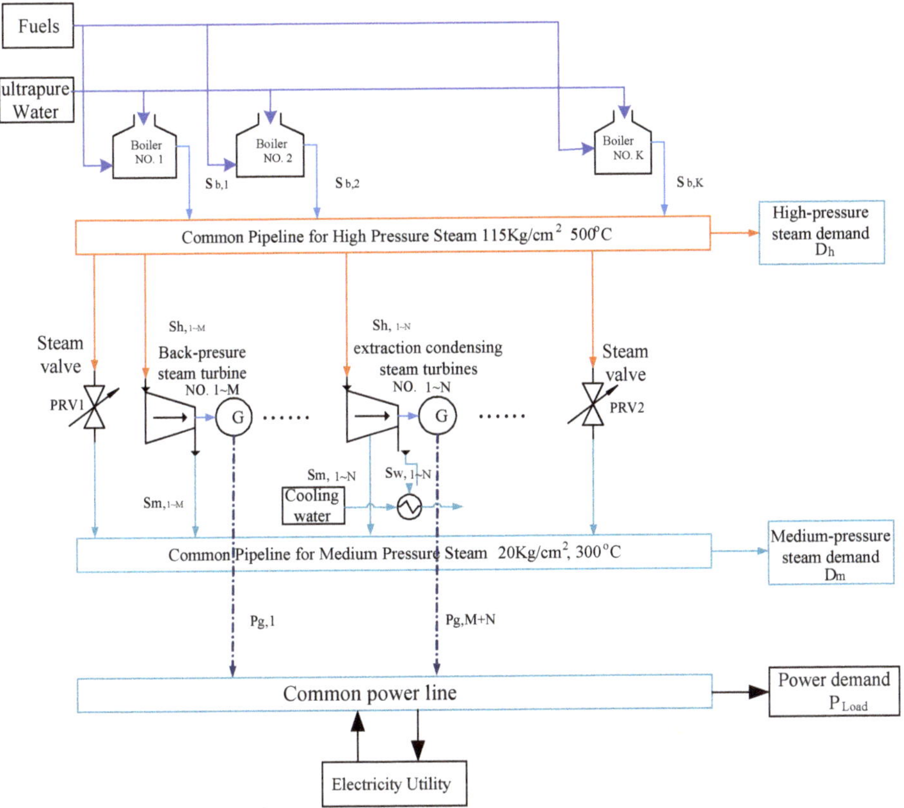

Figure 1. Energy flow of a cogeneration system.

2.1. Operation Models of Boilers

By using the Least-Square Support Vector Machine (LSSVM) [33], the operation models of boilers can be calculated from boiler operational records as shown in Figure 2. The temperature, pressure, fuel consumption, and steam generation for a real cogeneration system are measured from the operational data of boilers. The LSSVM is used for model training, and the input layer data are transferred to the output layer through the LSSVM. In general, using the Radial Basis Function Network (RBFN) kernel function, $K(x, y) = e^{(-\sigma^2 |x-y|^2)}$, can yield a good prediction of the LSSVM model. Therefore, we adopt the RBFN kernel function as the kernel function of the LSSVM model. The error is calculated by using Mean Absolute Percentage Error (MAPE) as shown in Equation (1):

$$\text{MAPE} = \frac{1}{T} \sum_{t=1}^{T} \frac{\left| S_t^A - S_t^F \right|}{S_t^A} \times 100\% \tag{1}$$

where S_t^A is t actual operation data to be predicted S_t^F is the t operation data constructed with LSSVM, and T is total training time.

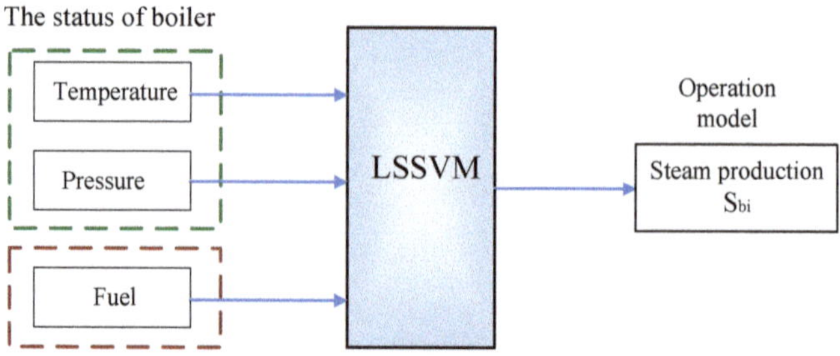

Figure 2. Operation model of boilers using the LSSVM.

2.2. Operation Model of Steam Turbines

The steam from back-pressure steam turbines flows to the single inlet. After the high temperature and high pressure steam enters the steam turbine, the pressure reduces, the volume expands, and the temperature reduces. The steam in the outlet end is the 20 kg/300 °C medium pressure steam required for the process. The operation of the back-pressure steam turbine relies on the relationship between steam flow at the outlet and the generated electricity; the operation model of the back-pressure steam turbine is constructed by LSSVM, as shown in Figure 3.

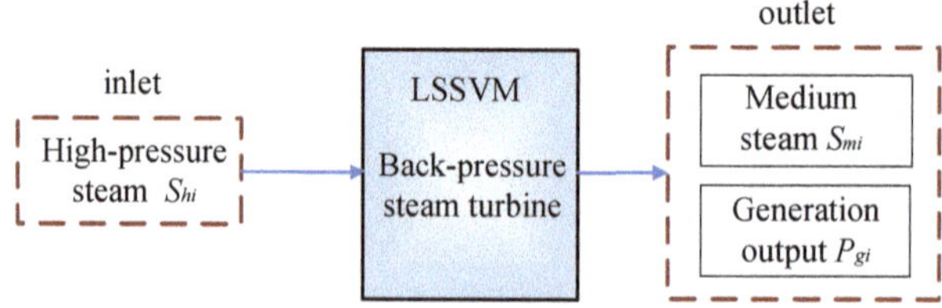

Figure 3. Operation model of the back-pressure steam turbine using the LSSVM.

Extraction condensing turbines are different from the backpressure turbines, meaning the extraction condensing turbines have a single steam inlet. In the multi sections, after extraction of the middle/low pressure steam and exhaust of the steam at the final section, the condensing turbines are shown as in Figure 1. LSSVM is used to construct the generated electricity functions between the process steam outlet flow and the steam flow at the condensing section. The operation model constructed by LSSVM is shown in Figure 4.

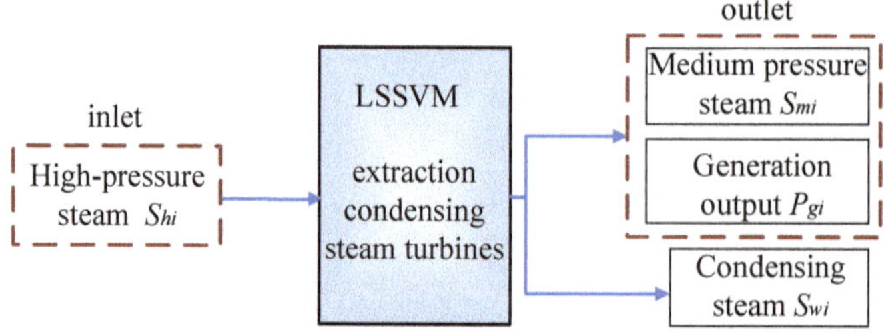

Figure 4. Operation model of the extraction condensing steam turbines using LSSVM.

2.3. Emission Model

CO_2 emission models may be defined based upon the amount of fuel consumed. The model of emission for CO_2 is formulated by the IPCC [34] as:

$$E_{co2,i}(t) = H\big(P_{gi}(t)\big) \times 4.1868 \times \frac{44}{12} \times CEP_i \times COR_i \tag{2}$$

$H\big(P_{g,i}(t)\big) = d_i + e_i P_{gi}(t) + f_i P_{gi}^2(t) + g_i P_{gi}^3(t)$, approximated by the three order function gives the thermal conductivity of each kind of unit. d_i, e_i, f_i, g_i are the coefficients of the emission of unit i. CEP_i is the carbon emission parameter of unit i (21.1 kg-C/GJ for oil, 25.8 kg-C/GJ for coal, 15.3 kg-C/GJ for natural gas) and COR_i is the carbon oxidizing rate of unit i (0.99 for oil, 0.98 for coal, 0.995 for natural gas).

2.4. Objective Function and Constraints

The purpose of the proposed scheme is to maximize profit while satisfying operational constraints. The objective function, including profit from steam sold, profit from electricity sold, fuel costs, emissions costs, wheeling costs, and water costs, is formulated in Equation (3):

$$\max TC(\cdot) = \sum_{t=1}^{T} \big[TC_{steam}(t) + TC_{elect}(t) - TC_{fuel}(t) - TC_{emiss}(t) - TC_{tran}(t) - TC_{water}(t) \big] \tag{3}$$

(1) The profit from thermal steam sold:

$$TC_{steam}(t) = S_{out}(t) \times Steam_{cost} \tag{4}$$

(2) The profit from electricity sold:

$$TC_{elect}(t) = TOU(t) \times P_{tie}(t) \tag{5}$$

(3) The cost of fuel:

$$TC_{fuel}(t) = \sum_{i=1}^{K} F_{bi}(S_{bi}(t)) \times Fuel\_cos\,t \tag{6}$$

(4) Emission costs:

$$TC_{emiss}(t) = C_C \times \sum_{i=1}^{K} E_{CO2,i}(t) \tag{7}$$

(5) Wheeling costs:

$$TC_{tran}(t) = WUC \times P_{tie}(t) \tag{8}$$

(6) The cost of pure water:

$$TC_{water}(t) = WC \times W_b(t) \tag{9}$$

C_c: The charged emission price for CO_2.(NT\$235/ton) [35]; $F_{bi}(S_{bi}(t))$: consumed enthalpy of the i steam boiler at t hour; $Fuel\_cost$: The fuel cost(NT\$/MBTU) for coal, gas, and oil; $S_{out}(t)$: The thermal steam sold at time t (ton/h); $Steam_{cost}$: The price of thermal steam sold (NT\$/ton); $P_{tie}(t)$: The electricity sold to utility at time t; $TOU(t)$: Time-of-Use rate (NT\$/KWH), as shown in Table 1 [32]; $Wb(t)$: The water used by boilers at time t (ton/h); WC: The cost of water (NT\$/ton); WUC: the wheeling cost (NT\$/MWH).

Table 1. The time-of-use rates of Taipower Company.

	Electricity Sale Rate (NT\$/KWH)		Utility Buy-Back Rate NT\$/KWH
	Level 1	Level 2	
Peak Period	3.04	2.7480	3.04
Semi-peak Period	1.83	1.5767	1.83
Off-peak Period	0.69	0.4729	0.69

Level 1: power exported at under 20% rated capacity; Level 2: power exported at over 20% rated capacity.

The operation constraints are considered as follows:

(1) Power balance in the power system:

$$\sum_{i=1}^{M+N} P_{gi}(t) - P_{ite}(t) - P_{load}(t) = 0 \tag{10}$$

(2) Steam balance for boilers, turbines, sold, and industrial processes:

$$\sum_{i=1}^{K} S_{bi}(t) - D_h(t) - \sum_{i=1}^{M+N} S_{hi}(t) = 0 \tag{11}$$

$$\sum_{i=1}^{M+N} S_{hi}(t) - \sum_{i=1}^{M+N} S_{mi}(t) - \sum_{i=1}^{N} S_{wi}(t) = 0 \tag{12}$$

$$\sum_{i=1}^{M+N} S_{mi}(t) - D_m(t) - S_{out}(t) = 0 \tag{13}$$

(3) Operation constraints for boilers, steam turbines and power generation:

$$S_{bi}^{min} \leq S_{bi}(t) \leq S_{bi}^{max} \quad , \quad i = 1,2,3,\ldots, K \tag{14}$$

$$S_{hi}^{min} \leq S_{hi}(t) \leq S_{hi}^{max} \quad , \quad i = 1,2,3,\ldots, M+N \tag{15}$$

$$S_{mi}^{min} \leq S_{mi}(t) \leq S_{mi}^{max} \quad , \quad i = 1,2,3,\ldots, M+N \tag{16}$$

$$S_{wi}^{min} \leq S_{wi}(t) \leq S_{wi}^{max} \quad , \quad i = 1,2,3,\ldots, N \tag{17}$$

$$P_{gi}^{min} \leq P_{gi}(t) \leq P_{gi}^{max} \quad , \quad i = 1,2,3,\ldots, M+N \tag{18}$$

$D_h(t)$, $D_m(t)$: The high/medium pressure steam demands of industry at time t (T/h); $P_{load}(t)$: The load of cogeneration system at time t; S_{bi}^{min}, S_{bi}^{max}: Minimal/maximal limits of steam for boiler i; S_{hi}^{min}, S_{hi}^{max}: Minimal/maximal limits of high pressure steam for turbine i; S_{mi}^{min}, S_{mi}^{max}: Minimal/maximal limits of medium pressure extraction steam for turbine i; S_{wi}^{min}, S_{wi}^{max}: Minimal/maximal limits of medium pressure exhausted steam for turbine i; P_{gi}^{min}, P_{gi}^{max}: Minimal/maximal limits of the generated power for turbine i.

3. Solution Algorithm

This study proposes an EACO, which combines the ACO and Genetic Algorithm (GA), in order to achieve the optimal energy trading dispatch of a cogeneration system. Crossover and mutation mechanisms are integrated into the ACO procedure, and serve to generate offspring in order to escape from the local optimum. The EACO procedure applied in the energy trading dispatch of a cogeneration system is described as follows.

(1) Input Data

Input data includes high/medium steam demand, internal load, plant type, plant capacity, and number of plants.

(2) Set EACO Parameters

EACO parameters include the population of ants (k), the number of generations (G), initial pheromone ($\tau_0 = 0.1$), the relative influence of the pheromone trail ($\alpha = 1$), the relative influence of the heuristic information ($\beta = 2$), and the pheromone evaporation rate ($\rho = 0.5$).

(3) Initialized Individuals

$\mathcal{R}_s^i = \left\{ S_b^i, S_m^i, S_w^i \right\}$ is an individual, $i = 1, 2, \ldots, k$. k, which is the number of ants, is set to 30 in our study. s is the number of parameters. All individuals are set between the minimal and maximal limits with a uniform distribution as shown in Equation (19). The fitness score of each $\mathcal{R}_s^i$ is obtained by calculating the objective function ($TC(\cdot)$) by considering Equations (10)~(18). The fitness values were arranged in descending order from the maximum ($TC(\mathcal{R}_s^i)_{max}$) to the minimum ($TC(\mathcal{R}_s^i)_{min}$).

$$\mathcal{R}_s^i = \mathcal{R}_{smin}^i + \text{Rand} \times \left(\mathcal{R}_{smax}^i - \mathcal{R}_{smin}^i \right) \tag{19}$$

Rand: The uniform random number in (0,1).

(4) The State Transition Rule

The state-based ants are generated according to the level of pheromone and constrained conditions. The transition probability for the $k - ant$ from one state s to the next j is at the $t - th$ interval, as given in Equation (20):

$$P_{t,sj}^k(g) = \begin{cases} \dfrac{[\tau_{t,sj}(g)]^\alpha \times [\eta_{t,sj}(g)]^\beta}{\sum_{l \in N_t^k(s)} [\tau_{t,sl}(g)]^\alpha \times [\eta_{t,sl}(g)]^\beta} \, , & \text{if } j \in N_t^k(s) \\[4mm] 0 & , \quad \text{others} \end{cases} \tag{20}$$

where $\eta_{t,sj}(g)$ and $\eta_{t,sl}(g)$ are the inverse of the edge distance at the $g - th$ generation, which are defined as Equations (21) and (22):

$$\eta_{t,sl}(g) = \frac{1}{\left| TC\left(\mathcal{R}_{t,s} \right) - TC\left(\mathcal{R}_{t,optimal} \right) \right|} \tag{21}$$

$$\eta_{t,sl}(g) = \frac{1}{\left| TC\left(\mathcal{R}_{t,s} \right) - TC\left(\mathcal{R}_{t,l} \right) \right|} \, , \quad l \in N_t^k(s) \tag{22}$$

$TC(\cdot)$ is the objective function as given in Equation (3). $TC\left(\mathcal{R}_{t,s} \right)$ and $TC\left(\mathcal{R}_{t,l} \right)$ are the score of the $s - th$ and $l - th$ individuals at the $t - th$ interval, and $TC\left(\mathcal{R}_{t,optimal} \right)$ is the optimal fitness score at the $t - th$ interval. $N_t^k(s)$ is the number of feasible individuals at the $t - th$ interval.

$\tau_{t,sj}(g)$ and $\tau_{t,sl}(g)$ are the pheromone intensity on edge (s, j) and edge (s, l) at the $g - th$ generation. Ant k positioned on state s chooses to move to the next state by taking account of $\tau_{t,sl}$ and $\eta_{t,sl}$. When the value of $\tau_{t,sl}$ increases, this indicates there has been a lot of traffic on this path and it is therefore more desirable in order to reach the optimal solution. When the value of $\eta_{t,sl}$ increases, it indicates that the current state should have a higher probability. Each stage contains several states, while the order of states selected at each stage can be combined as an achievable path deemed to be a feasible solution to the problem.

(5) Ant Reproduction

New ants are generated by the scheme of crossover and mutation. Crossover is a structured recombination operation that exchanges two individual ants. Mutation is the occasional random alteration of an individual. The crossover and mutation scheme is described as follows:

(i)   Randomly select two parents, and generate offspring by assigning a Control Variable $(CV(g))$

    (a)   If $rang > CV(g)$: Mutation is used;

    (b)   If $rang > CV(g)$: Crossover is used. *Rand*: the uniform random in $[0, 1]$; $CV(g)$: the control variable between 0.1 to 0.9. The initial value set to 0.5; g: the current iteration number.

(ii)   If $TC(\mathcal{R}_{t,optimal})$ comes from crossover used, the control variable $CV(g+1)$ will be decreased as shown in Equation (23):

$$CV(g+1) = CV(g) - RP \tag{23}$$

where $RP = \frac{|CV(g)-CV(g-1)|}{2}$ is the regulated parameter. When RP is added in the crossover process, the higher probability for crossover operation will produce the next offspring.

(iii)   If $TC(\mathcal{R}_{t,optimal})$ comes from mutation, the control parameter $CV(g+1)$ will be increased as shown in Equation (24):

$$CV(g+1) = CV(g) + D \tag{24}$$

Similarly, when RP is added in the mutation process, the higher probability for mutation operation will produce the next offspring.

(iv)   If $TC_{min}(g-1) = TC_{min}(g)$, the control variable needs to hold back. If $CV(g) > CV(g-1)$, we have:

$$CV(g+1) = CV(g) - D \tag{25}$$

otherwise,

$$CV(g+1) = CV(g) + D \tag{26}$$

The crossover operator proceeds to exchange two individual ants by random. $\mathcal{R}_{t,s}$ and $\mathcal{R}_{t,l}$, which are the $s-th$ and $l-th$ individual ants at the $t-th$ interval, are exchanged by the crossover operator. The mutation operation is carried out to produce another individual ant. Each individual ant is mutated and created to a new individual ant by using (27).

$$\mathcal{R}_{t,s}^{j+1} = \mathcal{R}_{t,s}^{j} + N(0,\sigma^2) \tag{27}$$

$N(0,\sigma^2)$ represents a Gaussian random variable with mean 0 and variable σ^2. σ^2 can be calculated by:

$$\sigma = \beta \times \left(\mathcal{R}_{t,s,\,max}^{j} - \mathcal{R}_{t,s,\,min}^{j}\right) \times \frac{TC(\cdot)^k}{TC(\cdot)_{max}^k} \tag{28}$$

β, which is a mutation factor at the j-th generation, is set within $[0, 1]$.

(6)   Update the Pheromone

While building a solution to the problem, the pheromone of a visited route can be dynamically adjusted by Equation (29). This process is called the "local pheromone-updating rule":

$$\tau_{t,sj}^{k+1} = (1-\rho)\tau_{t,sj}^{k} + \Delta\tau_{t,sj}^{k} \tag{29}$$

ρ is the constant of pheromone intensity $(0 \le \rho \le 1)$ and $\Delta\tau_{t,sj}^{k}$ is the deviation of pheromone intensity on edge (s, j) at the $t-th$ interval, as shown in Equation (30):

$$\Delta\tau_{t,st}^{k} = \begin{cases} Q/e_{t,sj}^{k} \quad, \quad the\ path(s,j)\ for\ k-th\ ant \\[2em] 0 \qquad\qquad, \quad other \end{cases} \tag{30}$$

Q is the release rate of pheromone $(0 \le Q \le 1)$ and $\Delta e_{t,sj}^{k}$ is the path error (s, j) for the k-th ant.

(7) Stopping Rule

If a pre-specified stopping condition is satisfied, the run must be stopped and the results outputted; otherwise, return to step 4. In this study, the stopping rule is set at 300 generations.

4. Case Study

The proposed algorithm was tested with three back-pressure steam turbines, four extraction condenser steam turbines and seven steam boilers using gas, oil, and coal as fuel. Steam generation was measured in the field. All facilities including generators, boilers, and steam turbines have their capacity limitation. The limits of all facilities are introduced in Tables 2 and 3.

Table 2. Maximal and minimal limits of boiler flows and generators.

Unit No.	Fuel Type	Min. (ton)	Max. (ton)	Unit No.	Min. (MW)	Max. (MW)
Boiler1	Gas	68	137.5	Gen1	4.1	10
Boiler2	Gas	52	120	Gen2	4.9	10
Boiler3	Gas	60	137.5	Gen3	4.4	10
Boiler4	Oil	52	100	Gen4	15.6	50
Boiler5	Coal	127	250	Gen5	20	100
Boiler6	Coal	84	280	Gen6	20	100
Boiler7	Coal	90	300	Gen7	20	100

Table 3. Steam output limits.

Unit No.	Unit Type	Min. (ton/h)	Max. (ton/h)
M_{m1}	Back-pressure	75	120
M_{m2}	Back-pressure	55	140
M_{m3}	Back-pressure	50	80
M_{m4}	Extraction condenser steam	30	150
M_{m5}	Extraction condenser steam	30	150
M_{m6}	Extraction condenser steam	30	150
M_{m7}	Extraction condenser steam	30	150
M_{w4}	Extraction condenser steam	20	200
M_{w5}	Extraction condenser steam	25	300
M_{w6}	Extraction condenser steam	25	300
M_{w7}	Extraction condenser steam	25	300

4.1. Modeling Tests for Boilers and Steam Turbines

In this paper, the operation data of the boilers in the cogeneration system are recorded during the periods of each normal working day. The number of operation data samples for each boiler is 60; these are used to establish the operational models of boilers 1–7. The error results are shown in Table 4, and the example for operation models of boilers 1–2 are shown in Figure 5.

Table 4. The error results for the operation model of boilers.

Unit No.	Number. of Operation Data	LS Error (%)	LSSVM Error (%)
Boiler1	60	4.0999	3.1537
Boiler2	60	3.0520	2.0379
Boiler3	60	3.1001	1.6948
Boiler4	60	2.9621	1.8252
Boiler5	60	2.6386	1.7779
Boiler6	60	2.6326	1.7169
Boiler7	60	2.7150	1.5325

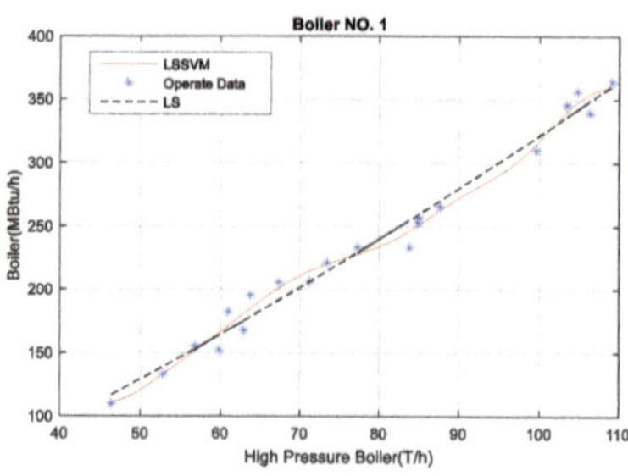
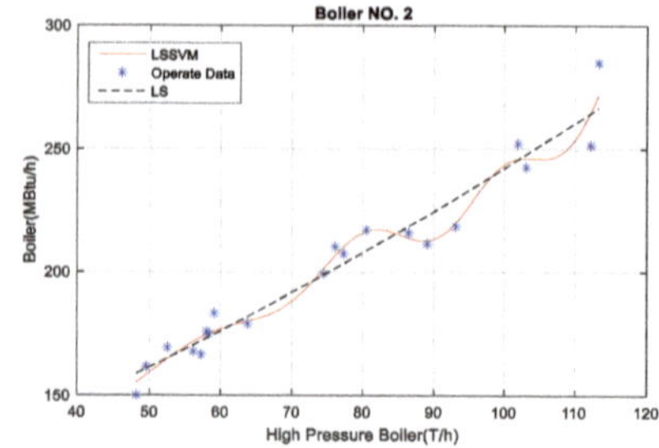

Figure 5. Example of the operation models of boilers 1–2.

In this paper, the operation data of the backpressure steam turbines (ST1~3) and extraction condensing steam turbines (ST4~7) in the cogeneration system are recorded and used by LS and LSSVM to establish the operation models of the steam turbines (ST1~7). The error results are shown in Table 5, and the example for the operation curves of steam turbines ST1 and ST6 are shown in Figure 6. The error results for the operation model of steam turbines is from 1.3916% to 2.1475%. It can be proved that the accuracy is reliable.

Table 5. The error results for the operation model of steam turbines.

Unit No.	Number. of Operation Data	LS Error (%)	LSSVM Error (%)
ST1	54	2.2081	1.5324
ST2	54	2.1404	1.4494
ST3	90	2.1960	1.7109
ST4	90	2.2494	1.9637
ST5	90	2.6757	2.1475
ST6	135	2.0326	1.5379
ST7	135	7.4892	1.3916

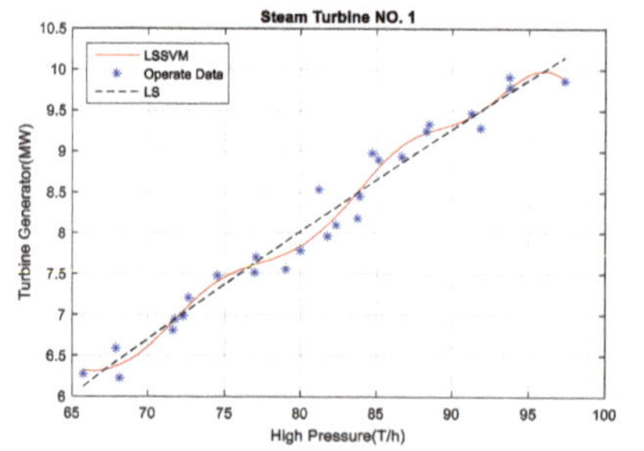
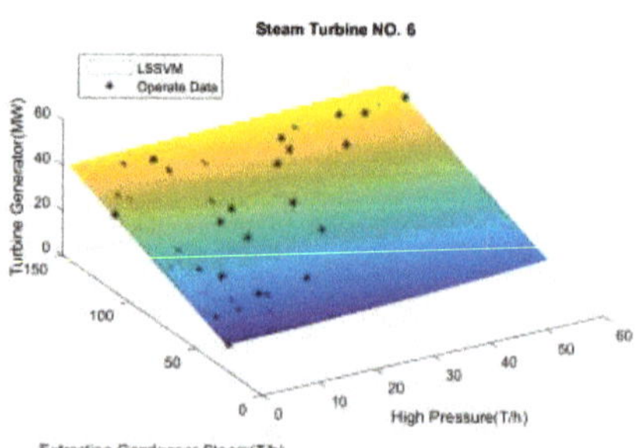

Figure 6. Example for the operation models of steam turbines (ST1/ST6).

4.2. Results at Different TOU Intervals

EACO was used to solve the energy dispatch of the cogeneration systems under a liberalized market. In the study system, the internal load was 30 MW and the steam demands of the industry for high- and medium-pressure were 60 T/H and 600 T/H, respectively. Table 6 shows the simulation results at the different TOU intervals. Table 6 indicates that all constraints were met. Since the electricity price is higher at the peak period, the operation strategy of cogeneration systems sold more power to the utility in order to achieve greater benefit. During off-peak periods, the cogeneration systems sell more thermal steam power to attain further benefits. It can be shown that the operation dispatch of the cogeneration systems during the peak period generated 273.321 MW and sold about 243.321 MW to the utility. During the peak period, all generators produced their highest outputs, while tending to sell the thermal steam to industries during semi-peak and off-peak periods. The TOU rate plays an important role in the economic dispatch of a cogeneration system.

Table 6. Dispatch results at different TOU intervals.

TOU	Peak Period	Semi-Peak Period	Off-Peak Period
P_{g1} (MW)	5.322	4.100	6.235
P_{g2} (MW)	6.704	5.182	5.061
P_{g3} (MW)	5.355	6.175	6.362
P_{g4} (MW)	50.000	50.000	50.000
P_{g5} (MW)	70.212	33.103	32.389
P_{g6} (MW)	49.873	33.967	21.210
P_{g7} (MW)	85.856	31.445	31.445
Total Generation (MW)	273.321	163.972	152.703
P_{tie} (MW)	243.321	133.972	122.703
S_w (ton/h)	521.944	129.219	98.716
S_{out} (ton/h)	0.000	535.781	566.284

Table 7 gives a profit analysis of cogeneration systems at different TOU intervals. The profits during peak period, semi-peak period, and off-peak peroid are 267,678.05, 13,606.32, and −121,047.33, respectively. The profit is lost during the off-peak period. As the cogeneration systems sell more power to utilities during the peak period, greater profits are realized. In the off-peak period, cogeneration systems sell more thermal steam to industries, thereby optimizing the energy trading dispatch. From Table 7, it is noted that the TOU rate influences the profits of the cogeneration system.

Table 7. Profit analysis of cogeneration systems at different TOU intervals. Unit: NT$/H.

TOU Item	Peak Period	Semi-Peak Period	Off-Peak Period
The profit for thermal steam sold	0.00	364,331.01	385,073.34
The profit from electricity sold	690,837.19	230,485.22	74,525.76
The cost of fuel	350,910.68	512,287.60	512,287.60
Emissions cost	199,499.69	208,611.45	208,611.45
Wheeling cost	12,166.03	6698.62	6135.14
The cost for pure water	13,200.00	13,200.00	13,200.00
Profit	267,678.05	13,606.32	−121,047.33

4.3. Convergence Test

Table 8 shows the comparisons of EP, GA, PSO, ACO, and EACO during different periods. An IBM PC with a P-IV2.0 GHz CPU and 512 MB SDRAM was used for this test. From this, the improvement of the EACO over other algorithms is clear. Figure 7 shows the convergent characteristics of EP, GA, PSO, ACO, and EACO during the peak period. The average execution times for EACO and ACO were only 1.85 s and 1.67 s, respectively. Although the executed performance of EACO was subtle, it showed the capacity of EACO to explore a solution more likely to offer maximum benefit.

Table 8. Comparison of EP, GA, PSO, ACO, and EACO algorithms. Unit: NT$/H.

Algorithm	Peak Period	Semi-Peak Period	Off-Peak Period
EP	263,849.60	12,813.87	− 123,289.77
GA	265,247.33	13,104.46	−122,999.72
PSO	266,904.29	13,489.53	−122,030.83
ACO	266,151.07	13,104.46	−122,192.54
EACO	267,678.05	13,606.32	−121,047.33

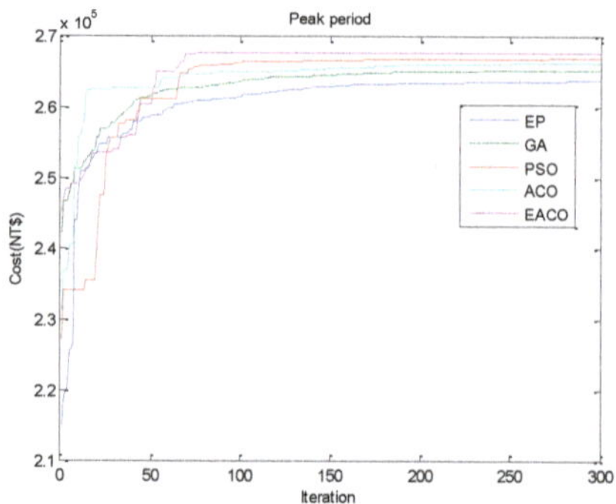

Figure 7. Convergence characteristics of EP, GA, PSO, ACO, and EACO during the peak period.

4.4. Robustness Test

Four algorithms including EP, GA, PSO, and ACO, were also tested at the peak period; the results are shown in Table 9. Each algorithm was executed using 100 trials during the robustness test. The number of optimum and average converged profit for EACO are 87 and NT$261,618.29. It is seen that EACO offers a greater ability to attain maximum profits and a higher probability of finding the best solution.

Table 9. Robustness test for EP, GA, PSO, ACO, and EACO algorithms in a peak period.

Algorithm	Maximal Converged Profit (NT$)	Minimal Converged Profit (NT$)	Average Converged Profit (NT$)	Average Number of Generations to Converge	No. of Trials Reaching Optimum	Average Execution Time (s)
EP	263,849.60	243,046.09	257,971.45	196	48	1.5341
GA	265,247.33	245,041.34	260,068.28	237	54	2.2153
PSO	266,904.29	254,865.39	262,555.59	173	57	1.6124
ACO	266,151.07	252,420.70	261,618.29	213	66	1.6741
EACO	267,678.05	259,030.88	264,500.37	184	87	1.8476

4.5. The Influence of Carbon Price

Table 10 shows the impact of various carbon prices experienced during the peak period. Table 10 suggests that if carbon prices are higher, CO_2 emissions will decrease. Similarly, due to the fuel type for extraction-condenser turbines being oil or coal, power generation will be reduced depending upon the carbon price. The purpose of various carbon prices here is to illustrate the tradoff between profit and emission costs, and also to show that generators may more economically dispatch trade electricity or CO_2 emission to find better profit.

Table 10. The impact of various carbon prices during the peak period.

Carbon Price (NT$)	Back-Pressure Turbine Generation (MW)	Extraction-Condenser Turbine Generation (MW)	Profit (NT$)	Emission Cost (NT$)
0	18.12	254.55	322,925.81	0.00
400	18.00	249.85	241,469.31	74,714.14
800	18.11	212.03	188,316.27	96,104.88
1200	17.52	172.72	142,085.10	99,394.94
1600	17.01	154.71	111,351.34	100,419.06
2000	15.36	149.44	94,250.70	105,522.53
2400	18.17	135.76	74,312.66	108,103.17
2800	18.11	135.76	56,215.64	123,906.39
3200	15.36	138.73	39,297.50	138,744.85

Figure 8 shows the profit-emission tradeoff curves during the peak period. Figure 8 provides diversified alternatives for decision makers, showing the effects of various carbon prices. Replaced with the maximal allowable emission as a constraint, an appropriate decision can be chosen to satisfy the desired level of profit and emission costs.

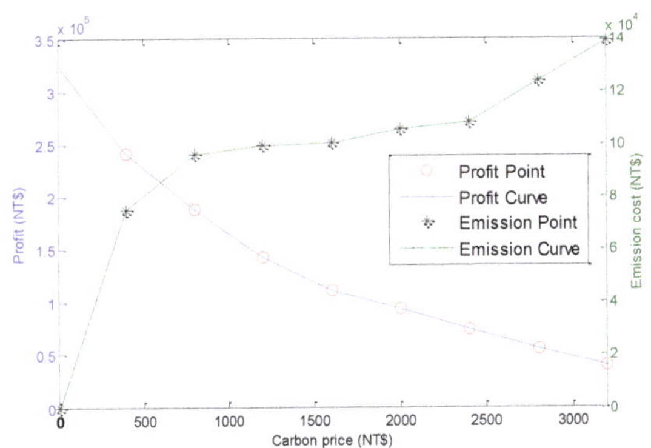

Figure 8. Profit-emission tradeoff curves during the peak period.

5. Conclusions

In this paper, an EACO is proposed to maximize the profit of energy trading using a cogeneration system. The objective function is formulated based upon a maximal profit model, which includes profit from steam sold, profit from electricity sold, fuel costs, CO_2 emission costs, wheeling costs, and water costs. By considering the various carbon prices, the profits of the energy trading dispatches are evaluated while considering three different TOU scenarios. The effectiveness of the EACO is demonstrated and simulated on a real cogeneration system. Our analysis points to expectations of the TOU rate or carbon price for the energy trading dispatch. With the advantages of both heuristic ideals and ACO, EACO has threefold conventional ideals: the complicated problem is solvable, with a better performance than ACO, and the more likelihood to get a global optimum than heuristic methods. The results indicate that both provide good tools for determining the optimum energy trading operation of a cogeneration system. This shows that the tradeoff between investment cost and environmental protection can be clearly predetermined in the liberty market. EACO also has great potential to be further applied to many ill-conditioned problems in power system planning and operations.

Author Contributions: W.-M.L. generalized the novel algorithms and designed the system planning projects. C.-Y.Y. used system parameters and models for the simulation test. M.-T.T. performed the editing and experimental model results. H.-S.H. written the orginal draft preparation. C.-S.T. provided hardware tools and system model related materials. All the authors were involved in exploring system validation and results, and permitting the benefits of the published document.

Funding: This research was funded by Ministry of Science and Technology of R.O.C. grant number MOST 107-2221-E-230-008-MY3.

Conflicts of Interest: The authors declare no conflict of interest.

References

1. Aluisio, B.; Dicorato, M.; Forte, G.; Trovato, M. An optimization procedure for Microgrid day-ahead operation in the presence of CIP facilities. *Sustain. Energy Grids Netw.* **2017**, *11*, 34–45. [CrossRef]
2. Motevasel, M.; Seifi, A.R.; Niknam, T. Multi-objective energy management of CHP (combined heat and power)-based micro-grid. *Energy* **2013**, *51*, 123–136. [CrossRef]
3. Basu, A.K.; Bhattacharya, A.; Chowdhury, S.; Chowdhury, S.P. Planned Scheduling for Economic Power Sharing in a CHP-Based Micro-Grid. *IEEE Trans. Power Syst.* **2012**, *27*, 30–38. [CrossRef]

4. Isa, N.M.; Tan, C.W.; Yatim, A. A comprehensive review of cogeneration system in a microgrid: A perspective from architecture and operating system. *Renew. Sustain. Energy Rev.* **2018**, *81*, 2236–2263. [CrossRef]

5. Farghal, S.; El-Dewieny, R.; Riad, A. Optimum operation of cogeneration plants with energy purchase facilities. *IEE Proc. C Gener. Transm. Distrib.* **1987**, *134*, 313–319. [CrossRef]

6. Rooijers, F.; Van Amerongen, R. Static economic dispatch for co-generation systems. *IEEE Trans. Power Syst.* **1994**, *9*, 1392–1398. [CrossRef]

7. Al Asmar, J.; Lahoud, C.; Brouche, M. Decision-making strategy for cogeneration power systems integration in the Lebanese electricity grid. *Energy Procedia* **2017**, *119*, 801–805. [CrossRef]

8. Skorek-Osikowska, A.; Remiorz, L.; Bartela, Ł.; Kotowicz, J. Potential for the use of micro-cogeneration prosumer systems based on the Stirling engine with an example in the Polish market. *Energy* **2017**, *133*, 46–61. [CrossRef]

9. Asano, H.; Sagai, S.; Imamura, E.; Ito, K.; Yokoyama, R. Impacts of time-of-use rates on the optimal sizing and operation of cogeneration systems. *IEEE Trans. Power Syst.* **1992**, *7*, 1444–1450. [CrossRef]

10. Lee, C.-H.; Huang, S.-C.; Chang, C.-A.; Chen, B.-K. Operation of Steam Turbines under Blade Failures during the Summer Peak Load Periods. *Energies* **2014**, *7*, 7415–7433. [CrossRef]

11. Gambini, M.; Vellini, M.; Stilo, T.; Manno, M.; Bellocchi, S. High-Efficiency Cogeneration Systems: The Case of the Paper Industry in Italy. *Energies* **2019**, *12*, 335. [CrossRef]

12. Jayakumar, N.; Subramanian, S.; Ganesan, S.; Elanchezhian, E. Grey wolf optimization for combined heat and power dispatch with cogeneration systems. *Int. J. Electr. Power Energy Syst.* **2016**, *74*, 252–264. [CrossRef]

13. Nguyen, T.T.; Vo, D.N.; Dinh, B.H. Cuckoo search algorithm for combined heat and power economic dispatch. *Int. J. Electr. Power Energy Syst.* **2016**, *81*, 204–214. [CrossRef]

14. Tsay, M.-T.; Lin, W.-M.; Lee, J.-L. Application of evolutionary programming for economic dispatch of cogeneration systems under emission constraints. *Int. J. Electr. Power Energy Syst.* **2001**, *23*, 805–812. [CrossRef]

15. Tsay, M.-T.; Lin, W.-M.; Lee, J.-L. Interactive best-compromise approach for operation dispatch of cogeneration systems. *IEE Proc. Gener. Transm. Distrib.* **2001**, *148*, 326. [CrossRef]

16. Carpaneto, E.; Chicco, G.; Mancarella, P.; Russo, A. Cogeneration planning under uncertainty: Part I: Multiple time frame approach. *Appl. Energy* **2011**, *88*, 1059–1067. [CrossRef]

17. Thorin, E.; Brand, H.; Weber, C. Long-term optimization of cogeneration systems in a competitive market environment. *Appl. Energy* **2005**, *81*, 152–169. [CrossRef]

18. Chen, S.-L.; Tsay, M.-T.; Gow, H.-J. Scheduling of cogeneration plants considering electricity wheeling using enhanced immune algorithm. *Int. J. Electr. Power Energy Syst.* **2005**, *27*, 31–38. [CrossRef]

19. Yusta, J.; Jesus, P.D.O.-D.; Khodr, H.; Jesus, P.D.O.-D. Optimal energy exchange of an industrial cogeneration in a day-ahead electricity market. *Electr. Power Syst. Res.* **2008**, *78*, 1764–1772. [CrossRef]

20. Safder, U.; Ifaei, P.; Yoo, C. Multi-objective optimization and flexibility analysis of a cogeneration system using thermorisk and thermoeconomic analyses. *Energy Convers. Manag.* **2018**, *166*, 602–636. [CrossRef]

21. He, L.; Lu, Z.; Pan, L.; Zhao, H.; Li, X.; Zhang, J. Optimal Economic and Emission Dispatch of a Microgrid with a Combined Heat and Power System. *Energies* **2019**, *12*, 604. [CrossRef]

22. Chang, H.-H. Genetic algorithms and non-intrusive energy management system based economic dispatch for cogeneration units. *Energy* **2011**, *36*, 181–190. [CrossRef]

23. Basu, M. Artificial immune system for combined heat and power economic dispatch. *Int. J. Electr. Power Energy Syst.* **2012**, *43*, 1–5. [CrossRef]

24. Gambardella, L.; Dorigo, M. Ant colony system: A cooperative learning approach to the traveling salesman problem. *IEEE Trans. Evol. Comput.* **1997**, *1*, 53–66.

25. Mullen, R.J.; Monekosso, D.; Barman, S.; Remagnino, P. A review of ant algorithms. *Expert Syst. Appl.* **2009**, *36*, 9608–9617. [CrossRef]

26. Saravuth, P.; Issarachai, N.; Waree, K. Ant colony optimisation for economic dispatch problem with non-smooth cost functions. *Int. J. Electr. Power Energy Syst.* **2010**, *32*, 478–487.

27. Priyadarshi, N.; Ramachandaramurthy, V.K.; Padmanaban, S.; Azam, F. An Ant Colony Optimized MPPT for Standalone Hybrid PV-Wind Power System with Single Cuk Converter. *Energies* **2019**, *12*, 167. [CrossRef]

28. Church, C.; Morsi, W.; El-Hawary, M.; Diduch, C.; Chang, L. Voltage collapse detection using Ant Colony Optimization for smart grid applications. *Electr. Power Syst. Res.* **2011**, *81*, 1723–1730. [CrossRef]

29. Zhou, J.; Wang, C.; Li, Y.; Wang, P.; Li, C.; Lu, P.; Mo, L. A multi-objective multi-population ant colony optimization for economic emission dispatch considering power system security. *Appl. Math. Model.* **2017**, *45*, 684–704. [CrossRef]

30. Kıran, M.S.; Özceylan, E.; Gunduz, M.; Paksoy, T. A novel hybrid approach based on Particle Swarm Optimization and Ant Colony Algorithm to forecast energy demand of Turkey. *Energy Convers. Manag.* **2012**, *53*, 75–83. [CrossRef]

31. Goldberg, D.E. *Genetic Algorithm in Search, Optimization and Machine Learning*; Addison Wesley: Boston, MA, USA, 1989.

32. Taiwan Power Company (TPC). *Time-of-Use Rate for Cogeneration Plants. The electricity Rates Structure for Taipower Company*; Taiwan Power Company (TPC): Taipei, Taiwan, 2017.

33. Suykens, J.A.K.; Vandewalle, J. Least square support vector machine. *Neural Process. Lett.* **1999**, *9*, 293–300. [CrossRef]

34. Intergovernmental Penal on Climate Change. Available online: http://www.ipcc.com/ (accessed on 8 April 2017).

35. Nord Pool Power Exchanger. Available online: http://www.nasdaqomx.com/commodities (accessed on 11 April 2017).

Article

Numerical Simulation of Crystalline Silicon Heterojunction Solar Cells with Different p-Type a-SiO$_x$ Window Layer

Chia-Hsun Hsu [1], Xiao-Ying Zhang [1], Hai-Jun Lin [1], Shui-Yang Lien [1,2,*], Yun-Shao Cho [2] and Chang-Sin Ye [3]

[1] School of Opto-electronic and Communication Engineering, Xiamen University of Technology, Xiamen 361024, China

[2] Department of Materials Science and Engineering, Da-Yeh University, Chunghwa 51595, Taiwan

[3] Metal Industries Research & Development Centre Opto-Electronics System Section, Kaohsiung 81160, Taiwan

* Correspondence: sylien@xmut.edu.cn

Received: 5 May 2019; Accepted: 23 June 2019; Published: 2 July 2019

Abstract: In this study, p-type amorphous silicon oxide (a-SiO$_x$) films are deposited using a radio-frequency inductively-coupled plasma chemical vapor deposition system. Effects of the CO$_2$ gas flow rate on film properties and crystalline silicon heterojunction (HJ) solar cell performance are investigated. The experimental results show that the band gap of the a-SiO$_x$ film can reach 2.1 eV at CO$_2$ flow rate of 10 standard cubic centimeters per minute (sccm), but the conductivity of the film deteriorates. In the device simulation, the transparent conducting oxide and contact resistance are not taken into account. The electrodes are assumed to be perfectly conductive and transparent. The simulation result shows that there is a tradeoff between the increase in the band gap and the reduction in conductivity at increasing CO$_2$ flow rate, and the balance occurs at the flow rate of six sccm, corresponding to a band gap of 1.95 eV, an oxygen content of 34%, and a conductivity of 3.3 S/cm. The best simulated conversion efficiency is 25.58%, with an open-circuit voltage of 741 mV, a short-circuit current density of 42.3 mA/cm^2, and a fill factor of 0.816%.

Keywords: heterojunction; crystalline silicon; solar cell; silicon oxide

1. Introduction

Crystalline silicon heterojunction (SHJ) solar cells have attracted significant attention in recent years, as they can provide high performance together with the prospect of low-cost fabrication and a decrease of silicon wafer thickness below 100 µm [1]. The advantage of the heterojunction between amorphous and crystalline silicon was first introduced into the so-called HIT concept (Hetero-junction with Intrinsic Thin-layer) by the former company SANYO (Osaka, Japan) (currently part of the company Panasonic) in 1992 [2]. The SHJ HIT solar cell is composed of a single thin crystalline silicon wafer surrounded by ultra-thin intrinsic amorphous silicon (a-Si:H) and n-type and p-type doped a-Si:H layers, which can be deposited at temperatures below 200 °C and thus can be used in the processing of thin wafers. On the two doped layers, transparent conducting oxide (TCO) layers and metal electrodes are formed with sputtering and screen-printing methods, respectively. The TCO layer on the top also works as an anti-reflection layer.

One of the key technologies of the SHJ solar cells is the interface defect density between the silicon wafer and a-Si:H layers. For n-type crystalline silicon-based SHJ, the p-type a-Si:H is used as an emitter, and it is well-known that the doped a-Si:H layers contain a significantly high defect density. The interface defects between the wafer and the doped a-Si:H is high, and thus a very thin intrinsic a-Si:H is required to be inserted at the interface. Considerable efforts have been made to reduce the carrier recombination rate at the a-Si:H/c-Si interface [3–6]. Due to the higher band gap of a-Si:H (about 1.7 eV) compared to that of c-Si, a carrier-selective emitter can be formed, which repels electrons by using a high barrier presented in the conduction band, while the holes can tunnel through the interface to the p-type a-Si:H emitter [7]. Typically, in n-type SHJ solar cells, the band mismatch is about 0.15–0.2 eV for the conduction band, and is about 0.45 eV for the valence band. The p-type emitter is also important for SHJ solar cells, which should have a low absorption coefficient as the light absorbed in this emitter would not contribute to the device current due to the high defects in the doped a-Si:H [8]. The emitter requires high conductivity in order to decrease the series resistance (R_s) and to increase the carrier collection. There is increasing interest in developing various materials to replace a-Si:H emitters, such as nanocrystalline silicon [9] to increase conductivity and silicon carbide [10,11] to increase the band gap.

In this study, p-type amorphous silicon oxide (a-SiO$_x$) films are prepared using inductively-coupled plasma chemical vapor deposition (ICPCVD). The a-SiO$_x$:H thin films having low light absorption, a wide band gap, and high conductivity are suitable for the use as a window layer of the SHJ solar cells. The objective of this work is to deposit a-SiO$_x$:H films with different band gaps and conductivity by varying the carbon dioxide (CO_2) gas flow rate and to investigate their effects on SHJ solar cell performance.

2. Materials and Methods

Boron-doped a-SiO$_x$ films were deposited using a 13.56 MHz ICPCVD system using a gas mixture of B_2H_6, SiH_4 and CO_2. The substrate temperature was kept at 200 °C. The CO_2 gas flow rate was varied from 2 to 10 standard cubic centimeters per minute (sccm). The power density was 0.03 W/cm^2. Detailed deposition parameters are summarized in Table 1. The films were deposited on flat glass substrates for optical, electrical, and structural characterization. The band gap (E_g) of the films was obtained by using a Tauc plot using the following equation [12]:

$$(\alpha h v)^{\frac{1}{n}} = A(h v - E_g) \tag{1}$$

where α is the absorption coefficient determined from transmission and reflection spectra measured by a UV-visible spectrometer, $h v$ is the photon energy, n is 1/2 for direct band gap materials or 2 for indirect band gap materials, and A is the band tailing parameter. The plotting of $(\alpha h v)^{(1/n)}$ versus $h v$ gives a straight line in a certain region. The extrapolation of this straight line intercepts the $(h v)$-axis to give E_g. The atomic ratio of oxygen to silicon of the a-SiO$_x$ films was measured by using X-ray photoelectron spectroscopy (XPS). The conductivity of the films was measured in a perpendicular direction of the films. PC1D simulation software (version 5.9, University of New South Wales, Sydney, Australia) was used. The device structure was p-type a-SiO$_x$ (15 nm)/intrinsic a-Si:H (5 nm)/n-type c-Si (100 μm)/n-type a-Si:H (15 nm). The front and rear textured surfaces, with depths of 3 μm and angles of 54.74°, were applied. The device scheme is shown in Figure 1. The minimum set of the simulation parameters is summarized in Table 2, where N_a is acceptor concentration, N_d is donor concentration, μ_n is electron mobility, μ_p is hole mobility, and S_{eff} is the surface recombination rate.

Table 1. Deposition parameters for a-SiO$_x$ films.

Parameter	Value
Substrate temperature (°C)	200
Pressure (Torr)	0.005
B_2H_6 flow rate (sccm)	0.5
Si(CH$_3$)$_3$ flow rate (sccm)	8
CO_2 (sccm)	2–10
Power density (W/cm^2)	0.03
Thickness (nm)	15

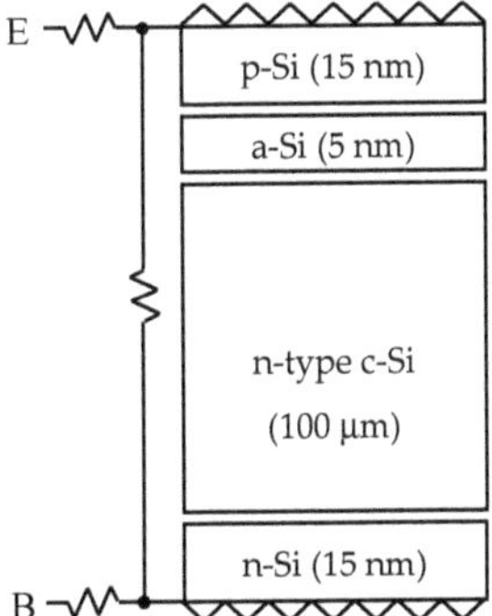

Figure 1. Device structure used for PC1D simulation.

Table 2. Minimum set of simulation parameters for heterojunction solar cells.

Device Parameter		Value		
Light source		AM1.5G		
Texture		Front and rear (depth: 3 μm)		
Emitter conductivity (S/cm)		variable		
Internal shunt element		0.3 S		
Layer parameter	p a-SiO$_x$	i a-Si:H	n c-Si	n a-Si:H
E_g (eV)	variable	1.7	1.12	1.7
N_a (cm^{-3})	10^{19}	0	0	0
N_d (cm^{-3})	0	0	10^{16}	10^{18}
μ_n (cm^2/Vs)	5	5	1200	5
μ_p (cm^2/Vs)	1	1	400	1
S_{eff} (cm/s)	10^4	0	0	10^4

3. Results

Figure 2 shows the O/Si atomic ratio and band gap of the a-SiO$_x$ films deposited at a different CO_2 flow rate. It can be seen that the O/Si ratio is about 14.7% at the CO_2 flow rate of 2 sccm. The ratio increases by increasing the CO_2 flow rate. The ratio reaches 44% when the flow rate increases to 10 sccm. This result indicates that the oxygen content in the films increases when the flow rate increases. For a typical a-Si:H film deposited by plasma chemical vapor deposition, the band gap of the films is about 1.6–1.7 eV depending on hydrogen incorporation [13]. The low CO_2 gas flow rate of 2 sccm leads to an O/Si ratio of 14.7%, and this oxygen incorporation increased the band gap of the film to 1.78 eV, which is slightly higher than a-Si:H films (about 1.7 eV). At a CO_2 flow rate of 6 sccm, the film with O/Si ratio of 34.3% corresponds to a band gap of 1.95 eV. Further increasing the flow rate to 10 sccm, or O/Si ratio up to 44.1%, gives a band gap of 2.1 eV. When a higher band gap material is used for an emitter of n-type SHJ solar cells, the mismatch at the conduction band and valence band could help formation of a selective contact. When a wider band gap p-type material is in contact with n-type c-Si, the band at the interface of the c-Si side will bend downward. The Fermi-level at the interface region is, then, much closer to the valence band compared to the conduction band, and this forms a minority carrier inversion layer that could significantly lower the interface recombination rate and eventually enhance the open-circuit voltage (V_{oc}) [14]. Furthermore, the V_{oc} of an n-type SHJ can be expressed as [15]:

$$V_{oc} = \frac{\varphi_D}{q} - \frac{AkT}{q} \ln\left(\frac{qN_c S_n}{J_{sc}}\right) \tag{2}$$

where φ_B is the effective barrier height for recombination, A is the ideality factor, q is the electron charge, k is the Boltzmann constant, T is the absolute temperature, N_c is the effective densities of state of the conduction band, S_n is the surface recombination rate for electrons, and J_{sc} is the short-circuit current density. The φ_B is the distance from the Fermi level to conduction band at the interface. Thus, a higher band gap emitter could in turn have more band bending that increases φ_B and eventually V_{oc}. The a-SiO$_x$ films with the highest band gap of 2.1 eV are expected to have a higher V_{oc} among others. In addition, the high band gap will also allow more incident light to enter the device, and this may enhance J_{sc}.

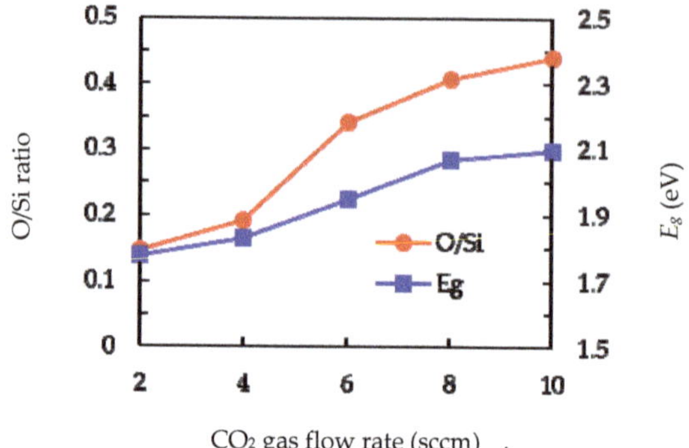

Figure 2. Atomic ratio of O/Si and band gap of the films deposited with different CO$_2$ flow rate. Standard cubic centimeters per minute (sccm).

Figure 3 shows the perpendicular conductivity of the a-SiO$_x$ films deposited at a different CO$_2$ gas flow rate. It can be seen that the conductivity reduces from 4.1 to 2.5 S/cm when the CO$_2$ flow rate increases from 2 to 10 sccm. The increase of the oxygen content in the films reduces the conductivity. The O/Si ratio of the films increases as the CO$_2$ flow rate increases. The film structure would be relatively close to SiO$_2$, and thus the conductivity would decrease. The overall series resistance of SHJ devices should be the sum of the resistances of indium tin oxide (ITO), the p-type emitter, the n-type c-Si, and the n-type a-Si:H (resistances of metal contacts are neglected). The perpendicular conductivity of the p-type emitter can be used as an indicator to evaluate the series resistance contributed by the p-type emitter. The lower oxygen content films have a high conductivity, and thus are expected to have a low series resistance and, ultimately, a higher fill factor (FF). Therefore, there is a tradeoff between the gain in V_{oc} brought by the increased band gap and the loss in FF due to the decreased conductivity at an increasing CO$_2$ flow rate.

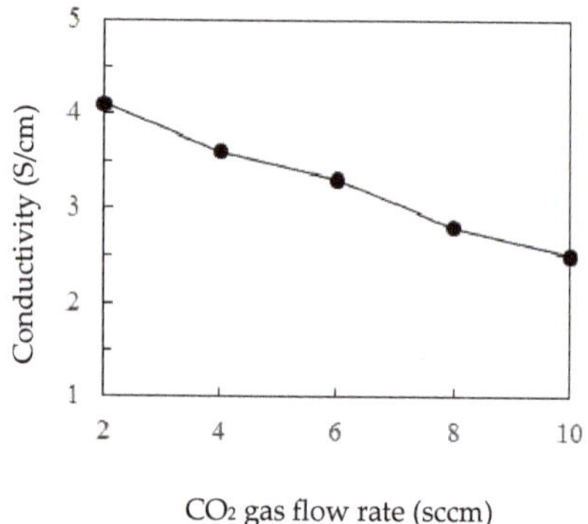

Figure 3. Conductivity of a-SiO$_x$ films as a function of CO$_2$ flow rate.

To determine the best film for the p-type emitter among the a-SiO$_x$ films with different CO$_2$ flow rates, the properties of the films are input into simulation software. The absorption coefficient spectra of the films were analyzed using the transmittance spectra. The reflectance spectra are input according to the measured data for the structure of ITO/p-type a-SiO$_x$/intrinsic a-Si:H/n-type c-S/n-type a-Si:H structure, where the ITO was deposited using a DC magnetron sputter system. We prepared an intrinsic a-Si:H thin film by plasma enhanced chemical vapor deposition using H$_2$ and SiH$_4$ on the both sides of c-Si. The intrinsic a-Si:H layer had a hydrogen content of about 10% and a band gap of 1.7 eV. The measured minority carrier lifetime (using Sinton WCT120) was 344 µs, corresponding to a surface recombination rate of 0.15 cm/s. In the simulation, by assuming the ideal passivation of the intrinsic a-Si:H on c-Si for simplification, the surface recombination rate of the c-Si interface is not considered. The deposition of different p-type a-SiO$_x$:H layers might not change the i-layer properties, and thus the hydrogen and oxygen in the p-layer would not significantly affect the c-Si surface's passivation quality. Therefore, the simulation may present an overestimated result, but the trend and influence of the p-type emitter layers on device performance are still validated. Figure 4 shows the external quantum efficiency (EQE) in a wavelength range of 300–550 nm for SHJ solar cells with different p-type emitters. The difference between the curves with wavelengths higher than 550 nm is neglected, so the spectral response in the long wavelength region is not shown. It is also noted that as the light is incident from the p-type emitter side, the impact of the emitter is mainly at the short-wavelengths. It can be seen that the EQE response increases with increasing the CO$_2$ flow rate. This enhancement is mainly explained by the increased band gap and accordingly reduced absorption of the light in the emitter region. Note that the light absorbed in the emitter would not contribute to the device current. The relationship between J_{sc} and EQE is given by

$$J_{sc} = \frac{q}{hc} \int \text{EQE}(\lambda) \cdot P_{\text{AM1.5G}}(\lambda) d\lambda \tag{3}$$

where h is the Planck constant, λ is the wavelength, and $P(\lambda)$ is the sun spectra at the AM1.5G condition. Therefore, a higher EQE response gives a higher J_{sc}. This evidences the expectation that higher band gap emitter can improve the J_{sc} of the device. It is noted that when the CO$_2$ flow rate is greater than 6 sccm, corresponding to a band gap of 1.95 eV, the EQE curves are similar. This might indicate that light absorption in emitter is already too low to gain further improvement.

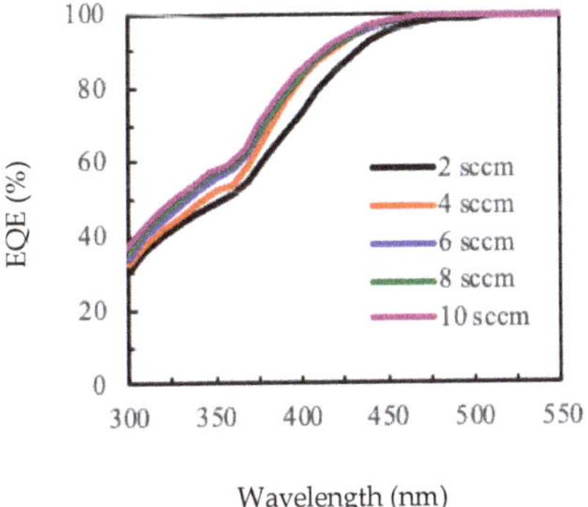

Figure 4. Simulated external quantum efficiency (EQE) spectra in 300–550 nm for crystalline silicon heterojunction (SHJ) solar cells with different p-type emitters.

The simulated current density-voltage (J-V) curves of the SHJ solar cells with different p-type emitters are shown in Figure 5. The extracted photovoltaic external parameters, such as V_{oc}, J_{sc}, FF and conversion efficiency (η), are listed in Table 3. It can be seen that the J_{sc} increases from 41.63 to 42.31 mA/cm^2 when the CO$_2$ flow rate increases from 2 to 8 sccm. The FF, on the other hand, reduces from 0.825 to 0.785 after increasing the CO_2 flow rate from 2 to 8 sccm. It is noted that one main factor affecting the solar cell's FF is R_s, which is related to the conductivity in each layer, carrier transport through the junction, and the contact resistivity of the TCO/p-layer/i-layer stack to c-Si. The carrier transport is influenced by i and p layer thickness, their band gap, the activation energy of doped layer, and work function of the TCO layer [16]. In this simulation, the TCO layer is not considered. The front and rear electrodes are assumed to be perfectly conductive and transparent. The contact resistivity of the TCO/p-layer/i-layer stack to c-Si is also not considered. The total R_s is, therefore, mainly influenced by p-layer conductivity. The case of 10 sccm shows a nearly 5% drop in FF compared to the case of 2 sccm. The reduction in FF is mainly attributed to the decreased p-layer conductivity due to the increase in the oxygen content of the emitter layer. In contrast, J_{sc} increases by 1.6%, which is much lower than the drop of FF. The V_{oc} increases from 730 to 745 mV when the CO_2 flow rate increases from 2 to 8 sccm. The solar cell V_{oc} might be affected by Si surface passivation and the band bending at the a-Si/c-Si interfaces. The latter is related to the band gap, activation energy, and work function of TCO layer. The TCO layer is not considered in the simulation. The activation energies of the p-layers deposited at the CO_2 flow rate of 2, 4, 6, 8, and 10 sccm are 52, 54, 58, 67, and 73 meV, respectively. Overall, the value of η increase from 25.07% to 25.58% by increasing the CO_2 flow rate from 2 to 6 sccm, and further increasing the flow rate from 6 to 10 sccm leads to a reduction in η from 25.58% to 24.74%. From this result, it can be seen that the gain in J_{sc} and V_{oc} due to using a large band gap emitter will possibly be offset by the loss in FF. The emitter properties should, therefore, be carefully controlled. In the present study, solar cells featuring an a-SiO$_x$ emitter with a band gap of 1.95 eV have the best η.

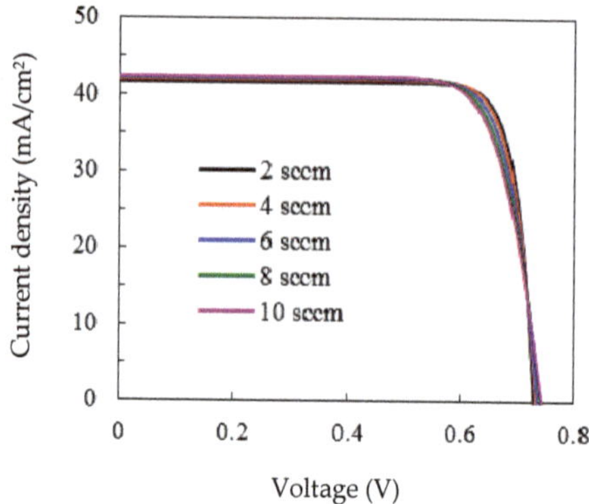

Figure 5. Simulated J-V curves of SHJ solar cells with different p-type emitters.

Table 3. External photovoltaic parameters for SHJ solar cells with different p-type emitters.

CO_2 (sccm)	V_{oc} (mV)	J_{sc} (mA/cm²)	FF	η (%)
2	730	41.63	0.825	25.07
4	734	41.92	0.822	25.29
6	741	42.3	0.816	25.58
8	743	42.31	0.796	25.02
10	745	42.31	0.785	24.74

4. Conclusions

In this work, different p-type a-SiO$_x$ films were prepared using ICPCVD for use as an emitter of n-type SHJ solar cells. The CO_2 flow rate is varied, and the film structure changed from a-Si to a-SiO$_x$, corresponding to a change in band gap from 1.78 to 2.1 eV. However, the perpendicular conductivity decreased when the oxygen content in the films increased. For the device simulation, the TCO layer was not taken into account. The front and rear electrodes are assumed to be perfectly conductive and transparent, and the contact resistance is considered to not be influenced by changing the p-type layer. The simulation result shows that the gain in J_{sc} saturates when the band gap reaches 1.9 eV. In addition, using a wide band gap p-type emitter for an n-type SHJ solar cell can enhance the V_{oc} up to 745 mV and J_{sc} to 42.31 mA/cm^2, but these improvements might be possibly offset by the decrease in FF. Finally, the optimal a-SiO$_x$ is that with a band gap of 1.95 eV and conductivity of 3.3 S/cm. The best conversion efficiency is 25.58%, which is 0.5% abs higher than that with a band gap similar to conventional p-type a-Si:H emitters.

Author Contributions: Conceptualization, Methodology, Investigation and Software, C.-H.H.; Validation and Formal Analysis, X.-Y.Z., H.-J.L., Y.-S.C. and C.-S.Y.; Funding Acquisition, X.-Y.Z. and S.-Y.L.; Writing-Original Draft Preparation, C.-H.H. and S.-Y.L.; Writing-Review & Editing, C.-H.H and S.-Y.L.; Supervision, S.-Y.L.

Funding: This work is sponsored by the science and technology project of Xiamen (No. 3502Z20183054) and the Science and Technology Program of the Educational Office of Fujian Province (No. JT180432).

Conflicts of Interest: The authors declare no conflict of interest.

References

1. Taguchi, M.; Tsunomura, Y.; Inoue, H.; Taira, S.; Nakashima, T.; Baba, T.; Sakata, H.; Maruyama, E. High-Efficiency HIT Solar Cell on Thin (<100 μm) Silicon Wafer. In Proceedings of the 24th European Photovoltaic Solar Energy Conference, Hamburg, Germany, 21–25 September 2009; pp. 1690–1693.

2. Tanaka, M.; Taguchi, M.; Matsuyama, T.; Sawada, T.; Tsuda, S.; Nakano, S.; Hanafusa, H.; Kuwano, Y. Development of New a-Si/c-Si Heterojunction Solar Cells: ACJ-HIT (Artificially Constructed Junction-Heterojunction with Intrinsic Thin-Layer). *Jpn. J. Appl. Phys.* **1992**, *31*, 3518. [CrossRef]

3. Schulze, T.F.; Korte, L.; Ruske, F.; Rech, B. Band lineup in amorphous/crystalline silicon heterojunctions and the impact of hydrogen microstructure and topological disorder. *Phys. Rev. B* **2011**, *83*, 165314. [CrossRef]

4. Descoeudres, A.; Holman, Z.; Barraud, L.; Morel, S.; De Wolf, S.; Ballif, C. >21% Efficient Silicon Heterojunction Solar Cells on n- and p-Type Wafers Compared. *IEEE J. Photovolt.* **2013**, *3*, 83–89. [CrossRef]

5. De Wolf, S.; Kondo, M. Abruptness of a-Si:H/c-Si interface revealed by carrier lifetime measurements. *Appl. Phys. Lett.* **2007**, *90*, 042111. [CrossRef]

6. Kinoshita, T.; Fujishima, D.; Yano, A.; Ogane, A.; Tohoda, S.; Matsuyama, K.; Nakamura, Y.; Tokuoka, N.; Kanno, H.; Sakata, H.; et al. The approaches for high efficiency HIT™ solar cell with very thin (<100 μm) silicon wafer over 23%. In Proceedings of the 26th European Photovoltaic Solar Energy Conference and Exhibition, Hamburg, Germany, 5–9 September 2011; pp. 871–874.

7. Taguchi, M.; Yano, A.; Tohoda, S.; Matsuyama, K.; Nakamura, Y.; Nishiwaki, T.; Fujita, K.; Maruyama, E. 24.7% Record Efficiency HIT Solar Cell on Thin Silicon Wafer. *IEEE J. Photovolt.* **2014**, *4*, 96–99. [CrossRef]

8. Battaglia, C.; Cuevas, A.; De Wolf, S. High-efficiency crystalline silicon solar cells: Status and perspectives. *Energy Environ. Sci.* **2016**, *9*, 1552–1576. [CrossRef]

9. Wurfel, U.; Cuevas, A.; Würfel, P. Charge Carrier Separation in Solar Cells. *IEEE J. Photovolt.* **2015**, *5*, 461. [CrossRef]

10. Moldovon, A.; Feldmann, F.; Zimmer, M.; Rentsch, J.; Benick, J.; Hermle, M. Tunnel oxide passivated carrier-selective contacts based on ultra-thin SiO$_2$ layers. Sol. Energy Mater. *Sol. Cells* **2015**, *142*, 123–127. [CrossRef]

11. Upadhyaya, A.D.; Ok, Y.W.; Chang, E.; Upadhyaya, V.; Madani, K.; Tate, K.; Rounsaville, B.; Choi, C.J.; Chandrasekaran, V.; Yelundur, V.; et al. Ion-Implanted Screen-Printed n-Type Solar Cell With Tunnel Oxide Passivated Back Contact. *IEEE J. Photovolt.* **2016**, *6*, 153. [CrossRef]

12. Tauc, J. Optical properties and electronic structure of amorphous Ge and Si. *Mater. Res. Bull.* **1968**, *3*, 37–46. [CrossRef]

13. Scherg-Kurmes, H.; Körner, S.; Ring, S.; Klaus, M.; Korte, L.; Ruske, F.; Schlatmann, R.; Rech, B.; Szyszka, B. High mobility In$_2$O$_3$:H as contact layer for a-Si:H/c-Si heterojunction and μc-Si:H thin film solar cells. *Thin Solid Films* **2015**, *594*, 316–322. [CrossRef]

14. Tomasi, A.; Sahli, F.; Seif, J.P.; Fanni, L.; Agut, S.M.N.; Geissbuehler, J.; Paviet-Salomon, B.; Nicolay, S.; Barraud, L.; Niesen, B.; et al. Transparent Electrodes in Silicon Heterojunction Solar Cells: Influence on Contact Passivation. *IEEE J. Photovolt.* **2016**, *6*, 17. [CrossRef]

15. Römer, U.; Peibst, R.; Ohrdes, T.; Lim, B.; Krügener, J.; Wietler, T.; Brendel, R. Ion Implantation for Poly-Si Passivated Back-Junction Back-Contacted Solar Cells. *IEEE J. Photovolt.* **2015**, *5*, 507–514. [CrossRef]
16. Procel, P.; Yang, G.; Isabella, O.; Zeman, M. Theoretical evaluation of contact stack for high efficiency IBC-SHJ solar cells. *Sol. Energy Mater. Sol. Cells* **2018**, *186*, 66–77. [CrossRef]

MDPI
St. Alban-Anlage 66
4052 Basel
Switzerland
Tel. +41 61 683 77 34
Fax +41 61 302 89 18
www.mdpi.com

Energies Editorial Office
E-mail: energies@mdpi.com
www.mdpi.com/journal/energies